MANUAL OF MILITARY TRAINING

BY

CAPTAIN JAMES A. MOSS
UNITED STATES ARMY

Intended, primarily, for use in connection with the instruction and training of Cadets in our military schools, and, of COMPANY officers of the Organized Militia, and, secondarily, as a guide to COMPANY officers of the Regular Army, the aim being to make efficient fighting COMPANIES and to qualify our Cadets and Militia officers for the duty and responsibilities of COMPANY officers of Volunteers.

Fredonia Books
Amsterdam, The Netherlands

Manual of Military Training

by
Captain James A. Moss

ISBN: 1-58963-722-4

Reprinted from the 1914 edition

Fredonia Books
Amsterdam, The Netherlands
http://www.fredoniabooks.com

In order to make original editions of historical works available to scholars at an economical price, this facsimile of the original edition of 1914 is reproduced from the best available copy and has been digitally enhanced to improve legibility, but the text remains unaltered to retain historical authenticity.

PREFATORY

The COMPANY is the *basic fighting tactical unit,*—it is the foundation rock upon which an army is built—and the fighting efficiency of a COMPANY is based on systematic and thorough training.

This book is a presentation of MILITARY TRAINING as manifested in the training and instruction of a COMPANY. The book contains everything pertaining to the training and instruction of COMPANY officers, noncommissioned officers and privates, and the officer who masters its contents and who makes his COMPANY proficient in the subjects embodied herein, will be in every way qualified, *without the assistance of a single other book,* to command with credit and satisfaction, in peace and in war, a COMPANY that will be an *efficient fighting weapon.*

The book is divided as follows:

PART I. DRILLS, EXERCISES, GUARD DUTY, TARGET PRACTICE, CEREMONIES AND INSPECTIONS.

PART II. MISCELLANEOUS SUBJECTS PERTAINING TO COMPANY TRAINING AND INSTRUCTION.

PART III. COMPANY FIELD TRAINING.

A program of instruction and training covering a given period can readily be arranged by looking over the TABLE OF CONTENTS, and selecting therefrom such subjects as it is desired to use, the number and kind, and the time to be devoted to each, depending upon the time available, and climatic and other conditions. It is suggested that, for the sake of variety, in drawing up a program of instruction and training, when practicable, a part of each day or a part of each drill time, be devoted to theoretical work and a part to practical work. For example, the theoretical work could be carried on in the forenoon and the practical work in the afternoon, or the theoretical work could be carried on from, say, 8 to 9:30 a. m., and the practical work from 9 or 9:30 to 10:30 or 11 a. m.

Attention is invited to the completeness of the Index, whereby one is enabled to locate at once any point covered in the book.

INDEX

(The numbers refer to paragraphs.)

[5]

[7]

[9]

DRILL REGULATIONS (Cont'd)

DRILL REGULATIONS (Cont'd)

[13]

[14]

[15]

[16]

[17]

TABLE OF CONTENTS
PART I
DRILLS, EXERCISES, GUARD DUTY, TARGET PRACTICE, CERE-
MONIES AND INSPECTIONS

[20]

[21]

PART III

COMPANY FIELD TRAINING

**IN THE ATTACK, THE DEFENSE, THE SERVICE OF SECURITY, THE
SERVICE OF INFORMATION, NIGHT OPERATIONS, INTRENCH-
MENTS, OBSTACLES, FIELD FIRING, CAMPING, AND
INDIVIDUAL COOKING**

[22]

[23]

[24]

PART I

DRILLS, EXERCISES, GUARD DUTY, TAR GET PRACTICE, CEREMONIES AND INSPECTIONS

CHAPTER I

INFANTRY DRILL REGULATIONS

(To include Changes No. 8, September 3, 1914.)

DEFINITIONS

(The numbers following the paragraphs are those of the Drill Regulations, and references in the text to certain paragraph numbers refer to these numbers and not to the numbers preceding the paragraphs.)

(NOTE.—Company drills naturally become monotonous. The monotony, however, can be greatly reduced by repeating the drills under varying circumstances. In the manual of arms, for instance, the company may be brought to open ranks and the officers and sergeants directed to superintend the drill in the front and rear ranks. As the men make mistakes they are fallen out and drilled nearby by an officer or noncommissioned officer. Or, the company may be divided into squads, each squad leader drilling his squad, falling out the men as they make mistakes, the men thus fallen out reporting to a designated officer or noncommissioned officer for drill. The men who have drilled the longest in the different squads are then formed into one squad and drilled and fallen out in like manner. The variety thus introduced stimulates a spirit of interest and rivalry that robs the drill of much of its monotony.

It is thought the instruction of a company in drill is best attained by placing special stress on squad drill. The noncommissioned officers should be thoroughly instructed, practically and theoretically, by one of the company officers and then be required to instruct their squads. The squads are then united and drilled in the school of the company.—Author)

DEFINITIONS

1. **Alignment:** A straight line upon which several elements are formed, or are to be formed; or the dressing of several elements upon a straight line.

FIG 1

NOTE.—The line A-B, on which a body of troops is formed or is to be formed, or the act of dressing a body of troops on the line, is called an alignment.—Author.

2. **Base:** The element on which a movement is regulated.

3. **Battle sight:** The position of the rear sight when the leaf is laid down.

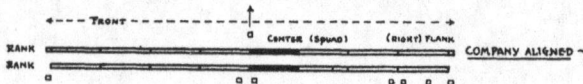

FIG 2

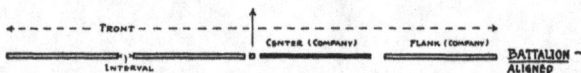

FIG 3

4. **Center:** The middle point or element of a command. (See Figs. 2, 3 and 5.)

5. **Column:** A formation in which the elements are placed one be hind another. (See Figs. 4, 5, 6 and 7.)

6. **Deploy:** To extend the front. In general to change from column to line, or from close order to extended order.

7. **Depth:** The space from head to rear of any formation, including the leading and rear elements. The depth of a man is assumed to be 12 inches. (See Figs. 4, 5, 6 and 7.)

8. **Distance:** Space between elements in the direction of depth. Distance is measured from the back of the man in front to the breast of the man in rear. The distance between ranks is 40 inches in both line and column. (See Figs. 4, 5, 6 and 7.)

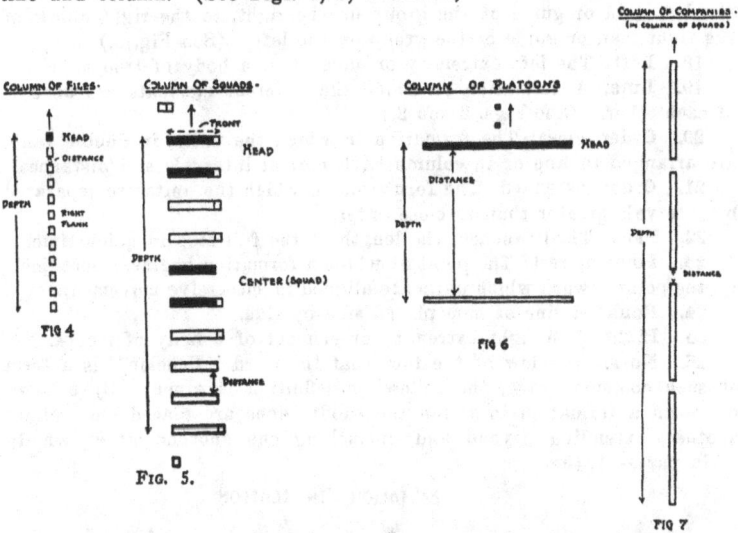

9. **Element:** A file, squad, platoon, company, or larger body, forming part of a still larger body.

10. **File:** Two men, the front-rank man and the corresponding man of the rear rank. The front-rank man is the **file leader.** A file which has no rear-rank man is a **blank file.** The term file applies also to a single man in a single-rank formation.

11. **File closers:** Such officers and noncommissioned officers of a company as are posted in rear of the line. For convenience, all men posted in the line of file closers.

12. **Flank:** The right or left of a command in line or in column; also the element on the right or left of the line. (See Figs. 2, 3 and 4.)

[27]

13. Formation: Arrangement of the elements of a command. The placing of all fractions in their order in line, in column, or for battle.

14. Front: The space, in width, occupied by an element, either in line or in column. The front of a man is assumed to be 22 inches. Front also denotes the direction of the enemy. (See Figs. 2, 3 and 5).

15. Guide: An officer, noncommissioned officer, or private upon whom the command or elements thereof regulates its march.

16. Head: The leading element of a column. (See Figs. 4, 5 and 6.)

17. Interval: Space between elements of the same line. The interval between men in ranks is 4 inches and is measured from elbow to elbow. Between companies, squads, etc., it is measured from the left elbow of the left man or guide of the group on the right, to the right elbow of the right man or guide of the group on the left. (See Fig. 3.)

18. Left: The left extremity or element of a body of troops.

19. Line: A formation in which the different elements are abreast of each other. (See Figs. 2 and 3.)

20. Order, close: The formation in which the units, in double rank, are arranged in line or in column with normal intervals and distances.

21. Order, extended: The formation in which the units are separated by intervals greater than in close order.

22. Pace: Thirty inches; the length of the full step in quick time.

23. Point of rest: The point at which a formation begins. Specifically, the point toward which units are aligned in successive movements.

24. Rank: A line of men placed side by side.

25. Right: The right extremity or element of a body of troops.

26. Note. In view of the fact that the word "Echelon" is a term of such common usage, the following definition is given: By echelon we mean a formation in which the subdivisions are placed one behind another, extending beyond and unmasking one another either wholly or in part.—Author.

BATTALION IN ECHELON

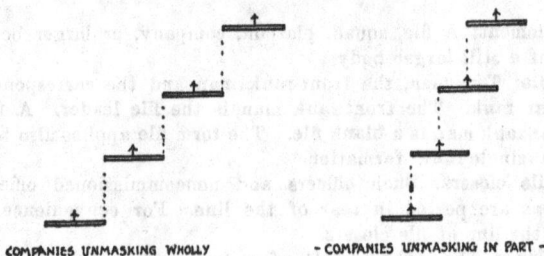

COMPANIES UNMASKING WHOLLY   - COMPANIES UNMASKING IN PART -

INTRODUCTION

(The numbers following the paragraphs are those of the Drill Regulations.—Author.)

27. Success in battle is the ultimate object of all military training; success may be looked for only when the training is intelligent and thorough. (1)

28. Commanding officers are accountable for the proper training of their respective organizations within the limits prescribed by regulations and orders.

The excellence of an organization is judged by its field efficiency. The field efficiency of an organization depends primarily upon its effectiveness as a whole. Thoroughness and uniformity in the training of the units of an organization are indispensable to the efficiency of the whole; it is by such means alone that the requisite teamwork may be developed. (2)

29. Simple movements and elastic formations are essential to correct training for battle. (3)

30. The Drill Regulations are furnished as a guide. They provide the principles for training and for increasing the probability of success in battle.

In the interpretation of the regulations, the spirit must be sought. Quibbling over the minutiæ of form is indicative of failure to grasp the spirit. (4)

31. The principles of combat are considered in Part II of these regulations. They are treated in the various schools included in Part I, only to the extent necessary to indicate the functions of the various commanders and the division of responsibility between them. The amplification necessary to a proper understanding of their application is to be sought in Part II. (5)

32. The following important distinctions must be observed:

(a) Drills executed at **attention** and the ceremonies are **disciplinary exercises** designed to teach precise and soldierly movement, and to inculcate that prompt and subconscious obedience which is essential to proper military control. To this end, smartness and precision should be exacted in the execution of every detail. Such drills should be frequent, but short.

33. (b) The purpose of **extended order drill** is to teach the **mechanism** of deployment, of the firings, and, in general, of the employment of troops in combat. Such drills are in the nature of disciplinary exercises and should be frequent, thorough, and exact in order to habituate men to the firm control of their leaders. Extended order drill is executed **at ease**. The company is the largest unit which executes extended order drill.

34. (c) **Field exercises** are for instruction in the duties incident to campaign. Assumed situations are employed. Each exercise should conclude with a discussion, on the ground, of the exercise and principles involved.

35. (d) The **combat exercise, a form of field exercise** of the company, battalion, and larger units, consists of the **application of tactical principles** to assumed situations, employing in the execution the appropriate formations and movements of close and extended order.

Combat exercises must simulate, as far as possible, the battle conditions assumed. In order to familiarize both officers and men with such conditions, companies and battalions will frequently be consolidated to provide war-strength organizations. Officers and noncommissioned officers not required to complete the full quota of the units participating are assigned as observers or umpires.

The firing line can rarely be controlled by the voice alone: thorough training to insure the proper use of prescribed signals is necessary.

The exercise should be followed by a brief drill at attention in order to restore smartness and control. (6)

36. In field exercises the enemy is said to be **imaginary** when his position and force are merely assumed; **outlined** when his position and force are indicated by a few men; **represented** when a body of troops acts as such. (7)

General Rules for Drills and Formations

37. When the **preparatory** command consists of more than one part, its elements are arranged as follows:

(1) For movements to be executed successively by the subdivisions or elements of an organization: (a) Description of the movement; (b) how executed, or on what element executed.

(2) For movements to be executed simultaneously by the subdivisions of an organization: (a) The designation of the subdivisions; (b) the movement to be executed. (8)

38. Movements that may be executed toward either flank are explained as toward but one flank, it being necessary to substitute the word ''left'' for ''right,'' and the reverse, to have the explanation of the corresponding movement toward the other flank. The commands are given for the execution of the movements toward either flank. The substitute word of the command is placed within parentheses. (9)

39. Any movement may be executed either from the halt or when marching, unless otherwise prescribed. If at a halt, the command for movements involving marching need not be prefaced by **forward,** as 1. **Column right (left),** 2. **MARCH.** (10)

40. Any movement not specially excepted may be executed in double time.

If at a halt, or if marching in quick time, the command **double time** precedes the command of execution. (11)

41. In successive movements executed in double time the leading or base unit marches in **quick time** when not otherwise prescribed; the other units march in **double time** to their places in the formation ordered and then conform to the gait of the leading or base unit. If marching in double time, the command **double time** is omitted. The leading or base unit marches in quick time; the other units continue at double time to their places in the formation ordered and then conform to the gait of the leading or base unit. (12)

42. To hasten the execution of a movement begun in quick time, the command: 1. **Double time**, 2. **MARCH**, is given. The leading or base unit continues to march in quick time, or remains at halt if already halted; the other units complete the execution of the movement in double time and then conform to the gait of the leading or base unit. (13)

43. To stay the execution of a movement when marching, for the correction of errors, the command: 1. **In place**, 2. **HALT**, is given. All halt and stand fast, without changing the position of the pieces. To resume the movement the command: 1. **Resume**, 2. **MARCH**, is given. (14)

44. To revoke a preparatory command, or, being at a halt, to begin anew a movement improperly begun, the command, **as you were**, is given, at which the movement ceases and the former position is resumed. (15)

45. Unless otherwise announced the guide of a company or subdivision of a company in line is **right**; of a battalion in line or line of subdivisions or of a deployed line, **center**; of a rank in column of squads, toward the side of the guide of the company.

To march with guide other than as prescribed above, or to change the guide: **Guide (right, left, or center).**

In successive formations into line, the guide is toward the point of rest; in platoons or larger subdivisions it is so announced.

The announcement of the guide, when given in connection with a movement, follows the command of execution for that movement. Exception: 1. **As skirmishers, guide right (left** or **center), 2. MARCH.** (16)

46. **The turn on the fixed pivot** by subdivisions is used in all formations from line into column and the reverse.

The turn on the moving pivot is used by subdivisions of a column in executing changes of direction. (17)

47. Partial changes of direction may be executed:

By interpolating in the preparatory command the word **half.** as **Column half right (left),** or **Right (left) half turn.** A change of direction of 45° is executed.

By the command: **INCLINE TO THE RIGHT (LEFT)**. The guide, or guiding element, moves in the indicated direction and the remainder of the command conforms. This movement effects slight changes of direction. (18)

48. The designations **line of platoons, line of companies, line of battalions**, etc., refer to the formations in which the platoons, companies, battalions, etc., each in column of squads, are in line. (19)

49. Full distance in column of subdivisions is such that in forming line to the right or left the subdivisions will have their proper intervals.

In column of subdivisions the guide of the leading subdivision is charged with the step and direction; the guides in rear preserve the trace, step, and distance. (20)

50. In close order, all details, detachments, and other bodies of troops are habitually formed in double rank.

To insure uniformity of interval between files when falling in, and in alignments, each man places the palm of the left hand upon the hip, fingers pointing downward. In the first case the hand is dropped by the side when the next man on the left has his interval; in the second case, at the command **front.** (21)

51. The posts of officers, noncommissioned officers, special units (such as band or machine-gun company), etc., in the various formations of the company, battalion, or regiment, are shown in plates.

In all changes from one formation to another involving a change of post on the part of any of these, posts are promptly taken by the most convenient route as soon as practicable after the command of execution for the movement; officers and noncommissioned officers who have prescribed duties in connection with the movement ordered, take their new posts when such duties are completed.

As instructors, officers and noncommissioned officers go wherever their presence is necessary. As file closers it is their duty to rectify mistakes and insure steadiness and promptness in the ranks. (22)

52. Except at ceremonies, the special units have no fixed places. They take places as directed; in the absence of directions, they conform as nearly as practicable to the plates, and in subsequent movements maintain their relative positions with respect to the flank or end of the command on which they were originally posted. (23)

53. General, field and staff officers are habitually mounted. The staff of an officer forms in single rank 3 paces in rear of .him, the right of the rank extending 1 pace to the right of a point directly in rear

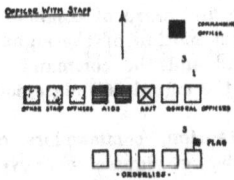
FIG 8

of him. Members of the staff are arranged in order from right to left as follows: General staff officers, adjutant, aids, other staff officers, arranged in each classification in order of rank, the senior on the right. The flag of the general officer and the orderlies are 3 paces in rear of the staff, the flag on the right. When necessary to reduce the front of the staff and orderlies, each line executes **twos right** or **fours right**, as explained in the Cavalry Drill Regulations, and follows the commander.

When not otherwise prescribed, staff officers draw and return saber with their chief. (24)

54. In making the about, an officer, mounted, habitually turns to the left.

When the commander faces to give commands, the staff, flag, and orderlies do not change position. (25)

55. For ceremonies, all mounted enlisted men of a regiment or smaller unit, except those belonging to the machine-gun organizations, are consolidated into a detachment; the senior present commands if no officer is in charge. The detachment is formed as a platoon or squad of cavalry in line or column of fours; noncommissioned staff officers are on the right or in the leading ranks. (27)

56. For ceremonies, such of the noncommissioned staff officers as are dismounted are formed 5 paces in rear of the color, in order of rank from right to left. In column of squads they march as file closers. (28)

Other than for ceremonies, noncommissioned staff officers and orderlies accompany their immediate chiefs unless otherwise directed. If mounted, the noncommissioned staff officers are ordinarily posted on the right or at the head of the orderlies. (29)

57. In all formations and movements a noncommissioned officer commanding a platoon or company carries his piece as the men do, if he is so armed, and takes the same post as an officer in like situation. When the command is formed in line for ceremonies, a noncommissioned officer commanding a company takes post on the right of the right guide after the company has been aligned. (30)

ORDERS, COMMANDS, AND SIGNALS

58. Commands only are employed in drill at **attention**. Otherwise either a **command, signal** or **order** is employed, as best suits the occasion, or one may be used in conjunction with another. (31)

59. Signals should be freely used in instruction, in order that officers and men may readily know them. In making arm signals the saber, rifle, or headdress may be held in the hand. (32)

60. Officers and men fix their attention at the first word of command, the first note of the bugle or whistle, or the first motion of the signal. A signal includes both the preparatory command and the command of execution; the movement commences as soon as the signal is understood, unless otherwise prescribed. (33)

61. Except in movements executed at **attention**, commanders or leaders of subdivisions repeat orders, commands, or signals whenever such repetition is deemed necessary to insure prompt and correct execution.

Officers, platoon leaders, guides, and musicians are equipped with whistles. Guides and musicians assist by repeating signals when necessary. Battalion and Company Commanders will use a whistle of different tone from that of the whistle used by platoon leaders and musicians. (34)

62. Prescribed signals are limited to such as are essential as a substitute for the voice under conditions which render the voice inadequate.

Before or during an engagement special signals may be agreed upon to facilitate the solution of such special difficulties as the particular situation is likely to develop, but it must be remembered that simplicity and certainty are indispensable qualities of a signal. (35)

Orders

63. In these regulations an **order** embraces instructions or directions given orally or in writing in terms suited to the particular occasion and not prescribed herein.

Orders are employed only when the **commands** prescribed herein do not sufficiently indicate the will of the commander.

Orders are more fully described in paragraphs 378 to 383, inclusive. (36)

Commands

64. In these regulations a **command** is the will of the commander expressed in the phraseology prescribed herein. (37)

There are two kinds of commands:

The **preparatory** command, such as **forward**, indicates the movement that is to be executed.

The command of **execution**, such as **MARCH, HALT**, or **ARMS**, causes the execution.

Preparatory commands are distinguished by bold face, those of **execution by CAPITALS**.

Where it is not mentioned in the text who gives the commands prescribed, they are to be given by the commander of the unit concerned.

The **preparatory** command should be given at such an interval of time before the command of execution as to admit of being properly understood; the command of **execution** should be given at the instant the movement is to commence.

The tone of command is animated, distinct, and of a loudness proportioned to the number of men for whom it is intended.

Each **preparatory** command is enunciated distinctly, with a rising inflection at the end, and in such manner that the command of **execution** may be more energetic.

The command of **execution** is firm in tone and brief. (38)

65. Majors and commanders of units larger than a battalion repeat such commands of their superiors as are to be executed by their units, facing their units for that purpose. The battalion is the largest unit that executes a movement at the command of execution of its commander. (39)

66. When giving commands to troops it is usually best to face toward them.

Indifference in giving commands must be avoided as it leads to laxity in execution. Commands should be given with spirit at all times. (40)

Bugle Signals

67. The authorized bugle signals are published in Part V of these regulations.

The following bugle signals may be used off the battle field, when not likely to convey information to the enemy:

Attention: Troops are brought to attention.

Attention to orders: Troops fix their attention.

Forward, MARCH: Used also to execute quick time from double time.

Double time, MARCH.

To the rear, MARCH: In close order, execute **squads right about**, **HALT.**

Assemble, MARCH.

The following bugle signals may be used on the battle field:

Fix bayonets.

Charge.

Assemble, March.

These signals are used only when intended for the entire firing line; hence they can be authorized only by the commander of a unit (for example, a regiment or brigade) which occupies a distinct section of the battle field. Exception: **Fix bayonet.** (See par. 318.)

The following bugle signals are used in exceptional cases on the battle field. Their principal uses are in field exercises and practice firing.

Commence firing: Officers charged with fire direction and control open fire as soon as practicable. When given to a firing line, the signal is equivalent to fire at will.

Cease firing: All parts of the line execute cease firing at once.

These signals are not used by units smaller than a regiment, except when such unit is independent or detached from its regiment. (41)

Whistle Signals

68. Attention to orders. A short blast of the whistle. This signal is used on the march or in combat when necessary to fix the attention of troops, or of their commanders or leaders, preparatory to giving commands, orders, or signals.

When the firing line is firing, each squad leader suspends firing and fixes his attention at a short blast of his platoon leader's whistle. The platoon leader's subsequent commands or signals are repeated and enforced by the squad leader. If a squad leader's attention is attracted by a whistle other than that of his platoon leader, or if there are no orders or commands to convey to his squad he resumes firing at once.

Suspend firing. A long blast of the whistle. (42)

Arm Signals

69. The following arm signals are prescribed. In making signals either arm may be used. Officers who receive signals on the firing line "repeat back" at once to prevent misunderstanding.

FIG 9

Forward, MARCH. Carry the hand to the shoulder; straighten and hold the arm horizontally, thrusting it in direction of march.

This signal is also used to execute quick time from double time.

FIG 10

HALT - ARM HELD STATIONARY

DOUBLE TIME - ARM MOVED UP AND DOWN SEVERAL TIMES

Halt. Carry the hand to the shoulder; thrust the hand upward and hold the arm vertically.

Double time, MARCH. Carry the hand to the shoulder; rapidly thrust the hand upward the full extent of the arm several times.

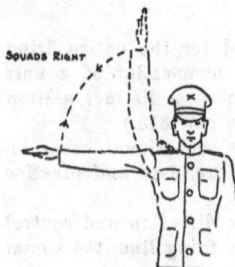

SQUADS RIGHT

FIG 11

Squads right, MARCH. Raise the arm laterally until horizontal; carry it to a vertical position above the head and swing it several times between the vertical and horizontal positions.

[36]

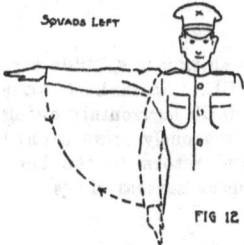

SQUADS LEFT

FIG 12

Squads left, MARCH. Raise the arm laterally until horizontal; carry it downward to the side and swing it several times between the downward and horizontal positions.

SQUADS RIGHT ABOUT
TO THE REAR

FIG 13

Squads right about, MARCH (if in close order) or, **To the rear, MARCH** (if in skirmish line). Extend the arm vertically above the head; carry it laterally downward to the side and swing it several times between the vertical and downward positions.

CHANGE DIRECTION

FIG 14

Change direction or Column right (left), MARCH. The hand on the side toward which the change of direction is to be made is carried across the body to the opposite shoulder, forearm horizontal; then swing in a horizontal plane, arm extended, pointing in the new direction.

AS SKIRMISHERS

FIG 15

As skirmishers, MARCH. Raise both arms laterally until horizontal.

As Skirmishers, Guide Center

FIG 16

As skirmishers, guide center, MARCH. Raise both arms laterally until horizontal; swing both simultaneously upward until vertical and return to the horizontal; repeat several times.

As Skirmishers, Guide Right

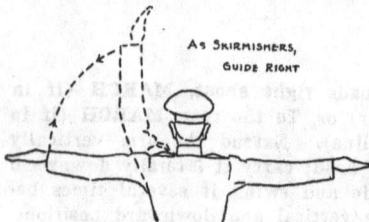

FIG 17

As skirmishers, guide right (left), MARCH. Raise both arms laterally until horizontal; hold the arm on the side of the guide steadily in the horizontal position; swing the other upward until vertical and return it to the horizontal; repeat several times.

Assemble

FIG 18

Assemble, MARCH. Raise the arm vertically to its full extent and describe horizontal circles.

To Announce Range – Battle Sight

FIG 19

Range, or **Change elevation.** To announce range, extend the arm toward the leaders or men for whom the signal is intended, fist closed; by keeping the fist closed battle sight is indicated;

Range 300 – Or Increase By 30°

FIG 20

by opening and closing the fist, expose thumb and fingers to a number equal to the hundreds of yards;

[38]

ADD 50 YARDS
(TOP VIEW)

FIG 21

to add 50 yards describe a short horizontal line with
forefinger.

DECREASE BY 300

FIG 22

To change elevation, indicate the **amount of**
increase or decrease by fingers as above; point upward
to indicate increase and downward to indicate decrease.

SUSPEND FIRING –
CEASE FIRING – SWING ARM UP AND
DOWN SEVERAL TIMES

FIG 23

Suspend firing. Raise and hold the forearm
steadily in a horizontal position in front of the
forehead, palm of the hand to the front.

Cease firing. Raise the forearm as in suspend firing and swing it up and down several times
in front of the face.

PLATOON

FIG. 24

Platoon. Extend the arm horizontally toward
the platoon leader; describe small circles with the
hand. (See par. 44.)

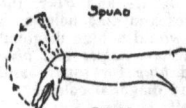

SQUAD

FIG 25

Squad. Extend the arm horizontally toward
the platoon leader; swing the hand up and down
from the wrist. (See par. 44.)

Rush. Same as double time. (43)

The signals **platoon** and **squad** are intended primarily for communication between the captain and his platoon leaders. The signal
platoon or **squad** indicates that the platoon commander is to cause the
signal which follows to be executed by platoon or squad. (44)

Flag Signals

70. The signal flags described below are carried by the company
musicians in the field.

In a regiment in which it is impracticable to make the permanent battalion division alphabetically, the flags of a battalion are as

shown; flags are assigned to the companies alphabetically, within their respective battalions, in the order given below.

First battalion:

     Company A. Red field, white square.

     Company B. Red field, blue square.

     Company C. Red field, white diagonals.

     Company D. Red field, blue diagonals.

Second battalion:

     Company E. White field, red square.

     Company F. White field, blue square.

     Company G. White field, red diagonals.

     Company H. White field, blue diagonals.

Third battalion:

     Company I. Blue field, red square.

     Company K. Blue field, white square.

     Company L. Blue field, red diagonals.

     Company M. Blue field, white diagonals.   (45)

Note.—An analysis of the above system of signal flags will show·—

1. The color of the field indicates the battalion, and the colors run in the order that is so natural to us all, viz: *Red, White* and *Blue.* Hence, *red* field indicates the *first* battalion; *white* field, the *second,* *blue* field, the *third.*

2. The *squares* indicate the first two companies of each battalion, and the *diagonals,* the second two. Hence,

Companies			Indicated by
A B	E F	I K	Squares
C D	G H	L M	Diagonals

3. The colors of the squares and diagonals in combination with those of the fields, run in the order that is so natural to us all, viz: *Red, White* and *Blue,* the color of any given field being, of course, omitted from the squares and diagonals, as a white square for instance, would not show on a white field, nor would a blue diagonal show on a blue field. For example, with a *red* field we would have *white* and *blue* for the square and diagonal colors, with a *white* field, *red* and *blue* for square and diagonal colors, with a *blue* field *red* and *white* for the square and diagonal colors.

4. From what has been said, the following table explains itself:

Battalion	Field	Co.	Squares	Diagonals
First	Red	A B C D	White Blue	 White Blue
Second	White	E F G H	Red Blue	 Red Blue
Third	Blue	I K L M	Red White	 Red White

Note how the square and diagonal colors always follow in the natural order of *red, white,* and *blue,* with the color of the field omitted.—Author.

71. In addition to their use in visual signaling, these flags serve to mark the assembly point of the company when disorganized by combat, and to mark the location of the company in bivouac and elsewhere, when such use is desirable. (46)

72. For communication between the firing line and the reserve or commander in rear, the subjoined signals (Signal Corps codes) are prescribed and should be memorized. In transmission, their concealment from the enemy's view should be insured. In the absence of signal flags, the headdress or other substitute may be used.

Letter of alphabet.	If signaled from the rear to the firing line.	If signaled from the firing line to the rear.
AAA	Ammunition going forward	Ammunition required.
CCC	Charge (mandatory at all times)..	Am about to charge if no instructions to the contrary.
C F	Cease firing	Cease firing.
D T	Double time or "rush"	Double time or "rush."
F	Commence firing	Am about to commence firing if no instructions to the contrary.
F B	Fix bayonets	
H H H ...	Halt	
L	Left	Left.
N	Negative	Negative.
O	Move forward	Am about to move forward.
Q	What is the range?	What is the range?
R	Range	Range.
R T	Right	Right.
S S S	Support going forward	Support needed.
S U F	Suspend firing	Suspend firing.
T	Target	Target.
Y	Affirmative	Affirmative. (47)

SCHOOL OF THE SOLDIER

73. The instructor explains briefly each movement, first executing it himself if practicable. He requires the recruits to take the proper positions unassisted and does not touch them for the purpose of correcting them, except when they are unable to correct themselves. He avoids keeping them too long at the same movement, although each should be understood before passing to another. He exacts by degrees the desired precision and uniformity. (48)

74. In order that all may advance as rapidly as their abilities permit, the recruits are grouped according to proficiency as instruction progresses. Those who lack aptitude and quickness are separated from the others and placed under experienced drill masters. (49)

Instruction Without Arms

For preliminary instruction a number of recruits, usually not exceeding three or four, are formed as a squad in single rank. (50)

Position of the Soldier, or Attention

75. Heels on the same line and as near each other as the conformation of the man permits.

Feet turned out equally and forming an angle of about 45°.

Knees straight without stiffness.

Hips level and drawn back slightly; body erect and resting equally on hips; chest lifted and arched; shoulders square and falling equally.

Arms and hands hanging naturally, thumb along the seam of the trousers.

Head erect and squarely to the front, chin drawn in so that the axis of the head and neck is vertical; eyes straight to the front.

Weight of the body resting equally upon the heels and balls of the feet. (51)

The Rests

76. Being at a halt, the commands are: **FALL OUT; REST; AT EASE**; and, 1. Parade, 2. **REST**.

At the command **fall out**, the men may leave the ranks, but are required to remain in the immediate vicinity. They resume their former places, at attention, at the command **fall in**.

At the command **rest** each man keeps one foot in place, but is not required to preserve silence or immobility.

At the command **at ease** each man keeps one foot in place and is required to preserve silence but not immobility. (52)

77. 1. Parade, 2. **REST**. Carry the right foot 6 inches straight to the rear, left knee slightly bent; clasp the hands, without constraint, in front of the center of the body, fingers joined, left hand uppermost, left thumb clasped by the thumb and forefinger of the right hand; preserve silence and steadiness of position. (53)

78. To resume the attention: 1. Squad, 2. **ATTENTION**.
The men take the position of the soldier. (54)

Eyes Right or Left.

79. 1. Eyes, 2. **RIGHT (LEFT)**, 3. **FRONT**.
At the command **right**, turn the head to the right oblique, eyes fixed on the line of eyes of the men in, or supposed to be in, the same rank. At the command **front**, turn the head and eyes to the front. (55)

FIG. 26.

Facings

80. To the flank: 1. **Right (left)**, 2. **FACE**.

Raise slightly the left heel and right toe; face to the right, turning on the right heel, assisted by a slight pressure on the ball of the left foot; place the left foot by the side of the right. Left face is executed on the left heel in the corresponding manner.

Right (left) half face is executed similarly, facing 45°.

"To face in marching" and advance, turn on the ball of either foot and step off with the other foot in the new line of direction; to face in marching without gaining ground in the new direction, turn on the ball of either foot and mark time. (56)

81. To the rear: 1. **About**, 2. **FACE**.

Carry the toe of the right foot about a half foot-length to the rear and slightly to the left of the left heel without changing the position of the left foot; face to the rear, turning to the right on the left heel and right toe; place the right heel by the side of the left. (57)

Salute with the Hand

82. 1. **Hand**, 2. **SALUTE**.

Raise the right hand smartly till the tip of the forefinger touches the lower part of the headdress above the right eye, thumb and fingers extended and joined, palm to the left, forearm inclined at about 45°, hand and wrist straight; at the same time look toward the person saluted. **(TWO)** Drop the arm smartly by the side.

For rules governing salutes, see "Honors and Salutes," pars. 758-765. (58)

Steps and Marchings

83. All steps and marchings executed from a halt, except right step, begin with the left foot. (59)

84. The length of the full step in quick time is 30 inches, measured from heel to heel, and the cadence is at the rate of 120 steps per minute.

The length of the full step in double time is 36 inches; the cadence is at the rate of 180 steps per minute.

The instructor, when necessary, indicates the cadence of the step by calling one, two, three, four, or left, right, the instant the left and right foot, respectively, should be planted. (60)

[43]

85. All steps and marchings and movements involving march are executed in **quick time** unless the squad be marching in **double time**, or **double time** be added to the command; in the latter case **double time is** added to the preparatory command. Example: 1. **Squad right, double time, 2. MARCH** (School of the Squad). (61)

Quick Time

86. Being at a halt, to march forward in quick time: 1. **Forward,** 2. **MARCH.**

At the command **forward**, shift the weight of the body to the right leg, left knee straight.

At the command **march**, move the left foot smartly straight forward 30 inches from the right, sole near the ground, and plant it without shock; next in like manner, advance the right foot and plant it as above; continue the march. The arms swing naturally. (62)

87. Being at a halt, or in march in quick time, to march in double time: 1. **Double time**, 2. **MARCH.**

If at a halt, at the first command shift the weight of the body to the right leg. At the command **march**, raise the forearms, fingers closed, to a horizontal position along the waist line; take up an easy run with the step and cadence of double time, allowing a natural swinging motion to the arms.

If marching in quick time, at the command **march**, given as either foot strikes the ground, take one step in quick time, and then step off in double time. (63)

To resume the quick time: 1. **Quick time**, 2. **MARCH.**

At the command **march**, given as either foot strikes the ground, advance and plant the other foot in double time; resume the quick time, dropping the hands by the sides. (64)

To Mark Time

88. Being in march: 1. **Mark time**, 2. **MARCH.**

At the command **march**, given as either foot strikes the ground, advance and plant the other foot; bring up the foot in rear and continue the cadence by alternately raising each foot about 2 inches and planting it on line with the other.

Being at a halt, at the command **march**, raise and plant the feet as described above. (65)

The Half Step

89. 1. **Half step**, 2. **MARCH.**

Take steps of 15 inches in quick time, 18 inches in double time. (66)

Forward, half step, halt, and **mark time** may be executed one from the other in quick or double time.

To resume the full step from half step or mark time: 1. **Forward,** 2. **MARCH.** (67)

[44]

Side Step

90. Being at a halt or mark time: 1. **Right (left) step**, 2. **MARCH**.
Carry and plant the right foot 15 inches to the right; bring the left foot beside it and continue the movement in the cadence of quick time.

The side step is used for short distances only and is not executed in double time.

If at order arms, the side step is executed **at trail** without command. (68)

Back Step

91. Being at a halt or mark time: 1. **Backward**, 2. **MARCH**.
Take steps of 15 inches straight to the rear.

The back step is used for short distances only and is not executed in double time.

If at order arms, the back step is executed **at trail** without command. (69)

To Halt

92. To arrest the march in quick or double time: 1. **Squad**, 2. **HALT**.
At the command halt, given as either foot strikes the ground, plant the other foot as in marching; raise and place the first foot by the side of the other. If in double time, drop the hands by the sides. (70)

To March by the Flank

93. Being in march: 1. **By the right (left) flank**, 2. **MARCH**.
At the command march, given as the right foot strikes the ground, advance and plant the left foot, then face to the right in marching and step off in the new direction with the right foot. (71)

To March to the Rear

94. Being in march: 1. **To the rear**, 2. **MARCH**.
At the command march, given as the right foot strikes the ground, advance and plant the left foot; turn to the right about on the balls of both feet and immediately step off with the left foot.

If marching in double time, turn to the right about, taking four steps in place, keeping the cadence, and then step off with the left foot. (72)

Change Step

95. Being in march: 1. **Change step**, 2. **MARCH**.
At the command march, given as the right foot strikes the ground, advance and plant the left foot; plant the toe of the right foot near the heel of the left and step off with the left foot.

The change on the right foot is similarly executed, the command march being given as the left foot strikes the ground. (73)

Manual of Arms

96. As soon as practicable the recruit is taught the use, nomenclature and care of his rifle; when fair progress has been made in the instruction without arms, he is taught the manual of arms; instruction without arms and that with arms alternate. (74)

97. The following rules govern the carrying of the piece:

First. The piece is not carried with cartridges in either the chamber or the magazine except when specifically ordered. When so loaded, or supposed to be loaded, it is habitually carried locked; that is, with the safety lock turned to the "safe." At all other times it is carried unlocked, with the trigger pulled. (See par. 336).

98. Second. Whenever troops are formed under arms, pieces are immediately inspected at the commands: 1. Inspection, 2. ARMS; 3. Order (Right shoulder, port), 4. ARMS.

A similar inspection is made immediately before dismissal.

If cartridges are found in the chamber or magazine they are removed and placed in the belt.

99. Third. The cut-off is kept turned "off" except when cartridges are actually used.

100. Fourth. The bayonet is not fixed except in bayonet exercise. on guard, or for combat.

101. Fifth. Fall in is executed with the piece at the order arms. Fall out, rest, and at ease are executed as without arms. On resuming attention the position of order arms is taken.

102. Sixth. If at the order, unless otherwise prescribed, the piece is brought to the right shoulder at the command march, the three motions corresponding with the first three steps. Movements may be executed at the trail by prefacing the preparatory command with the words at trail; as, 1. At trail, forward, 2. MARCH; the trail is taken at the command march.

When the facings, alignments, open and close ranks, taking interval or distance, and assemblings are executed from the order, raise the piece to the trail while in motion and resume the order on halting.

103. Seventh. The piece is brought to the order on halting. The execution of the order begins when the halt is completed.

104. Eighth. A disengaged hand in double time is held as when without arms. (75)

105. The following rules govern the execution of the manual of arms:

First. In all positions of the left hand at the balance (center of gravity, bayonet unfixed) the thumb clasps the piece; the sling is included in the grasp of the hand.

Second. In all positions of the piece "diagonally across the body" the position of the piece, left arm and hand are the same as in port arms.

Third. In resuming the order from any position in the manual, the motion next to the last concludes with the butt of the piece about 3 inches from the ground, barrel to the rear, the left hand above and near the right, steadying the piece, fingers extended and joined, forearm and wrist straight and inclining downward, all fingers of the right hand grasping the piece. To complete the order, lower the piece gently to the ground with the right hand, drop the left quickly by the side, and take the position of order arms.

Allowing the piece to drop through the right hand to the ground, or other similar abuse of the rifle to produce effect in executing the manual, is prohibited.

Fourth. The cadence of the motions is that of quick time; the recruits are first required to give their whole attention to the details of the motions, the cadence being gradually acquired as they become accustomed to handling their pieces. The instructor may require them to count aloud in cadence with the motions.

Fifth. The manual is taught at a halt and the movements are for the purpose of instruction, divided into motions and executed in detail; in this case the command of execution determines the prompt execution of the first motion, and the commands, two, three, four, that of the other motions.

To execute the movements in detail, the instructor first cautions: By the numbers; all movements divided into motions are then executed as above explained until he cautions: Without the numbers; or commands movements other than those in the manual of arms.

Sixth. Whenever circumstances require, the regular positions of the manual of arms and the firings may be ordered without regard to the previous position of the piece.

Under exceptional conditions of weather or fatigue the rifle may be carried in any manner directed. (76)

ORDER ARMS - STANDING

FIG 30

106. Position of order arms standing: The butt rests evenly on the ground, barrel to the rear, toe of the butt on a line with toe of, and touching, the right shoe, arms and hands hanging naturally, right hand holding the piece between the thumb and fingers. (77)

PRESENT ARMS

FIG 31.

107. Being at order arms: 1. **Present,** 2. **ARMS.**
With the right hand carry the piece in front of the center of the body, barrel to the rear and vertical, grasp it with the left hand at the balance, forearm horizontal and resting against the body. **(TWO)** Grasp the small of the stock with the right hand. (78)

108. Being at order arms: 1. **Port,** 2. **ARMS.**
With the right hand raise and throw the piece diagonally across the body, grasp it smartly with both hands; the right, palm down, at the small of the stock; the left, palm up, at the balance; barrel up, sloping to the left and crossing opposite the junction of the neck with the left shoulder; right forearm horizontal; left forearm resting against the body; the piece in a vertical plane parallel to the front. (79)

109. Being at present arms: 1. **Port,** 2. **ARMS.**

Carry the piece diagonally across the body and take the position of port arms. (80)

110. Being at port arms: 1. **Present,** 2. **ARMS.**

Carry the piece to a vertical position in front of the center of the body and take the position of present arms. (81)

111. Being at present or port arms: 1. **Order,** 2. **ARMS.**

Let go with the right hand; lower and carry the piece to the right with the left hand; regrasp it with the right hand just above the lower band; let go with the left hand, and take the next to the last position in coming to the order. **(TWO)** Complete the order. (82)

112. Being at order arms: 1. **Right shoulder,** 2. **ARMS.**

With the right hand raise and throw the piece diagonally across the body; carry the right hand quickly to the butt, embracing it, the heel between the first two fingers. **(TWO)** Without changing the grasp of the right hand, place the piece on the right shoulder, barrel up and inclined at an angle of about 45° from the horizontal, trigger guard in the hollow of the shoulder, right elbow near the side, the piece in a vertical plane perpendicular to the front; carry the left hand, thumb and fingers extended and joined, to the small of the stock, tip of the forefinger touching the cocking piece, wrist straight and elbow down. **(THREE)** Drop the left hand by the side. (83)

113. Being at right shoulder arms: 1. **Order,** 2. **ARMS.**

Press the butt down quickly and throw the piece diagonally across the body, the right hand retaining the grasp of the butt. **(TWO), (THREE)** Execute order arms as described from port arms. (84)

114. Being at port arms: 1. **Right shoulder,** 2. **ARMS.**

Change the right hand to the butt. **(TWO), (THREE)** As in right shoulder arms from order arms. (85)

115. Being at right shoulder arms: 1. **Port,** 2. **ARMS.**

Press the butt down quickly and throw the piece diagonally across the body, the right hand retaining its grasp of the butt. **(TWO)** Change the right hand to the small of the stock. (86)

116. Being at right shoulder arms: 1. **Present,** 2. **ARMS.**

Execute port arms. **(THREE)** execute present arms. (87)

117. Being at present arms: 1. **Right shoulder,** 2. **ARMS.**

Execute port arms. **(TWO), (THREE), (FOUR)** Execute right shoulder arms as from port arms. (88)

118. Being at port arms: 1. **Left shoulder**, 2. **ARMS**.
Carry the piece with the right hand and place it on the left shoulder, barrel up, trigger guard in the hollow of the shoulder; at the same time grasp the butt with the left hand, heel between first and second fingers, thumb and fingers closed on the stock. **(TWO)** Drop the right hand by the side.

119. Being at left shoulder arms: 1. **Port**, 2. **ARMS**.
Grasp the piece with the right hand at the small of the stock. **(TWO)** Carry the piece to the right with the right hand, regrasp it with the left, and take the position of port arms.

Left shoulder arms may be ordered directly from the order, right shoulder or present, or the reverse. At the command **arms** execute **port arms** and continue in cadence to the position ordered. (89)

120. Being at order arms: 1. **Parade**, 2. **REST**.
Carry the right foot 6 inches straight to the rear, left knee slightly bent; carry the muzzle in front of the center of the body, barrel to the left; grasp the piece with the left hand just below the stacking swivel, and with the right hand below and against the left.

121. Being at parade rest: 1. **Squad**, 2. **ATTENTION**.
Resume the order, the left hand quitting the piece opposite the right hip. (90)

122. Being at order arms: 1. **Trail**, 2. **ARMS**.
Raise the piece, right arm slightly bent, and incline the muzzle forward so that the barrel makes an angle of about 30° with the vertical.

When it can be done without danger or inconvenience to others, the piece may be grasped at the balance and the muzzle lowered until the piece is horizontal; a similar position in the left hand may be used. (91)

[50]

123. Being at trail arms: 1. **Order**, 2. **ARMS**.
Lower the piece with the right hand and resume the order. (92)

Rifle Salute

124. Being at right shoulder arms: 1. **Rifle**, 2. **SALUTE**.
Carry the left hand smartly to the small of the stock, forearm horizontal, palm of hand down, thumb and fingers extended and joined, forefinger touching end of cocking piece; look toward the person saluted. **(TWO)** Drop left hand by the side; turn head and eyes to the front. (93)

Being at order or trail arms: 1. **Rifle**, 2. **SALUTE**.
Carry the left hand smartly to the right side, palm of the hand down, thumb and fingers extended and joined, forefinger against piece near the muzzle; look toward the person saluted. **(TWO)** Drop the left hand by the side; turn the head and eyes to the front.

For rules governing salutes, see "Honors and Salutes" (pars. 758-765). (94)

The Bayonet

125. Being at order arms: 1. **Fix**, 2. **BAYONET**.
If the bayonet scabbard is carried on the belt: Execute parade rest; grasp the bayonet with the right hand, back of hand toward the body; draw the bayonet from the scabbard and fix it on the barrel, glancing at the muzzle; resume the order.

If the bayonet is carried on the haversack: Draw the bayonet with the left hand and fix it in the most convenient manner. (95)

126. Being at order arms: 1. **Unfix**, 2. **BAYONET**.
If the bayonet scabbard is carried on the belt: Execute parade rest; grasp the handle of the bayonet firmly with the right hand, pressing the spring with the forefinger of the right hand; raise the bayonet until the handle is about 12 inches above the muzzle of the piece; drop the point to the left, back of the hand toward the body, and, glancing at the scabbard, return the bayonet, the blade passing between the left arm and the body; regrasp the piece with the right hand and resume the order.

If the bayonet scabbard is carried on the haversack: Take the bayonet from the rifle with the left hand and return it to the scabbard in the most convenient manner.

If marching or lying down, the bayonet is fixed and unfixed in the most expeditious and convenient manner and the piece returned to the original position.

Fix and **unfix** bayonet are executed with promptness and regularity but not in cadence. (96) (See Par. 337)

127. CHARGE BAYONET. Whether executed at halt or in motion, the bayonet is held toward the opponent as in the position of **guard** in the Manual for Bayonet Exercise.

Exercises for instruction in bayonet combat are prescribed in the Manual for Bayonet Exercise. (97)

The Inspection

128. Being at order arms: 1. **Inspection**, 2. **ARMS**.

At the second command take the position of port arms. **(TWO)** Seize the bolt handle with the thumb and forefinger of the right hand, turn the handle up, draw the bolt back, and glance at the chamber. Having found the chamber empty, or having emptied it, raise the head and eyes to the front. (98) (See Par. 338.)

129. Being at inspection arms: 1. **Order (Right shoulder, port)**, 2. **ARMS**.

At the preparatory command push the bolt forward, turn the handle down, pull the trigger, and resume **port arms**. At the command **arms**, complete the movement ordered. (99) (See Par. 339.)

To Dismiss the Squad

130. Being at halt: 1. **Inspection**, 2. **ARMS**, 3. **Port**, 4. **ARMS**, 5. **DISMISSED**. (100)

SCHOOL OF THE SQUAD

131. Soldiers are grouped into squads for purposes of instruction, discipline, control, and order. (101)

132. The squad proper consists of a corporal and seven privates.

The movements in the School of the Squad are designed to make the squad a fixed unit and to facilitate the control and movement of the company. If the number of men grouped is more than 3 and less than 12, they are formed as a squad of 4 files, the excess above 8 being posted as file closers. If the number grouped is greater than 11, 2 or more squads are formed and the group is termed a platoon.

For the instruction of recruits, these rules may be modified. (102)

133. The corporal is the squad leader, and when absent is replaced by a designated private. If no private is designated, the senior in length of service acts as leader.

The corporal, when in ranks, is posted as the left man in the front rank of the squad.

When the corporal leaves the ranks to lead his squad, his rear rank man steps into the front rank, and the file remains blank until the corporal returns to his place in ranks, when his rear rank man steps back into the rear rank. (103)

134. In battle officers and sergeants endeavor to preserve the integrity of squads; they designate new leaders to replace those disabled, organize new squads when necessary, and see that every man is placed in a squad.

Men are taught the necessity of remaining with the squad to which they belong and, in case it be broken up or they become separated therefrom, to attach themselves to the nearest squad and platoon leaders, whether these be of their own or of another organization. (104)

The squad executes the **halt, rests, facings, steps and marchings**, and the **manual of arms** as explained in the School of the Soldier. (105)

To Form the Squad

135. To form the squad the instructor places himself 3 paces in front of where the center is to be and commands: **FALL IN**.

The men assemble at attention, pieces at the order, and are arranged by the corporal in double rank, as nearly as practicable in order of height from right to left, each man dropping his left hand as soon as the man on his left has his interval. The rear rank forms with distance of 40 inches.

The instructor then commands: **COUNT OFF**.

At this command all except the right file execute **eyes right**, and beginning on the right, the men in each rank count **one, two, three, four**; each man turns his head and eyes to the front as he counts.

Pieces are then inspected. (106)

Alignments

136. To align the squad, the base file or files having been established: 1. **Right (Left)**, 2. **DRESS**, 3. **FRONT**.

At the command **dress** all men place the left hand upon the hip (whether dressing to the right or left); each man, except the base file, when on or near the new line executes **eyes right**, and, taking steps of 2 or 3 inches, places himself so that his right arm rests lightly against the arm of the man on his right, and so that his eyes and shoulders are in line with those of the men on his right; the rear rank men cover in file.

The instructor verifies the alignment of both ranks from the right flank and orders up or back such men as may be in rear, or in advance, of the line; only the men designated move.

At the command **front**, given when the ranks are aligned, each man turns his head and eyes to the front and drops his left hand by his side.

In the first drills the basis of the alignment is established on, or parallel to, the front of the squad; afterwards, in oblique directions.

Whenever the position of the base file or files necessitates a considerable movement by the squad, such movement will be executed by marching to the front or oblique, to the flank or backward, as the case may be, without other command, and at the trail. (107)

136a. To preserve the alignment when marching: **GUIDE RIGHT (LEFT)**.

The men preserve their intervals from the side of the guide, yielding to pressure from that side and resisting pressure from the opposite direction; they recover intervals, if lost, by gradually opening out or closing in; they recover alignment by slightly lengthening or shortening the step; the rear-rank men cover their file leaders at 40 inches.

In double rank, the front-rank man on the right, or designated flank, conducts the march; when marching faced to the flank, the leading man of the front rank is the guide. (108)

To Take Intervals and Distances

137. Being in line at a halt: 1. **Take interval**, 2. **To the right (left)**, 3. **MARCH**, 4. **Squad**, 5. **HALT**.

Being in line at a halt.

1. **Take interval**, 2. **To the right (left)**
At the second command the rear-rank men march backward 4 steps and halt;

3. MARCH
At the command **march** all face to the right and the leading man of each rank steps off; the other men step off in succession, each following the preceding man at 4 paces, rear-rank men marching abreast of their file leaders.

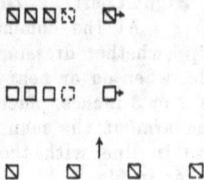

4. **Squad**, 5. **HALT**
At the command **halt**, given when all have their intervals, all halt and face to the front. (109)

[54]   FIG 40

(At Intervals)

(Assembly)

(Assembled)

137a. Being at intervals, to assemble the squad:

1. **Assemble, to the right (left)**, 2. **MARCH.**

The front-rank man on the right stands fast, the rear-rank man on the right closes to 40 inches. The other men face to the right, close by the shortest line, and face to the front. (110)

FIG 41

(At Distances)

(In Line)

FIG 42

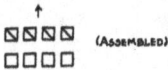

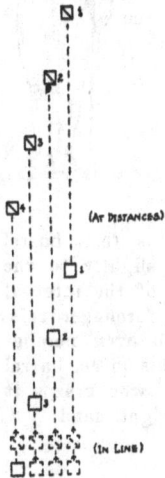

138. Being in line at a halt and having counted off: 1. **Take distance,** 2. **MARCH,** 3. **Squad,** 4. **HALT.**

At the command march No. 1 of the front rank moves straight to the front; Nos. 2, 3, and 4 of the front rank and Nos. 1, 2, 3, and 4 of the rear rank, in the order named, move straight to the front, each stepping off so as to follow the preceding man at 4 paces. The command halt is given when all have their distances.

In case more than one squad is in line, each squad executes the movement as above. The guide of each rank of numbers is right. (111)

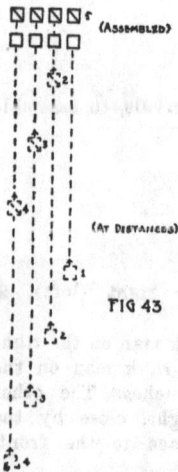

(ASSEMBLED)

(AT DISTANCES)

FIG 43

139. Being at distances, to assemble the squad: 1. **Assemble,** 2. **MARCH.**

No. 1 of the front rank stands fast; the other numbers move forward to their proper places in line. (112)

To Stack and Take Arms

140. Being in line at a halt: **STACK ARMS.** Each **even** number of the **front** rank grasps his piece with the left hand at the upper band

and rests the butt between his feet, barrel to the front, muzzle inclined slightly to the front and opposite the center of the interval on his right, the thumb and forefinger raising the stacking swivel; each **even** number of the **rear** rank then passes his piece, barrel to the rear, to his file leader, who grasps it between the bands with his right hand

and throws the butt about 2 feet in advance
of that of his own piece and opposite the
right of the interval, the right hand slipping
to the upper band, the thumb and forefinger
raising the stacking swivel, which he engages
with that of his own piece;

each **odd** number of the **front** rank raises his piece with
the right hand, carries it well forward, barrel to the front;
the left hand, guiding the stacking swivel,

engages the lower hook of the swivel of his
own piece with the free hook of that of the
even number of the rear rank; he then turns
the barrel outward into the angle formed by
the other two pieces and lowers the butt to
the ground, to the right and against the toe
of his right shoe.

The stacks made, the loose pieces are laid on them by the **even** numbers
of the front rank.

When each man has finished handling pieces, he takes the position of the soldier. (113.)

141. Being in line behind the stacks: **TAKE ARMS.**

(See preceding illustration.)

The loose pieces are returned by the **even** numbers of the **front** rank; each **even** number of the front rank grasps his own piece with the left hand, the piece of his rear rank man with his right hand, grasping both between the bands; each **odd** number of the **front** rank grasps his piece in the same way with the right hand; disengages it by raising the butt from the ground and then turning the piece to the right, detaches it from the stack; each **even** number of the front rank disengages and detaches his piece by turning it to the left,

and, then passes the piece of his rear-rank man to him,

and all resume the order. (114)

Should any squad have Nos. 2 and 3 blank files, No. 1 rear rank takes the place of No. 2 rear rank in making and breaking the stack; the stacks made or broken, he resumes his post.

Pieces not used in making the stacks are termed **loose pieces.**

Pieces are never stacked with the bayonet fixed. (115)

The Oblique March

142. For the instruction of recruits, the squad being in column or correctly aligned, the instructor causes the squad to **face half right** or **half left,** points out to the men their relative positions, and explains that these are to be maintained in the oblique march. (116)

143. I. **Right (Left) oblique,** 2. **MARCH.**

Each man steps off in a direction 45° to the right of his original

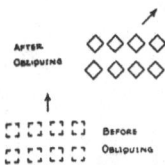

AFTER OBLIQUING

BEFORE OBLIQUING

FIG 44

front. He preserves his relative position, keeping his shoulders parallel to those of the guide (the man on the right front of the line or column), and so regulates his steps that the ranks remain parallel to their original front.

At the command **halt** the men halt faced to front.

To resume the original direction: 1. **Forward,** 2. **MARCH.**

The men half face to the left in marching and then move straight to the front.

If at **halfstep** or **mark time** while obliquing, the oblique march is resumed by the commands: 1. **Oblique,** 2. **MARCH.** (117)

To Turn on Moving Pivot

144. Being in line: 1. **Right (Left) turn,** 2. **MARCH.**

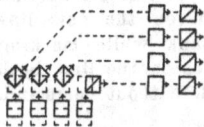

FIG 45

The movement is executed by each rank successively and on the same ground. At the second command, the pivot man of the front rank faces to the right in marching and takes the half step; the other men of the rank oblique to the right until opposite their places in line, then execute a second right oblique and take the half step on arriving abreast of the pivot man. All glance toward the marching flank while at half step and take the full step without command as the last man arrives on the line.

Right (Left) half turn is executed in a similar manner. The pivot man makes a half change of direction to the right and the other men make quarter changes in obliquing. (118)

To Turn on Fixed Pivot

145. Being in line, to turn and march: 1. **Squad right (left),** 2. **MARCH.**

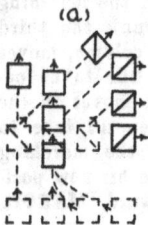

(a)

At the second command, the right flank man in the front rank faces to the right in marching and marks time; the other front rank men oblique to the right, place themselves abreast of the pivot, and mark time. In the rear rank the third man from the right, followed in column by the second and first, moves straight to the front.

(b)

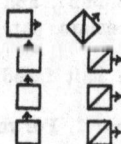

until in rear of his front-rank man,

(c)

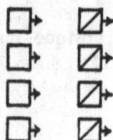

when all face to the right in marching and mark time;

See (a)

FIG 46.

the other number of the rear rank moves straight to the front four paces and places himself abreast of the man on his right. Men on the new line glance toward the marching flank while marking time and, as the last man arrives on the line, both ranks execute **forward, MARCH,** without command. (119)

146. Being in line, to turn and halt: 1. **Squad right (left),** 2. **MARCH,** 3. **Squad,** 4. **HALT.**

The third command is given immediately after the second. The turn is executed as prescribed in the preceding paragraph except that all men, on arriving on the new line, mark time until the fourth command is given, when all halt. The fourth command should be given as the last man arrives on the line. (120)

147. Being in line, to turn about and march: 1. **Squad right (left) about,** 2. **MARCH.**

At the second command, the front rank twice executes **squad right,** initiating the second **squad right** when the man on the marching flank has arrived abreast of the rank. In the rear rank the third man from the right, followed by the second and first in column, moves straight to the front until on the prolongation of the line to be occupied by the rear rank; changes direction to the right; moves in the new direction until in rear of his front-rank man, when all face to the right in marching, mark time, and glance toward the marching flank. The fourth man marches on the left of the third to his new position; as he arrives on the line, both ranks execute **forward, MARCH,** without command. (121)

148. Being in line, to turn about and halt: 1. **Squad right (left) about,** 2. **MARCH,** 3. **Squad,** 4. **HALT.**

The third command is given immediately after the second. The turn is executed as prescribed in the preceding paragraph except that all men, on arriving on the new line, mark time until the fourth command is given, when all halt. The fourth command should be given as the last man arrives on the line. (122)

To Follow the Corporal

149. Being assembled or deployed, to march the squad without unnecessary commands, the corporal places himself in front of it and commands: **FOLLOW ME.**

If in line or skirmish line, No. 2 of the front rank follows in the trace of the corporal at about 3 paces; the other men conform to the movements of No. 2, guiding on him and maintaining their relative positions.

If in column, the head of the column follows the corporal. (123)

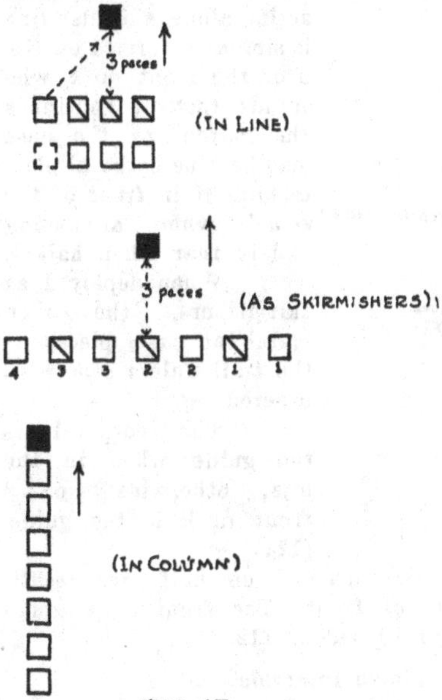

FIG 47

To Deploy as Skirmishers

150. Being in any formation, assembled: 1. **As skirmishers,** 2. **MARCH.**

The corporal places himself in front of the squad, if not already there. Moving at a run, the men place themselves abreast of the corporal at half-pace intervals, Nos. 1

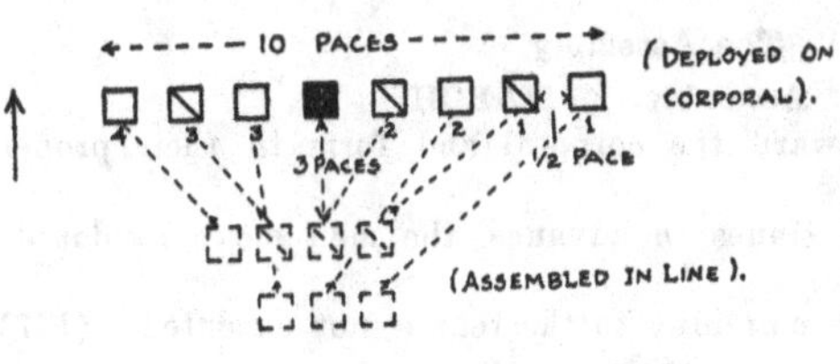

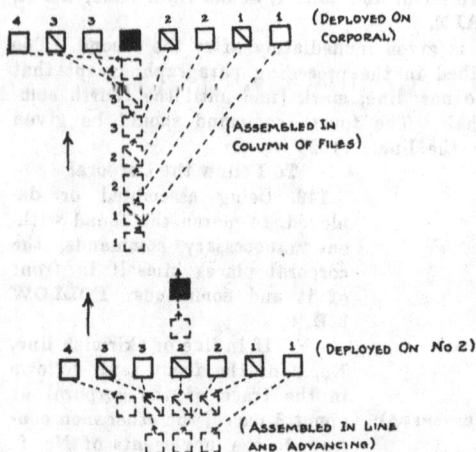

FIG 48.

and 2 on his right, Nos 3 and 4 on his left, rear-rank men on the right of their file leaders, extra men on the left of No. 1, all then conform to the corporal's gait.

When the squad is acting alone, skirmish line is similarly formed on No. 2 of the front rank, who stands fast or continues the march, as the case may be; the corporal plac-es himself in front of the squad when advancing and in rear when halted.

When deployed as skirmishers, the men march at ease, pieces at the trail unless otherwise ordered.

The corporal is the guide when in the line; otherwise No. 2 front rank is the guide. (124)

151. The normal interval between skirmishers is one-half pace, result-ing practically in one man per yard of front. The front of a squad thus deployed as skirmishers is about 10 paces. (125)

To Increase or Diminish Intervals

152. If assembled, and it is desired to deploy at greater than the normal interval; or if deployed, and it is desired to increase or decrease the interval: 1. As skirmishers, (so many) paces, 2. MARCH.

Intervals are taken at the indicated number of paces. If al-ready deployed, the men move by the flank toward or away from the guide. (126)

The Assembly

153. Being deployed: 1. Assemble. 2. MARCH.

The men move toward the corporal and form in their proper places.

If the corporal continues to advance, the men move in double time, form, and follow him.

The assembly while marching to the rear is not executed. (127)

[62]

Kneeling and Lying Down

154. If standing: **KNEEL.**

Half face to the right; carry the right toe about 1 foot to the left rear of the left heel;

kneel on right knee, sitting as nearly as possible on the right heel; left forearm across left thigh; piece remains in position of order arms, right hand grasping it above lower band. (128)

155. If standing or kneeling: **LIE DOWN.**

Kneel, but with right knee against left heel:

carry back the left foot and lie flat on the belly, inclining body about 35° to the right

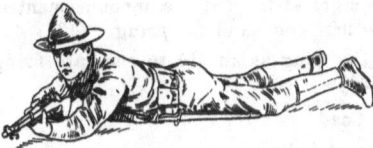

piece horizontal, barrel up, muzzle off the ground and pointed to the front; elbows on the ground; left hand at the balance, right hand grasping the small of the stock opposite the neck. This is the position of order arms, lying down. (129)

156. If kneeling or lying down: **RISE**.

If kneeling, stand up, faced to the front, on the ground marked by the left heel.

If lying down, raise body on both knees; stand up, faced to the front, on the ground marked by the knees. (130)

157. If lying down: **KNEEL**.

Raise the body on both knees; take the position of kneel. (131)

158. In double rank, the positions of kneeling and lying down are ordinarily used only for the better utilization of cover.

When deployed as skirmishers, a sitting position may be taken in lieu of the position kneeling. (132)

Loadings and Firings

159. The commands for loading and firing are the same whether standing, kneeling, or lying down. The firings are always executed at a halt.

When kneeling or lying down in double rank, the rear rank does not load, aim, or fire.

The instruction in firing will be preceded by a command for loading.

Loadings are executed in line and skirmish line only. (133)

160. Pieces having been ordered loaded are kept loaded without command until the command **unload**, or **inspection arms**, fresh clips being inserted when the magazine is exhausted. (134) (See Par. 340.)

160a. The aiming point or target is carefully pointed out. This may be done before or after announcing the sight setting. Both are indicated before giving the command for firing, but may be omitted when the target appears suddenly and is unmistakable; in such case battle sight is used if no sight setting is announced. (135)

161. The target or aiming point having been designated and the sight setting announced, such designation or announcement need not be repeated until a change of either or both is necessary.

Troops are trained to continue their fire upon the aiming point or target designated, and at the sight setting announced, until a change is ordered. (136)

162. If the men are not already in the position of load, that position is taken at the announcement of the sight setting; if the announcement is omitted, the position is taken at the first command for firing. (137)

163. When deployed, the use of the sling as an aid to accurate firing is discretionary with each man. (138)

To Load

164. Being in line or skirmish line at halt:
1. **With dummy (blank or ball) cartridges**, 2. **LOAD**.

[64]

At the command **load** each front-rank man or skirmisher faces half right and carries the right foot to the right, about 1 foot, to such position as will insure the greatest firmness and steadiness of the body; raises, or lowers, the piece and drops it into the left hand at the balance, the left thumb extended along the stock, muzzle at the height of the breast, and turns the cut-off up.

With the right hand he turns and draws the bolt back,

takes a loaded clip and inserts the end in the clip slots, places the thumb on the powder space of the top cartridge, the fingers extending around the piece and tips resting on the magazine floor plate;

forces the cartridges into the magazine by pressing down with the thumb; without removing the clip, thrusts the bolt home, turning down the handle; turns the safety lock to the "safe,"

[65]

164 (cont.)

and carries the hand to the small of the stock

Each rear rank man moves to the right front, takes a similar position opposite the interval to the right of his front rank man, muzzle of the piece extending beyond the front rank and loads.

A skirmish line may load while moving, the pieces being held as nearly as practicable in the position of load.

If kneeling or sitting, the position of the piece is similar; if kneeling, the left forearm rests on the left thigh;

if sitting the elbows are supported by the knees.

If lying down, the left hand steadies and supports the piece at the balance, the toe of the butt resting on the ground, the muzzle off the ground.

For reference, these positions (standing, kneeling, and lying down) are designated as that of **load**. (139). (See Par. 341.)

165. For instruction in loading: 1. **Simulate**, 2. **LOAD**.

Executed as above described except that the cut-off remains "off" and the handling of cartridges is simulated.

The recruits are first taught to **simulate** loading and firing; after a few lessons dummy cartridges may be used. Later, blank cartridges may be used. (140)

The rifle may be used as a single loader by turning the magazine "off." The magazine may be filled in whole or in part while "off" or "on" by pressing cartridges singly down and back until they are in the proper place. The use of the rifle as a single loader is, however, to be regarded as exceptional. (141) (See 342.)

To Unload

166. **UNLOAD.**

Take the position of load, turn the safety lock up and move bolt alternately back and forward until all the cartridges are ejected. After the last cartridge is ejected the chamber is closed by first thrusting the bolt slightly forward to free it from the stud holding it in place when the chamber is open, pressing the follower down and back to engage it under the bolt and then thrusting the bolt home; the trigger is pulled. The cartridges are then picked up, cleaned, and returned to the belt and the piece is brought to the order. (142) (See 343.)

To Set the Sight

167. **RANGE, ELEVEN HUNDRED (EIGHT-FIFTY, etc.), or BATTLE SIGHT.**

The sight is set at the elevation indicated. The instructor explains and verifies sight settings. (143)

To Fire by Volley

169. 1. **Ready,** 2. **AIM,** 3. **Squad,** 4. **FIRE.**

At the command **ready** turn the safety lock to the "ready";

at the command **aim** raise the piece with both hands and support the butt firmly against the hollow of the right shoulder, right thumb clasping the stock, barrel horizontal, left elbow well under the piece, right elbow as high as the shoulder; incline the head slightly forward and a little to the right, cheek against the stock,

left eye closed, right eye looking through the notch of the rear sight so as to perceive the object aimed at, second joint of the forefinger resting lightly against the front of the trigger and taking up the slack; top of front sight is carefully raised into, and held in, the line of sight.

Each rear-rank man aims through the interval to the right of his file leader and leans slightly forward to advance the muzzle of his piece beyond the front rank.

In aiming kneeling, the left elbow rests on the left knee, point of elbow in front of kneecap.

In aiming sitting, the elbows are supported by the knees.

In aiming, lying down, raise the piece with both hands; rest on both elbows and press the butt firmly against the right shoulder.

[69]

At the command **fire** press the finger against the trigger; fire without deranging the aim and without lowering or turning the piece; lower the piece in the position of **load** and load. (144)

To continue the firing: 1. **AIM**, 2. **Squad**, 3. **FIRE**.

Each command is executed as previously explained. **Load** (from magazine) is executed by drawing back and thrusting home the bolt with the right hand, leaving the safety lock at the "ready." (145)

To Fire at Will

169. FIRE AT WILL.

Each man, independently of the others, comes to the **ready**, aims carefully and deliberately at the aiming point or target, **fires, loads,** and continues the firing until ordered to **suspend** or **cease firing**. (146)

170. To increase (decrease) the rate of fire in progress the instructor shouts: **FASTER (SLOWER).**

Men are trained to fire at the rate of about three shots per minute at effective ranges and five or six at close ranges, devoting the minimum of time to loading and the maximum to deliberate aiming. To illustrate the necessity for deliberation, and to habituate men to combat conditions, small and comparatively indistinct targets, are designated. (147)

To Fire by Clip

171. CLIP FIRE.

Executed in the same manner as **fire at will**, except that each man, after having exhausted the cartridges then in the piece, **suspends firing.** (148) (See par. 344.)

To Suspend Firing

172. The instructor blows a **long blast** of the whistle and repeats same, if necessary, or commands: **SUSPEND FIRING.**

Firing stops; pieces are held, loaded and locked, in a position of readiness for instant resumption of firing, rear sights unchanged. The men continue to observe the target or aiming point, or the place at which the target disappeared, or at which it is expected to reappear.

This whistle signal may be used as a preliminary to **cease firing.** (149)

To Cease Firing

173. CEASE FIRING.

Firing stops; pieces not already there are brought to the position of load; those not loaded, are loaded; sights are laid, pieces are locked and brought to the order.

Cease firing is used for long pauses, to prepare for changes of position, or to steady the men. (150) (See par. 345.)

Commands for suspending or ceasing fire may be given at any time after the preparatory command for firing whether the firing has actually commenced or not. (151)

The Use Of Cover

174. The recruit should be given careful instruction in the individual use of cover.

It should be impressed upon him that, in taking advantage of natural cover, he must be able to fire easily and effectively upon the enemy; if advancing on an enemy, he must do so steadily and as rapidly as possible; he must conceal himself as much as possible while firing and while advancing. While setting his sight he should be under cover or lying prone. (152)

To teach him to fire easily and effectively, at the same time concealing himself from the view of the enemy, he is practiced in simulated firing in the prone, sitting, kneeling, and crouching positions, from behind hillocks, trees, heaps of earth or rocks, from depressions, gullies, ditches, doorways, or windows. He is taught to fire around the right side of his concealment whenever possible, or, when this is not possible, to rise enough to fire over the top of his concealment.

When these details are understood, he is required to select cover with reference to an assumed enemy and to place himself behind it in proper position for firing. (153)

The evil of remaining too long in one place, however good the concealment, should be explained. He should be taught to advance from cover to cover, selecting cover in advance before leaving his concealment.

It should be impressed upon him that a man running rapidly toward an enemy furnishes a poor target. He should be trained in springing from a prone position behind concealment, running at top speed to cover and throwing himself behind it. He should also be practiced in advancing from cover to cover by crawling, or by lying on the left side, rifle grasped in the right hand, and pushing himself forward with the right leg. (154)

He should be taught that, when fired on while acting independently, he should drop to the ground, seek cover, and then endeavor to locate his enemy. (155)

The instruction of the recruit in the use of cover is continued in the combat exercises of the company, but he must then be taught that the proper advance of the platoon or company and the effectiveness of its fire is of greater importance than the question of cover for individuals. He should also be taught that he may not move about or shift his position in the firing line except the better to see the target. (156)

Observation

175. The ability to use his eyes accurately is of great importance to the soldier. The recruit should be trained in observing his surrounding from positions and when on the march

He should be practiced in pointing out and naming military features of the ground; in distinguishing between living beings; in counting distant groups of objects or beings; in recognizing colors and forms. (157)

170. In the training of men in the mechanism of the firing line, they should be practiced in repeating to one another target and aiming point designations and in quickly locating and pointing out a designated target. They should be taught to distinguish, from a prone position, distant objects, particularly troops, both with the naked eye and with field glasses. Similarly, they should be trained in estimating distances. (158)

SCHOOL OF THE COMPANY

177. The captain is responsible for the theoretical and practical instruction of his officers and noncommissioned officers, not only in the duties of their respective grades, but in those of the next higher grades. (159)

178. The company in line is formed in double rank with the men arranged, as far as practicable, according to height from right to left, the tallest on the right.

The original division into squads is effected by the command: **COUNT OFF.** The squads, successively from the right, count off as in the School of the Squad, corporals placing themselves as Nos. 4 of the front rank. If the left squad contains less than six men, it is either increased to that number by transfers from other squads or is broken up and its members assigned to other squads and posted in the line of file closers. These squad organizations are maintained, by transfers if necessary, until the company becomes so reduced in numbers as to necessitate a new division into squads. No squad will contain less than six men. (160)

179. The company is further divided into two, three or four platoons, each consisting of not less than two nor more than four squads. In garrison or ceremonies the strength of platoons may exceed four squads. (161)

At the formation of the company the platoons or squads are numbered consecutively from right to left and these designations do not change.

For convenience in giving commands and for reference, the designations, **right, center, left,** when in line, and **leading, center, rear,** when in column, are applied to platoons or squads. These designations apply to the actual right, left, center, head, or rear, in whatever direction the company may be facing. The **center squad** is the middle or right middle squad of the company.

The designation "So-and-so's" squad or platoon may also be used. (162)

180. Platoons are assigned to the lieutenants and noncommissioned officers, in order of rank, as follows: 1, right; 2, left; 3, center (right center); 4, left center.

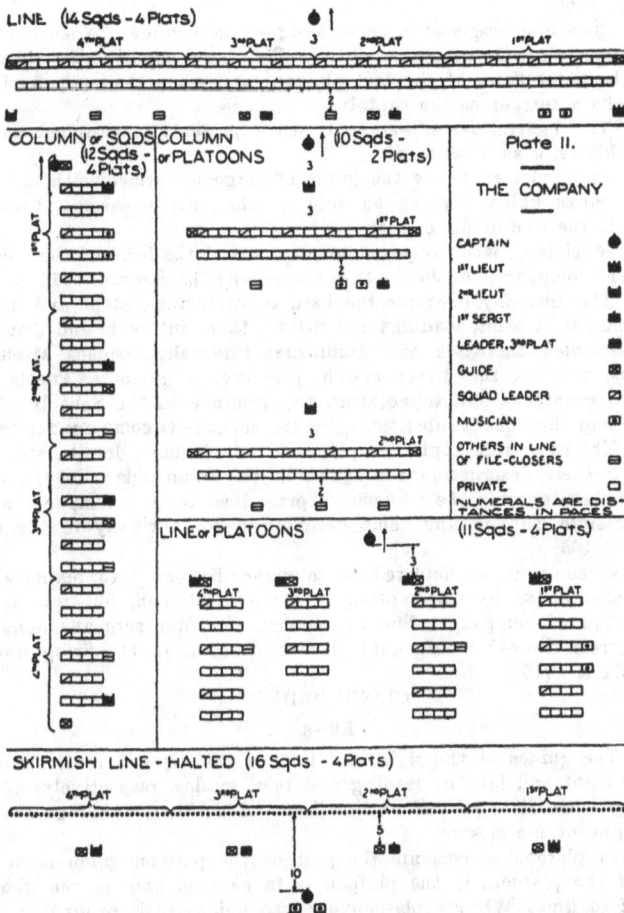

LINE (14 Sqds - 4 Plats)

COLUMN or SQDS (12 Sqds - 4 Plats)

COLUMN or PLATOONS (10 Sqds - 2 Plats)

Plate II.

THE COMPANY

CAPTAIN	
1ST LIEUT	
2ND LIEUT	
1ST SERGT	
LEADER, 3RD PLAT	
GUIDE	
SQUAD LEADER	
MUSICIAN	
OTHERS IN LINE OF FILE-CLOSERS	
PRIVATE	

NUMERALS ARE DISTANCES IN PACES

LINE or PLATOONS (11 Sqds - 4 Plats)

SKIRMISH LINE - HALTED (16 Sqds - 4 Plats)

The noncommissioned officers next in rank are assigned as guides, one to each platoon. If sergeants still remain, they are assigned to platoons as additional guides. When the platoon is deployed, its guide, or guides, accompany the platoon leader.

During battle, these assignments are not changed; vacancies are filled by noncommissioned officers of the platoon, or by the nearest available officers or noncommissioned officers arriving with reënforcing troops. (163)

181. The first sergeant is never assigned as a guide. When not commanding a platoon, he is posted as a file closer opposite the third file from the outer flank of the first platoon; and when the company is deployed he accompanies the captain.

The quartermaster sergeant, when present, is assigned according to his rank as a sergeant.

Enlisted men below the grade of sergeant, armed with the rifle, are in ranks unless serving as guides; when not so armed, they are posted in the line of file closers.

Musicians, when required to play, are at the head of the column. When the company is deployed, they accompany the captain. (164)

182. The company executes the **halt, rests, facings, steps** and **marchings, manual of arms, loadings** and **firings, takes intervals** and **distances** and **assembles, increases** and **diminishes intervals,** resumes **attention, obliques,** resumes the direct march, preserves alignments, **kneels, lies down, rises, stacks** and **takes arms,** as explained in the Schools of the Soldier and the Squad, substituting in the commands **company** for **squad.**

The same rule applies to platoons, detachments, details, etc., substituting their designation for **squad** in the commands. In the same manner these execute the movements prescribed for the company, whenever possible, substituting their designation for **company** in the commands. (165)

183. A company so depleted as to make division into platoons impracticable is led by the captain as a single platoon, but retains the designation of company. The lieutenants and first sergeant assist in fire control; the other sergeants place themselves in the firing line as skirmishers. (166)

CLOSE ORDER
Rules

184. The guides of the right and left, or leading and rear, platoons, are the right and left, or leading and rear, guides, respectively, of the company when it is in line or in column of squads. Other guides are in the line of file closers.

In platoon movements the post of the platoon guide is at the head of the platoon, if the platoon is in column, and on the guiding flank if in line. When a platoon has two guides their original assignment to flanks of the platoon does not change. (167)

The guides of a column of squads place themselves on the flank opposite the file closers. To change the guides and file closers to the other flank, the captain commands: 1. **File closers on left (right) flank;**

2. **MARCH.** The file closers dart through the column; the captain and guides change.

In column of squads, each rank preserves the alignment toward the side of the guide. (168)

185. Men in the line of file closers do not execute the loadings or firings.

Guides and enlisted men in the line of file closers execute the manual of arms during the drill unless specially excused, when they remain at the order. During ceremonies they execute all movements. (169)

186. In **taking intervals and distances,** unless otherwise directed, the right and left guides, at the first command, place themselves in the line of file closers, and, with them, take a distance of 4 paces from the rear rank. In taking intervals, at the command **march,** the file closers face to the flank and each steps off with the file nearest him. In **assembling** the guides and file closers resume their positions in line. (170)

187. In movements executed simultaneously by platoons (as **platoons right** or **platoons, column right),** platoon leaders repeat the preparatory command **(platoon right,** etc.), applicable to their respective platoons. The command of execution is given by the captain only. (171)

To Form the Company

188. At the sounding of the assembly the first sergeant takes position 6 paces in front of where the center of the company is to be, faces it, draws saber, and commands: **FALL IN.**

The right guide of the company places himself, facing to the front, where the right of the company is to rest, and at such point that the center of the company will be 6 paces from and opposite the first sergeant; the squads form in their proper places on the left of the right guide, superintended by the other sergeants, who then take their posts.

The first sergeant commands: **REPORT.** Remaining in position at the order, the squad leaders, in succession from the right, salute and report: **All present;** or, **Private(s) —— absent.** The first sergeant does not return the salutes of the squad leaders; he then commands: 1. **Inspection,** 2. **ARMS,** 3. Order, 4. **ARMS,** faces about, salutes the captain, reports: **Sir, all present or accounted for,** or the names of the unauthorized absentees, and, without command, takes his post.

If the company can not be formed by squads, the first sergeant commands: 1. **Inspection,** 2. **ARMS,** 3. **Right shoulder,** 4. **ARMS,** and calls the roll. Each man, as his name is called, answers **here** and executes order arms. The sergeant then effects the division into squads and reports the company as prescribed above.

[75]

The captain places himself 12 paces in front of the center of, and facing, the company in time to receive the report of the first sergeant, whose salute he returns, and then draws saber.

The lieutenants take their posts when the first sergeant has reported and draw saber with the captain. The company, if not under arms, is formed in like manner omitting reference to arms. (172)

189. For the instruction of platoon leaders and guides, the company, when small, may be formed in single rank. In this formation close order movements only are executed. The single rank executes all movements as explained for the front rank of a company. (173)

To Dismiss the Company

190. Being in line at a halt, the captain directs the first sergeant: **Dismiss the company.** The officers fall out; the first sergeant places himself faced to the front, 3 paces to the front and 2 paces from the nearest flank of the company, salutes, faces toward opposite flank of the company and commands: 1. Inspection, 2. **ARMS**, 3. Port, 4. **ARMS**, 5. **DISMISSED**. (174)

Alignments

191. The alignments are executed as prescribed in the School of the Squad, the guide being established instead of the flank file. The rear-rank man of the flank file keeps his head and eyes to the front and covers his file leader.

At each alignment the captain places himself in prolongation of the line, 2 paces from and facing the flank toward which the dress is made, verifies the alignment, and commands: **FRONT**.

Platoon leaders take a like position when required to verify alignments. (175)

Movements on the Fixed Pivot

192. Being in line, to turn the company: 1. **Company right (left)**, 2. **MARCH**, 3. Company, 4. **HALT**; or, 3. Forward, 4. **MARCH**.

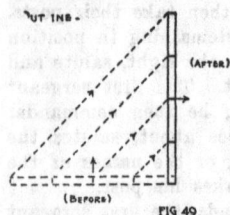

(AFTER)

(BEFORE)

FIG 49

For detail, see Fig. 46.

At the second command the right-flank man in the front rank faces to the right in marching and marks time; the other front-rank men oblique to the right, place themselves abreast of the pivot, and mark time; in the rear rank the third man from the right, followed in column by the second and first, moves straight to the front until in rear of his front-rank man, when all face to the right in marching and mark time; the remaining men of the rear rank move straight to the front 4 paces, oblique to the right, place themselves abreast of the third man, cover their file leaders, and mark time, the right guide steps back, takes post on the flank, and marks time.

The fourth command is given when the last man is 1 pace in rear of the new line.

The command **halt** may be given at any time after the movement begins; only those halt who are in the new position. Each of the others halts upon arriving on the line, aligns himself to the right, and executes front without command. (176)

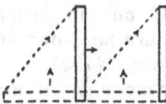

FROM LINE TO COLUMN OF PLATOONS

FIG 50

193. Being in line, to form column of platoons, or the reverse: 1. **Platoons right (left)**, 2. **MARCH**, 3. **Company**, 4. **HALT**; or, 3. **Forward**, 4. **MARCH**.

Executed by each platoon as described for the company.

Before forming line the captain sees that the guides on the flank toward which the movement is to be executed are covering. This is effected by previously announcing the guide to that flank. (177)

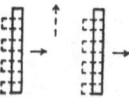

FROM LINE TO COLUMN OF SQUADS

FIG 51

FROM LINE OF PLATOONS TO COLUMN OF PLATOONS

194. Being in line, to form column of squads, or the reverse; or, being in line of platoons, to form column of platoons, or the reverse: 1. **Squads right (left)**, 2. **MARCH**; or, 1. **Squads right (left)**, 2. **MARCH**, 3. **Company**, 4. **HALT**.

Executed by each squad as described in the School of the Squad.

If the company or platoons be formed in line toward the side of the file closers, they dart through the column and take posts in rear of the company at the second command. If the column of squads be formed from line, the file closers take posts on the pivot flank, abreast of and 4 inches from the nearest rank. (178)

Movements on the Moving Pivot

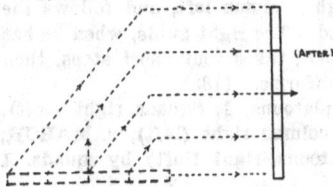

- OUTLINE -

(AFTER)

(BEFORE)   FIG 52

195. Being in line, to change direction: 1. **Right (Left) turn**, 2. **MARCH**, 3. **Forward**, 4. **MARCH**.

Executed as described in the School of the Squad, except that the men do not glance toward the marching flank and that all take the full step at the fourth command. The right guide is the pivot of the front rank. Each rear-rank man obliques on the same ground as his file leader. (179)

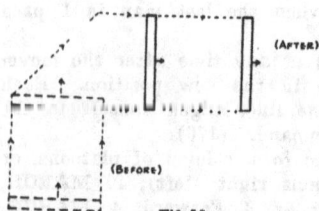

FIG 53

196. Being in column of platoons, to change direction: 1. **Column right (left).** 2. **MARCH.**

At the first command the leader of the leading platoon commands: **Right turn.** At the command **march** the leading platoon turns to the right on moving pivot; its leader commands: 1. **Forward,** 2. **MARCH,** on completion of the turn. Rear platoons march squarely up to the turning point of the leading platoon and turn at command of their leaders. (180)

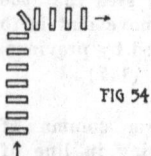

FIG 54

197. Being in column of squads, to change direction: 1. **Column right (left), 2. MARCH.**

At the second command the front rank of the leading squad turns to the right on moving pivot as in the School of the Squad; the other ranks, without command, turn successively on the same ground and in a similar manner. (181)

198. Being in column of squads, to form line of platoons or the reverse: 1. **Platoons, column right (left), 2. MARCH.**

Executed by each platoon as described for the company. (182)

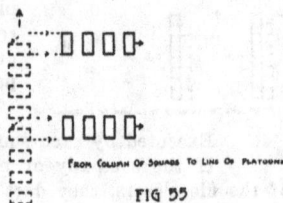

FROM COLUMN OF SQUADS TO LINE OF PLATOONS

FIG 55

199. Being in line, to form column of squads and change direction: 1. **Squads right (left), column right (left), 2. MARCH;** or, 1. **Right (Left) by squads, 2. MARCH.**

In the first case the right squad initiates the **column right** as soon as it has completed the **squad right.**

SQUADS RIGHT COLUMN RIGHT

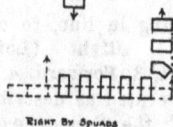

FIG 56

RIGHT BY SQUADS

In the second case, at the command **march,** the right squad marches **forward;** the remainder of the company executes **squads right, column left,** and follows the right squad. The right guide, when he has posted himself in front of the right squad, takes four short steps, then resumes the full step; the right squad conforms. (183)

200. Being in line, to form line of platoons: 1, **Squads right (left), platoons, column right (left), 2. MARCH;** or, 1. **Platoons, right (left) by squads, 2. MARCH.**

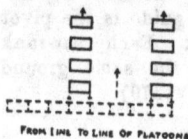

FIG 57

FROM LINE TO LINE OF PLATOONS

Executed by each platoon as described for the company in the preceding paragraph. (184)

[78]

Facing or Marching to the Rear

201. Being in line, line of platoons, or in column of platoons or squads, to face or march to the rear: 1. **Squads right (left) about**, 2. **MARCH**; or, 1. **Squads right (left) about**, 2. **MARCH**; 3. **Company**, 4. **HALT**.

Executed by each squad as described in the School of the Squad.

If the company or platoons be in column of squads, the file closers turn about toward the column, and take their posts; if in line, each darts through the nearest interval between squads. (185)

202. To march to the rear for a few paces: 1. **About**, 2. **FACE**, 3. **Forward**, 4. **MARCH**.

If in line, the guides place themselves in the rear rank, now the front rank; the file closers, on facing about, maintain their relative positions. No other movement is executed until the line is faced to the original front. (186)

On Right (Left) Into Line

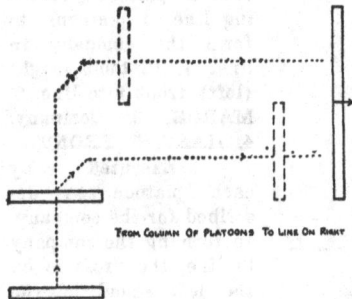

From Column Of Platoons To Line On Right

From Column Of Squads To Line On Right

FIG 57

203. Being in column of platoons or squads, to form line on right or left: 1. **On right (left) into line**, 2. **MARCH**, 3. **Company**, 4. **HALT**, 5. **FRONT**.

At the first command the leader of the leading unit commands: **Right turn.** The leaders of the other units command: **Forward**, if at a halt. At the second command the leading unit turns to the right on moving pivot. The command **halt** is given when the leading unit has advanced the desired distance in the new direction; it halts; its leader then commands: **Right dress.**

The units in rear continue to march straight to the front; each, when opposite the right of its place in line, executes **right turn** at the command of its leader; each is halted on the line at the command of its leader, who then commands: **Right dress.** All dress on the first unit in line.

If executed in double time, the leading squad marches in double time until halted. (187)

[79]

Front Into Line

204. Being in column of platoons or squads, to form line to the front: 1. **Right (Left) front into line,** 2. **MARCH,** 3. **Company,** 4. **HALT,** 5. **HALT.**

At the first command the leaders of the units in rear of the leading one command: **Right oblique.** If at a halt, the leader of the leading unit commands: **Forward.** At the second command the leading unit moves straight forward; the rear units oblique as indicated. The command halt is given when the leading unit has advanced the desired distance; it halts; its leader then commands: **Left dress.** Each of the rear units, when opposite its place in line, resumes the original direction at the command of its leader; each is halted on the line at the command of its leader, who then commands: **Left dress.** All dress on the first unit in line. (188)

205. Being in column of squads to form column of platoons, or being line of platoons, to form the company in line: 1. **Platoons, right (left) front into line,** 2. **MARCH,** 3. **Company,** 4. **HALT,** 5. **FRONT.**

Executed by each platoon as described for the company. In forming the company in line, the dress is on the left squad of the left platoon. If forming column of platoons, platoon leaders verify the alignment before taking their posts; the captain commands **front** when the alignments have been verified.

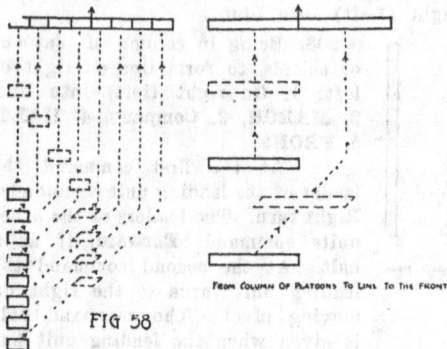

FIG 58

FROM COLUMN OF PLATOONS TO LINE TO THE FRONT

FROM COLUMN OF SQUADS TO LINE TO THE FRONT

When **front into line** is executed in double time the commands for halting and aligning are omitted and the guide is toward the side of the first unit in line. (189)

At Ease and Route Step

206. The column of squads is the habitual column of route, but **route step** and **at ease** are applicable to any marching formation. (190)

To march at route step: 1. **Route step,** 2. **MARCH.**

Sabers are carried at will or in the scabbard; the men carry their pieces at will, keeping the muzzles elevated; they are not required to preserve silence, nor to keep the step. The ranks cover and preserve their distance. If halted from route step, the men stand **at rest.** (191)

To march at ease: 1. **At ease**, 2. **MARCH**.

The company marches as in route step, except that silence is preserved; when halted, the men remain at ease. (192)

Marching at route step or at ease: 1. **Company**, 2. **ATTENTION**.

At the command **attention** the pieces are brought to the right shoulder and the cadenced step in quick time is resumed. (193)

To Diminish The Front of A Column of Squads

207. Being in column of squads: 1. **Right (left) by twos**, 2. **MARCH**.

At the command **march** all files except the two right files of the leading squad execute **in place halt**; the two left files of the leading squad oblique to the right when disengaged and follow the right files at the shortest practicable distance. The remaining squads follow successively in like manner. (194)

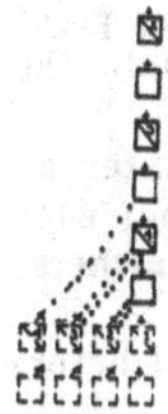

FIG 59

208. Being in column of squads or twos: 1. **Right (left) by file**, 2. **MARCH**.

At the command **march**, all files execute **in place halt** except the right file of the leading two or squad. The left file or files of the leading two or squad oblique successively to the right when disengaged and each follows the file on its right at the shortest practicable distance. The remaining twos or squads follow successively in like manner. (195)

FIG 60

Being in column of files or twos, to form column of squads; or, being in column of files, to form column of twos: 1. **Squads (Twos), right (left) front into line**, 2. **MARCH**.

At the command **march**, the leading file or files halt. The remainder of the squad, or two, obliques to the right and halts on line with the leading file or files. The remaining squads or twos close up and successively form in rear of the first in like manner.

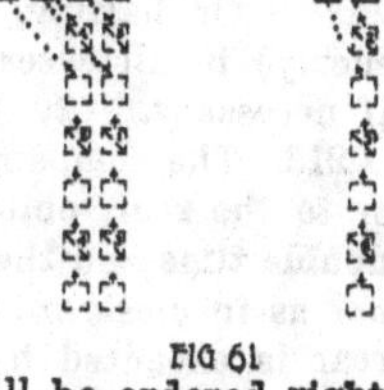

FIG 61

The movement described in this paragraph will be ordered **right** or **left**, so as to restore the files to their normal relative positions in the two or squad. (196)

The movements prescribed in the three preceding paragraphs are difficult of execution at attention and have no value as disciplinary exercises. (197)

Marching by twos or files can not be executed without serious delay and waste of road space. Every reasonable precaution will be taken to obviate the necessity for these formations. (198)

EXTENDED ORDER
Rules for Deployment

209. The command **guide right (left** or **center)** indicates the base squad for the deployment; if in line it designates the actual **right (left** or **center)** squad; if in column the command **guide right (left)** designates the **leading** squad, and the command **guide center** designates the **center** squad. After the deployment is completed, the guide is **center** without command, unless otherwise ordered. (199)

210. At the preparatory command for forming skirmish line, from either column of squads or line, each squad leader (except the leader of the base squad, when his squad does not advance), cautions his squad, **follow me** or **by the right (left) flank**, as the case may be; at the command **march**, he steps in front of his squad and leads it to its place in line. (200)

211. Having given the command for forming skirmish line, the captain, if necessary, indicates to the corporal of the base squad the point on which the squad is to march; the corporal habitually looks to the captain for such directions. (201)

The base squad is deployed as soon as it has sufficient interval The other squads are deployed as they arrive on the general line; each corporal halts in his place in line and commands or signals, **as skirmishers**; the squad deploys and halts abreast of him.

If tactical considerations demand it, the squad is deployed before arriving on the line. (202)

212. Deployed lines preserve a general alignment toward the guide. Within their respective fronts, individuals or units march so as best to secure cover or to facilitate the advance, but the general and orderly progress of the whole is paramount.

On halting, a deployed line faces to the front (direction of the enemy) in all cases and takes advantage of cover, the men lying down if necessary. (203)

213. The company in skirmish line **advances, halts** moves **by the flank,** or **to the rear, obliques,** resumes **the direct march,** passes from **quick** to **double time** and the reverse by the same commands and in a similar manner as in close order; if at a halt, the movement **by the flank** or **to the rear** is executed by the same commands as when marching. **Company right (left, half right, half left)** is executed as explained for the front rank, skirmish intervals being maintained. (204)

214. A platoon or other part of the company is deployed and marched in the same manner as the company, substituting in the commands, **platoon (detachment,** etc.**) for company.** (205)

Deployments

215. Being in line, to form skirmish line to the front: 1. **As skirmishers, guide right (left or center), 2. MARCH.**

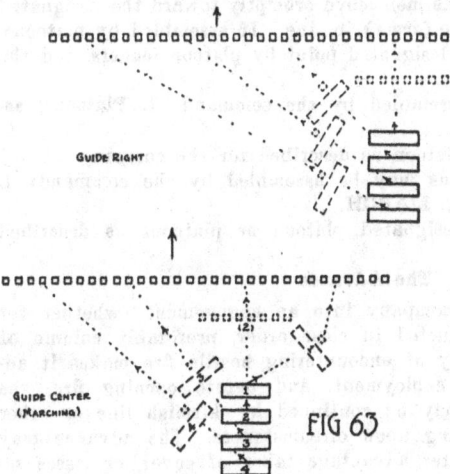

If marching, the corporal of the base squad moves straight to the front; when that squad has advanced the desired distance, the captain commands: 1. **Company,** 2. **HALT.** If the guide be **right (left),** the other corporals move to the **left (right)** front, and, in succession from the base, place their squads on the line; if the guide be center, the other corporals move to the **right** or **left** front, according as they are on the right or left of the center squad, and in succession from the center squad place their squads on the line.

FIG 62

If at a halt, the base squad is deployed without advancing; the other squads may be conducted to their proper places by the flank; interior squads may be moved when squads more distant from the base have gained comfortable marching distance. (206)

216. Being in column of squads, to form skirmish line to the front: 1. **As skirmishers, guide right (left or center), 2. MARCH.**

GUIDE RIGHT

GUIDE CENTER (MARCHING)

FIG 63

If marching, the corporal of the base squad deploys it and moves straight to the front; if at a halt, he deploys his squad without advancing. If the guide be **right (left),** the other corporals move to the **left (right) front,** and, in succession from the base, place their squads on the line; if the guide be **center,** the corporals in front of the center squad move to the right (if at a halt, to the right rear), the corporals

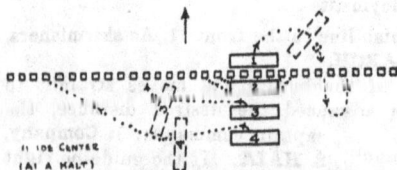

in rear of the center squad move to the left front, and each, in succession from the base, places his squad on the line.

The column of twos or files is deployed by the same commands and in like manner. (207)

217. The company in line or in column of squads may be deployed in an oblique direction by the same commands. The captain points out the desired direction; the corporal of the base squad moves in the direction indicated; the other corporals conform. (208)

218. To form skirmish line to the flank or rear the line or the column of squads is turned by squads to the flank or rear and then deployed as described. (209)

219. The intervals between men are increased or decreased as described in the School of the Squad, adding to the preparatory command, **quide right (left or center)** if necessary. (210)

The Assembly

220. The captain takes his post in front of, or designates, the element on which the company is to assemble and commands: 1. **Assemble,** 2. **MARCH.**

If in skirmish line the men move promptly toward the designated point and the company is re-formed in line. If assembled by platoons, these are conducted to the designated point by platoon leaders, and the company is re-formed in line.

Platoons may be assembled by the command: 1. **Platoons, assemble,** 2. **MARCH.**

Executed by each platoon as described for the company.

One or more platoons may be assembled by the command: 1. **Such platoon(s), assemble,** 2. **MARCH.**

Executed by the designated platoon or platoons as described for the company. (211)

The Advance

221. The advance of a company into an engagement (whether for attack or defense) is conducted in close order, preferably column of squads, until the probability of encountering hostile fire makes it advisable to deploy. After deployment, and before opening fire, the advance of the company may be continued in skirmish line or other suitable formation, depending upon circumstances. The advance may often be facilitated, or better advantage taken of cover, or losses reduced by the employment of the platoon or squad columns or by the use of a succession of thin lines. The selection of the method to be used is made by the captain or major, the choice depending upon conditions arising during the progress of the advance. If the deployment is found to be premature, it will generally be best to assemble the company and proceed in close order.

Patrols are used to provide the necessary security against surprise. (212)

222. Being in skirmish line: 1. **Platoon columns,** 2. **MARCH.**

The platoon leaders move forward through the center of their respective platoons; men to the right of the platoon leader march to the left and follow him in file; those to the left march in like manner to the right; each platoon leader thus conducts the march of his platoon in double column of files; platoon guides follow in rear of their respective platoons to insure prompt and orderly execution of the advance. (213)

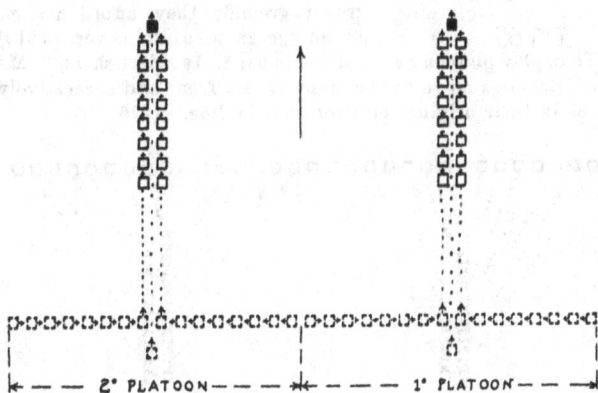

FIG 64

223. Being in skirmish line: 1. **Squad columns,** 2. **MARCH.**

Each squad leader moves to the front; the members of each squad oblique toward and follow their squad leader in single file at easy marching distances. (214)

224. Platoon columns are profitably used where the ground is so difficult or cover so limited as to make it desirable to take advantage of the few favorable routes; no two platoons should march within the area of burst of a single shrapnel.[1] **Squad** columns are of value principally in facilitating the advance over rough or brush-grown ground; they afford no material advantage in securing cover. (215)

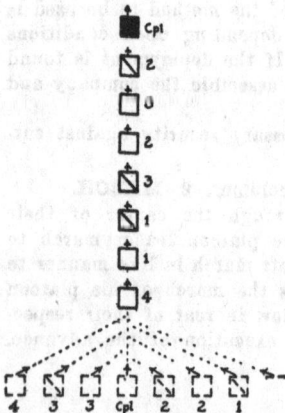

FIG 65.

225. To deploy platoon or squad columns: 1. **As skirmishers,** 2. **MARCH.**
Skirmishers move to the right or left front and successively place themselves in their original positions on the line. (216)

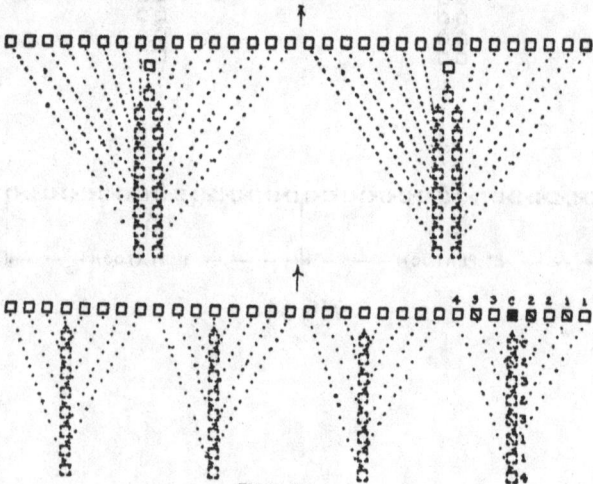

FIG 66.

226. Being in platoon or squad columns: 1. **Assemble,** 2. **MARCH.**

---

[1] Ordinarily about 20 yards wide.

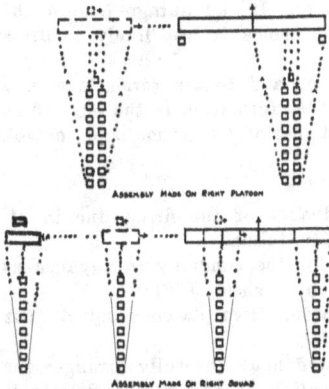

ASSEMBLY MADE ON RIGHT PLATOON

ASSEMBLY MADE ON RIGHT SQUAD

FIG 67

The platoon or squad leaders signal **assemble**. The men of each platoon or squad, as the case may be, advance and, moving to the right and left, take their proper places in line, each unit assembling on the leading element of the column and re-forming in line. The platoon or squad leaders conduct their units toward the element or point indicated by the captain, and to their places in line; the company is re-formed in line. (217)

227. Being in skirmish line, to advance by a succession of thin lines: 1. (Such numbers), forward, 2. MARCH.

The captain points out in advance the selected position in front of the line occupied. The designated number of each squad moves to the front; the line thus formed preserves the original intervals as nearly as practicable; when this line has advanced a suitable distance (generally from 100 to 250 yards, depending upon the terrain and the character of the hostile fire), a second is sent forward by similar commands, and so on at irregular distances until the whole line has advanced. Upon arriving at the indicated position, the first line is halted. Successive lines, upon arriving, halt on line with the first and the men take their proper places in the skirmish line.

Ordinarily each line is made up of one man per squad and the men of a squad are sent forward in order from right to left as deployed. The first line is led by the platoon leader of the right platoon, the second by the guide of the right platoon, and so on in order from right to left.

The advance is conducted in quick time unless conditions demand a faster gait.

The company having arrived at the indicated position, a further advance by the same means may be advisable. (218)

The advance in a succession of thin lines is used to cross a wide stretch swept, or likely to be swept, by artillery fire or heavy, long-range rifle fire which can not profitably be returned. Its purpose is the building up of a strong skirmish line preparatory to engaging in a fire fight. This method of advancing results in serious (though temporary) loss of control over the company. Its advantage lies in the fact that it offers a less definite target, hence is less likely to draw fire. (219)

The above are suggestions. Other and better formations may be devised to fit particular cases. The best formation is the one which advances the line farthest with the least loss of men, time, and control. (220)

The Fire Attack

228. The principles governing the advance of the firing line in attack are considered in the School of the Battalion.

When it becomes impracticable for the company to advance as a whole by ordinary means, it advances by rushes. (221)

Being in skirmish line: 1. **By platoon (two platoons, squad, four men, etc.), from the right (left)**, 2. **RUSH**.

The platoon leader on the indicated flank carefully arranges the details for a prompt and vigorous execution of the rush and puts it into effect as soon as practicable. If necessary, he designates the leader for the indicated fraction. When about to rush, he causes the men of the fraction to cease firing and to hold themselves flat, but in readiness to spring forward instantly. The leader of the rush (at the signal of the platoon leader, if the latter be not the leader of the rush) commands: **Follow me**, and running at top speed, leads the fraction to the new line, where he halts it and causes it to open fire. The leader of the rush selects the new line if it has not been previously designated.

The first fraction having established itself on the new line, the next like fraction is sent forward by its platoon leader, without further command of the captain, and so on, successively, until the entire company is on the line established by the first rush.

If more than one platoon is to join in one rush, the junior platoon leader conforms to the action of the senior.

A part of the line having advanced, the captain may increase or decrease the size of the fractions to complete the movement. (222)

When the company forms a part of the firing line, the rush of the company as a whole is conducted by the captain, as described for a platoon in the preceding paragraph. The captain leads the rush; platoon leaders lead their respective platoons; platoon guides follow the line to insure prompt and orderly execution of the advance. (223)

229. When the foregoing method of rushing, by running, becomes impracticable, any method of advance that **brings the attack closer to the enemy,** such as crawling, should be employed.

For regulations governing the charge, see paragraphs 318 and 319. (224)

(All rushes should be made with life and ginger, and all the men should start together. All rushes should be made under covering fire, and when a unit rushes forward the adjoining unit or units make up for the loss of fire thus caused by increasing the rate of their fire.

A unit commander about to rush forward, will not do so until he sees that the adjoining unit or units have started to give him the protection of their covering fire and, if necessary, he will call to them to do so. Each unit must be careful not to advance until the last unit that rushed forward has had time to take up an effective fire. When sights have to be adjusted at the conclusion of a rush, the men should do so in the prone position even though it be necessary for the men to kneel for firing. The same as the men who rush should start simultaneously from the prone position, so should they stop simultaneously, all men dropping down to the ground together, wherever they may be, at the command "Down", given by the unit commander when the leading men have reached the new position. The slower members who drop down in rear will crawl up to the line after the halt. So that the slower members may not be crowded out of the line, and also to prevent bunching, the faster men should leave room for them on the line.—Author.)

The Company in Support

230. To enable it to follow or reach the firing line, the support adopts suitable formations, following the principles explained in paragraphs 212-218.

The support should be kept assembled as long as practicable. If after deploying a favorable opportunity arises to hold it for some time in close formation, it should be reassembled. It is redeployed when necessary. (225)

The movements of the support as a whole and the dispatch of reënforcements from it to the firing line are controlled by the major.

A reënforcement of less than one platoon has little influence and will be avoided whenever practicable.

The captain of a company in support is constantly on the alert for the major's signals or commands. (226)

A reënforcement sent to the firing line joins it deployed as skirmishers. The leader of the reënforcement places it in an interval in the line, if one exists, and commands it thereafter as a unit. If no such suitable interval exists, the reënforcement is advanced with increased intervals between skirmishers; each man occupies the nearest interval in the firing line, and each then obeys the orders of the nearest squad leader and platoon leader. (227)

A reënforcement joins the firing line as quickly as possible without exhausting the men. (228)

The original platoon division of the companies in the firing line should be maintained and should not be broken up by the mingling of reënforcements.

Upon joining the firing line, officers and sergeants accompanying a reënforcement take over the duties of others of like grade who have been disabled, or distribute themselves so as best to exercise their normal functions. Conditions will vary and no rules can be prescribed. It is essential that all assist in mastering the increasing difficulties of control. (229)

The Company Acting Alone

231. In general, the company, when acting alone, is employed according to the principles applicable to the battalion acting alone; the

captain employs platoons as the major employs companies, making due allowance for the difference in strength.

The support may be smaller in proportion or may be dispensed with. (230)

The company must be well protected against surprise. Combat patrols on the flanks are specially important. Each leader of a flank platoon details a man to watch for the signals of the patrol or patrols on his flank. (231)

FIRE

232. Ordinarily pieces are loaded and extra ammunition is issued before the company deploys for combat.

In close order the company executes the firings at the command of the captain, who posts himself in rear of the center of the company.

Usually the firings in close order consist of saluting volleys only. (232)

233. When the company is deployed, the men execute the firings at the command of their platoon leaders; the latter give such commands as are necessary to carry out the captain's directions, and, from time to time, add such further commands as are necessary to continue, correct and control the fire ordered. (233)

234. The voice is generally inadequate for giving commands during fire and must be replaced by signals of such character that proper fire direction and control is assured. To attract attention, signals must usually be preceded by the whistle signal (short blast). A fraction of the firing line about to rush should, if practicable, avoid using the long blast signal as an aid to cease firing. Officers and men behind the firing line can not ordinarily move freely along the line, but must depend on mutual watchfulness and the proper use of the prescribed signals. All should post themselves so as to see their immediate superiors and subordinates. (234)

235. The musicians assist the captain by observing the enemy, the target, and the fire-effect, by transmitting commands or signals, and by watching for signals. (235)

236. Firing with blank cartridges at an outlined or represented enemy at distances less than 100 yards is prohibited. (236)

The effect of fire and the influence of the ground in relation thereto, and the individual and collective instruction in marksmanship, are treated in the *Small-Arms Firing Manual.* (237)

Ranges

237. For convenience of reference ranges are classified as follows:
0 to 600 yards, close range.
600 to 1,200 yards, effective range.
1,200 to 2,000 yards, long range.

2,000 yards and over, distant range. (238)

The distance to the target must be determined as accurately as possible and the sights set accordingly. Aside from training and morale, this is the most important single factor in securing effective fire at the longer ranges. (239)

Except in a deliberately prepared defensive position, the most accurate and only practicable method of determining the range will generally be to take the mean of several estimates.

Five or six officers or men, selected from the most accurate estimators in the company, are designated as **range finders** and are specially trained in estimating distances.

Whenever necessary and practicable, the captain assembles the range finders, points out the target to them, and adopts the mean of their estimates. The range finders then take their customary posts. (240)

Classes of Firing

238. Volley firing has limited application. In defense it may be used in the early stages of the action if the enemy presents a large, compact target. It may be used by troops executing **fire of position**. When the ground near the target is such that the strike of bullets can be seen from the firing line, **ranging volleys** may be used to correct the sight setting.

In combat, volley firing is executed habitually by platoon. (241)

Fire at will is the class of fire normally employed in attack or defense. (242)

Clip fire has limited application. It is principally used: 1. In the early stages of combat, to steady the men by habituating them to brief pauses in firing. 2. To produce a short burst of fire. (243)

The Target

239. Ordinarily the major will assign to the company an objective in attack or sector in defense; the company's target will lie within the limits so assigned. In the choice of target, tactical considerations are paramount; the nearest hostile troops within the objective or sector will thus be the usual target. This will ordinarily be the hostile firing line; troops in rear are ordinarily proper targets for artillery, machine guns, or, at times, infantry employing fire of position.

Change of target should not be made without excellent reasons therefor, such as the sudden appearance of hostile troops under conditions which make them more to be feared than the troops comprising the former target. (244)

240. The distribution of fire over the entire target is of special importance.

The captain allots a part of the target to each platoon, or each platoon leader takes as his target that part which corresponds to his position in the company. Men are so instructed that each fires on that part of the target which is directly opposite him. (245)

All parts of the target are equally important. Care must be exercised that the men do not slight its less visible parts. A section of the target not covered by fire represents a number of the enemy permitted to fire coolly and effectively. (246)

If the target can not be seen with the naked eye, platoon leaders select an object in front of or behind it, designate this as the aiming point, and direct a sight setting which will carry the cone of fire into the target. (247)

Fire Direction

241. When the company is large enough to be divided into platoons, it is impracticable for the captain to command it directly in combat. His efficiency in managing the firing line is measured by his ability to enforce his will through the platoon leaders. Having indicated clearly what he desires them to do, he avoids interfering except to correct serious errors or omissions. (248)

The captain **directs** the fire of the company or of designated platoons. He designates the target, and, when practicable, allots a part of the target to each platoon. Before beginning the fire action he determines the range, announces the sight setting, and indicates the class of fire to be employed and the time to open fire. Thereafter, he observes the fire effect, corrects material errors in sight setting, prevents exhaustion of the ammunition supply, and causes the distribution of such extra ammunition as may be received from the rear. (249)

Fire Control

242. In combat the platoon is the **fire unit**. From 20 to 35 rifles are as many as one leader can control effectively. (250)

Each platoon leader puts into execution the commands or directions of the captain, having first taken such precautions to insure correct sight setting and clear description of the target or aiming point as the situation permits or requires; thereafter, he gives such additional commands or directions as are necessary to exact compliance with the captain's will. He corrects the sight setting when necessary. He designates an aiming point when the target can not be seen with the naked eye. (251)

243. In general, **platoon leaders** observe the target and the effect of their fire and are on the alert for the captain's commands or signals; they observe and regulate the rate of fire. The **platoon guides** watch the firing line and check every breech of fire discipline. **Squad leaders** transmit commands and signals when necessary, observe the conduct of their squads and abate excitement, assist in enforcing fire discipline and participate in the firing. (252)

The best troops are those that submit longest to fire control. Loss of control is an evil which robs success of its greatest results. To avoid or delay such loss should be the constant aim of all.

Fire control implies the ability to stop firing, change the sight setting and target, and resume a well directed fire. (253)

Fire Discipline

244. "Fire discipline implies, besides a habit of obedience, a control of the rifle by the soldier, the result of training, which will enable him in action to make hits instead of misses. It embraces taking advantage of the ground; care in setting the sight and delivery of fire; constant attention to the orders of the leaders, and careful observation of the enemy; an increase of fire when the target is favorable, and a cessation of fire when the enemy disappears; economy of ammunition." (*Small-Arms Firing Manual.*)

In combat, shots which graze the enemy's trench or position and thus reduce the effectiveness of his fire have the approximate value of hits; such shots only, or actual hits, contribute toward fire superiority.

Fire discipline implies that, in a firing line without leaders, each man retains his presence of mind and directs effective fire upon the proper target. (254)

245. To create a correct appreciation of the requirements of fire discipline, men are taught that the rate of fire should be as rapid as is consistent with accurate aiming; that the rate will depend upon the visibility, proximity, and size of the target; and that the proper rate will ordinarily suggest itself to each trained man, usually rendering cautions or commands unnecessary.

In attack the highest rate of fire is employed at the halt preceding the assault, and in pursuing fire. (255)

246. In an advance by rushes, leaders of troops in firing positions are responsible for the delivery of heavy fire to cover the advance of each rushing fraction. Troops are trained to change slightly the direction of fire so as not to endanger the flanks of advanced portions of the firing line. (256)

247. In defense, when the target disappears behind cover platoon leaders suspend fire, prepare their platoons to fire upon the point where it is expected to reappear, and greet its reappearance instantly with vigorous fire. (257)

SCHOOL OF THE BATTALION

248. The battalion being purely a tactical unit, the major's duties are primarily those of an instructor in drill and tactics and of a tactical

commander. He is responsible for the theoretical and practical training of the battalion. He supervises the training of the companies of the battalion with a view to insuring the thoroughness and uniformity of their instruction.

In the instruction of the battalion as a whole, his efforts will be directed chiefly to the development of tactical efficiency, devoting only such time to the mechanism of drill and to the ceremonies as may be necessary in order to insure precision, smartness, and proper control. (258)

The movements explained herein are on the basis of a battalion of four companies; they may be executed by a battalion of two or more companies, not exceeding six. (259)

249. The companies are generally arranged from right to left according to the rank of the captains present at the formation. The arrangement of the companies may be varied by the major or higher commander.

After the battalion is formed, no cognizance is taken of the relative order of the companies. (260)

250. In whatever direction the battalion faces, the companies are designated numerically from right to left in line, and from head to rear in column, **first company, second company,** etc.

The terms **right** and **left** apply to actual right and left as the line faces; if the about by squads be executed when in line, the right company becomes the left company and the right center becomes the left center company.

The designation center company indicates the right center or the actual center company according as the number of companies is even or odd. (261)

251. The band and other special units, when attached to the battalion, take the same post with respect to it as if it were the nearest battalion shown in Plate IV. (262)

CLOSE ORDER
Rules

252. Captains repeat such preparatory commands as are to be immediately executed by their companies, as **forward, squads right,** etc.; the men execute the commands **march, halt,** etc., if applying to their companies, when given by the major. In movements executed in route step or at ease the captains repeat the command of execution, if necessary. Captains do not repeat the major's commands in executing the manual of arms, nor those commands which are not essential to the execution of a movement by their companies, as **column of squads, first company, squads right,** etc.

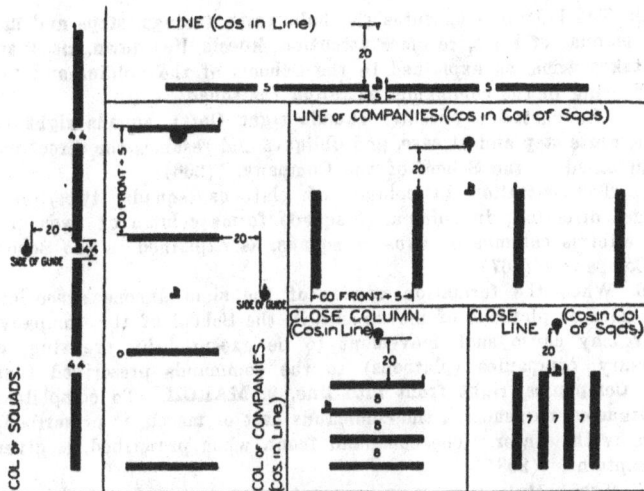

THE BATTALION

Plate III   MAJOR (WITH STAFF LTC) - ●   GUIDE AND DIRECTION - ↑   THE COLOR - ♦

NUMERALS ARE DISTANCES OR INTERVALS IN PACES

In giving commands or cautions captains may prefix the proper letter designations of their companies, as **A Company, HALT; B Company, squads right,** etc. (263)

253. At the command **guide center (right** or **left),** captains command: **Guide right** or **left,** according to the positions of their companies. **Guide center** designates the left guide of the center company. (264)

254. When the companies are to be dressed, captains place themselves on that flank toward which the dress is to be made, as follows:

The battalion in line: Beside the guide (or the flank file of the front rank, if the guide is not in line) and facing to the front.

The battalion in column of companies: Two paces from the guide, in prolongation of and facing down the line.

Each captain, after dressing his company, commands: **FRONT,** and takes his post.

The battalion being in line and unless otherwise prescribed, at the captain's command **dress,** or at the command **halt,** when it is prescribed that the company shall dress, the guide on the flank away from the point of rest, with his piece at right shoulder, dresses promptly on the captain and the companies beyond. During the dress he moves, if necessary, to the right and left only; the captain dresses the company on the line thus established. The guide takes the position of order arms at the command **front.** (265)

[95]

255. The battalion executes the halt, rests, facings, steps and marchings, manual of arms, resumes attention, kneels, lies down, rises, stacks and takes arms, as explained in the Schools of the Soldier and Squad, substituting in the commands battalion for squad.

The battalion executes squads right (left), squads right (left) about, route step and at ease, and obliques and resumes the direct march, as explained in the School of the Company. (266)

The battalion in column of platoons, squads, twos, or files changes direction; in column of squads forms column of twos or files and re-forms columns of twos or squads, as explained in the School of the Company. (267)

256. When the formation admits of the simultaneous execution by companies or platoons of movements in the School of the Company the major may cause such movement to be executed by prefixing, when necessary, companies (platoons) to the commands prescribed therein: as 1. Companies, right front into line, 2. MARCH. To complete such simultaneous movements, the commands halt or march, if prescribed, are given by the major. The command front, when prescribed, is given by the captains. (268)

257. The battalion as a unit executes the loadings and firings only in firing saluting volleys. The commands are as for the company, substituting battalion for company. At the first command for loading, captains take post in rear of the center of their respective companies. At the conclusion of the firing, the captains resume their posts in line.

On other occasions, when firing in close order is necessary, it is executed by company or other subdivision under instructions from the major. (269)

To Form the Battalion

258. For purposes other than ceremonies: The battalion is formed in column of squads. The companies having been formed, the adjutant posts himself so as to be facing the column, when formed, and 6 paces in front of the place to be occupied by the leading guide of the battalion; he draws saber; adjutant's call is sounded or the adjutant signals assemble.

The companies are formed, at attention, in column of squads in their proper order. Each captain, after halting his company, salutes the adjutant; the adjutant returns the salute and, when the last captain has saluted, faces the major and reports: Sir, the battalion is formed. He then joins the major. (270)

For ceremonies or when directed: The battalion is formed in line.

The companies having been formed, the adjutant posts himself so as to be 6 paces to the right of the right company when line is formed, and faces in the direction in which the line is to extend. He draws saber; **adjutant's call** is sounded; the band plays if present.

The right company is conducted by its captain so as to arrive from the rear, parallel to the line; its right and left guides precede it on the line by about 20 paces, taking post facing to the right at order arms, so that their elbows will be against the breasts of the right and left files of their company when it is dressed. The guides of the other companies successively prolong the line to the left in like manner and the companies approach their respective places in line as explained for the right company. The adjutant, from his post, causes the guides to cover.

When about 1 pace in rear of the line, each company is halted and dressed to the right against the arms of the guides.

The band, arriving from the rear, takes its place in line when the right company is halted; it ceases playing when the left company has halted.

When the guides of the left company have been posted, the adjutant, moving by the shortest route, takes post facing the battalion midway between the post of the major and the center of the battalion.

The major, staff, noncommissioned staff, and orderlies take their posts.

When all parts of the line have been dressed, and officers and others have reached their posts, the adjutant commands: 1. **Guides**, 2. **POSTS**, 3. **Present**, 4. **ARMS**. At the second command guides take their places in the line. The adjutant then turns about and reports to the major: **Sir, the battalion is formed**; the major directs the adjutant: **Take your post, Sir**; draws saber and brings the battalion to the order. The adjutant takes his post, passing to the right of the major. (271)

To Dismiss the Battalion

259. DISMISS YOUR COMPANIES.

Staff and noncommissioned staff officers fall out; each captain marches his company off and dismisses it. (272)

To Rectify the Alignment

260. Being in line at a halt, to align the battalion: 1. **Center (right or left)**, 2. **DRESS**.

The captains dress their companies successively toward the center (right or left) guide of the battalion, each as soon as the captain next toward the indicated guide commands: **Front**. The captains of the center companies (if the dress is center) dress them without waiting for each other. (273)

To give the battalion a new alignment: 1. **Guides center (right or left) company on the line**, 2. **Guides on the line**, 3. **Center (right or left)**, 4. **Dress**, 5. **Guides**, 6. **POSTS**.

At the first command, the designated guides places themselves on the line (par. 271) facing the center (right or left). The major establishes them in the direction he wishes to give the battalion.

At the second command, the guides of the other companies take posts, facing the center (right or left), so as to prolong the line.

At the command **dress**, each captain dresses his company to the flank toward which the guides of his company face.

At the command **posts**, given when all companies have completed the dress, the guides return to their posts. (274)

To Rectify the Column

261. Being in column of companies, or in close column, at a halt, if the guides do not cover or have not their proper distances, and it is desired to correct them, the major commands: 1. **Right (left)**, 2. **DRESS.**

Captains of companies in rear of the first place their right guides so as to cover at the proper distance; each captain aligns his company to the right and commands: **FRONT.** (275)

On Right (Left) into Line

262. Being in column of squads or companies: 1. **On right (left) into line**, 2. **MARCH**, 3. **Battalion**, 4. **HALT.**

Being in column of squads: At the first command, the captain of the leading company commands: **Squads right.** If at a halt each captain in rear commands: **Forward.** At the second command the leading company marches in line to the right; the companies in rear continue to march to the front and form successively on the left, each, when opposite its place, being marched in line to the right.

The fourth command is given when the first company has advanced the desired distance in the new direction; it halts and is dressed to the right by its captain; the others complete the movement, each being halted 1 pace in rear of the line established by the first company, and then dressed to the right.

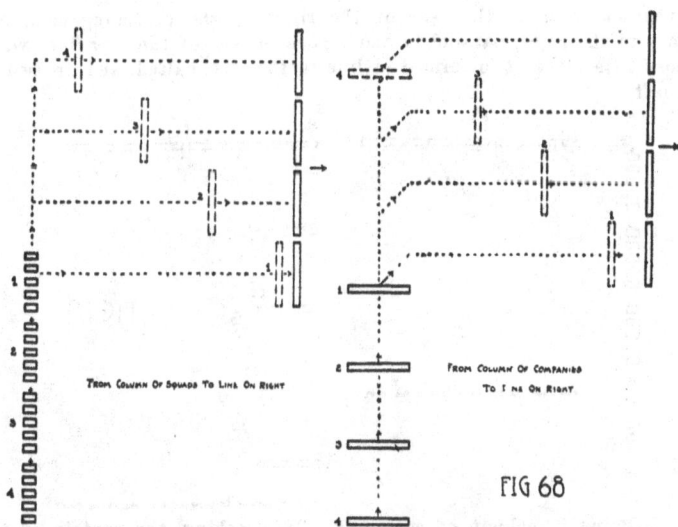

FROM COLUMN OF SQUADS TO LINE ON RIGHT

FROM COLUMN OF COMPANIES TO LINE ON RIGHT

FIG 68

Being in column of companies: At the first command, the captain of the first company commands: **Right turn.** If at a halt, each captain in rear commands: **Forward.** Each of the captains in rear of the leading company gives the command: 1. **Right turn,** in time to add, 2. **MARCH,** when his company arrives opposite the right of its place in line.

The fourth command is given and the movement completed as explained above.

Whether executed from column of squads or column of companies, each captain places himself so as to march beside the right guide after his company forms line or changes direction to the right.

If executed in double time, the leading company marches in double time until halted. (276)

Front into Line

263. Being in column of squads or companies: 1. **Right (left) front into line, 2. MARCH.**

Being in column of squads: At the first command, the captain of the leading company commands: **Column right;** the captains of the companies in rear, **column half right.** At the second command the leading company executes **column right,** and, as the last squad completes the change of direction, is formed in line to the left, halted, and dressed to the left. Each of the companies in rear is conducted by the most

264

convenient route to the rear of the right of the preceding company, thence to the right, parallel to and 1 pace in rear of the new line; when opposite its place, it is formed in line to the left, halted, and dressed to the left.

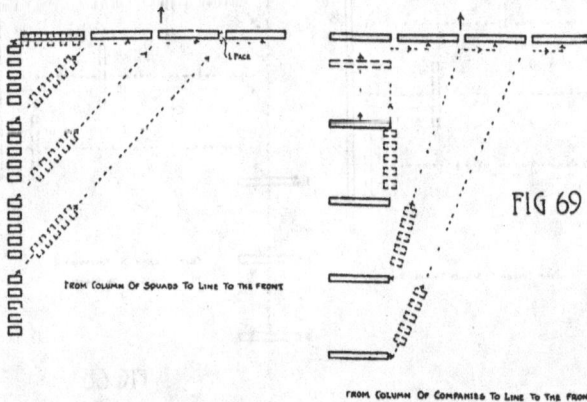

FROM COLUMN OF SQUADS TO LINE TO THE FRONT

FIG 69

FROM COLUMN OF COMPANIES TO LINE TO THE FRONT

Being in column of companies: If marching, the captain of the leading company gives the necessary commands to halt his company at the second command; if at a halt the leading company stands fast. At the first command, the captain of each company in rear commands: **Squads right**, or **Right by squads**, and after the second command conducts his company by the most convenient route to its place in line, as described above.

Whether executed from column of squads or column of companies, each captain halts when opposite or at the point where the left of his company is to rest. (277)

To Form Column of Companies Successively to the Right or Left

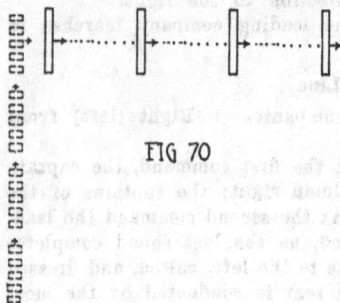

FIG 70

264. Being in column of squads: 1. **Column of companies, first company, squads right (left)**, 2. **MARCH.**

The leading company executes **squads right** and moves forward. The other companies move forward in column of squads and successively march in line to the right on the same ground as the leading company and in such manner that the guide covers the guide of the preceding company. (278)

[100]

To Form Column of Squads Successively to the Right or Left

265. Being in column of companies: 1. **Column of squads, first company, squads right (left), 2. MARCH.**

The leading company executes **squads right** and moves forward. The other companies move forward in column of companies and successively march in column of squads to the right on the same ground as the leading company. (279)

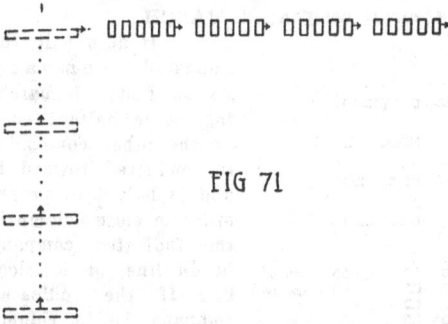

FIG 71

To Change Direction

266. Being in column of companies or close column: 1. **Column right (left), 2. MARCH.**

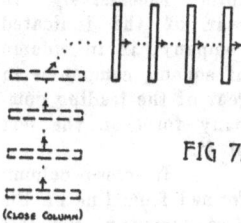

The captain of the first company commands: **Right turn.**

The leading company turns to the right on moving pivot, the captain adding: 1. **Forward, 2. MARCH,** upon its completion.

The other companies march squarely up to the turning point; each changes direction by the same commands and means as the first and in such manner that the guide covers the guide of the preceding company. (280)

FIG 72

(CLOSE COLUMN)

267. Being in line of companies or close line: 1. **Battalion right (left), 2. MARCH, 3. Battalion, 4. HALT.**

The right company changes direction to the right; the other companies are conducted by the shortest line to their places abreast of the first.

The fourth command is given when the right company has advanced the desired distance in the new direction; that company halts; the others halt successively upon arriving on the line. (281)

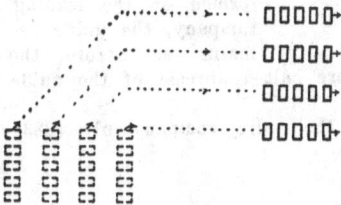

FIG 73

268. Being in column of squads, the battalion changes direction by the same commands and in the manner prescribed for the company. (282)

Mass Formations

269. Being in line, line of companies, column of companies or column of squads: 1. **Close on first (fourth) company, 2. MARCH.**

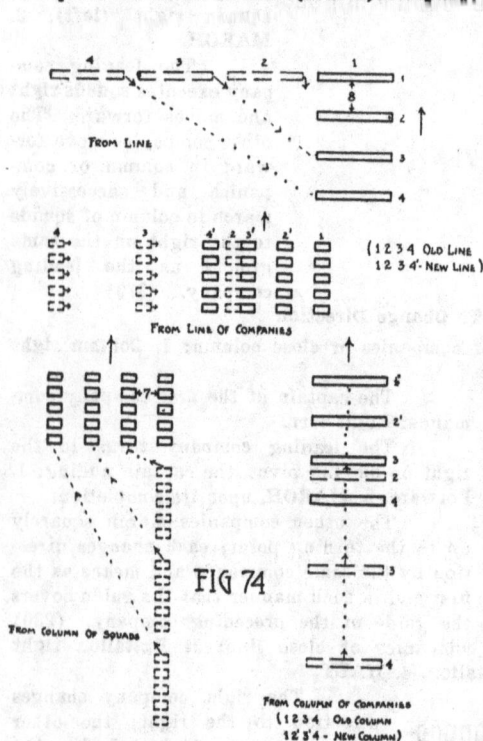

FROM LINE

FROM LINE OF COMPANIES

(1 2 3 4 OLD LINE
1 2 3 4'· NEW LINE)

FIG 74

FROM COLUMN OF SQUADS

FROM COLUMN OF COMPANIES
(1 2 3 4   OLD COLUMN
1 2' 3'4 - NEW COLUMN)

If at a halt, the indicated company stands fast; if marching, it is halted; each of the other companies is conducted toward it and is halted in proper order in close column if the indicated company be in line, or in close line if the indicated company be in column of squads.

If the battalion is in line, companies form successively in rear of the indicated company; if in column of squads, companies in rear of the leading company form on the left of it.

In close column formed from line on the first company, the left guides cover; formed on the fourth company, right guides cover. If formed on the leading company, the guide remains as before the formation. In close line, the guides are halted abreast of the guide of the leading company.

The battalion in column closes on the leading company only. (283)

To Extend the Mass

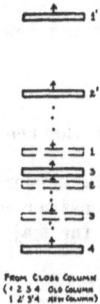

270. Being in close column or in close line: 1. **Extend on first (fourth) company,** 2. **MARCH.**

Being in close line: If at a halt, the indicated company stands fast; if marching, it halts; each of the other companies is conducted away from the indicated company and is halted in its proper order in line of companies.

Being in close column, the extension is made on the fourth company only. If marching, the leading company continues to march; companies in rear are halted and successively resume the march in time to follow at full distance. If at halt, the leading company marches; companies in rear successively march in time to follow at full distance.

FROM CLOSE COLUMN
(1 2 3 4 OLD COLUMN
1 2' 3'4 NEW COLUMN)

FIG 75

Close column is not extended in double time. (284)

271. Being in close column: 1. **Right (left) front into line,** 2. **MARCH.** Executed as from column of companies. (285)

272. Being in close column: 1. **Column of squads, first (fourth) company, squads right (left),** 2. **MARCH.**

The designated company marches in column of squads to the right. Each of the other companies executes the same movement in time to follow the preceding company in column. (286)

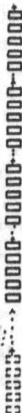

FIG 76

273. Being in close line: 1. **Column of squads, first (fourth) company, forward,** 2. **MARCH.**

The designated company moves forward. The other companies (halting if in march) successively take up the march and follow in column. (287)

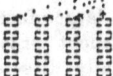

FROM CLOSE LINE

Route Step and At Ease

274. The battalion marches in **route step** and **at ease** as prescribed in the School of the Company. When marching in column of companies or platoons, the guides maintain the trace and distance.

In route marches the major marches at the head of the column; when necessary, the file closers may be directed to march at the head and rear of their companies. (288)

Assembly

275. The battalion being wholly or partially deployed, or the companies being separated: 1. **Assemble**, 2. **MARCH**.

The major places himself opposite to or designates the element or point on which the battalion is to assemble. Companies are assembled and marched to the indicated point. As the companies arrive the major or adjutant indicates the formation to be taken. (289)

COMBAT PRINCIPLES

Orders

276. The following references to orders are applicable to attack or defense. (290)

277. In extended order, the company is the largest unit to execute movements by prescribed commands or means. The major, assembling his captains if practicable, directs the disposition of the battalion by means of **tactical orders**. He controls its subsequent movements by such **orders** or **commands** as are suitable to the occasion. (291)

277a. In every disposition of the battalion for combat the major's order should give subordinates sufficient information of the enemy, of the position of supporting and neighboring troops, and of the object sought to enable them to conform intelligently to the general plan.

The order should then designate the companies which are to constitute the **firing line** and those which are to constitute the **support**. In attack, it should designate the direction or the objective, the order and front of the companies on the firing line, and should designate the right or left company as base company. In defense, it should describe the front of each company and, if necessary, the sector to be observed by each. (292)

When the battalion is operating alone, the major provides for the reconnaissance and protection of his flanks; if part of a larger force, the major makes similar provisions, when necessary, without orders from higher authority, unless such authority has specifically directed other suitable reconnaissance and protection. (293)

278. When the battalion is deployed upon the initiative of the major, he will indicate whether extra ammunition shall be issued; if deployed in pursuance of orders of higher authority, the major will cause the issue of extra ammunition, unless such authority has given directions to the contrary. (294)

Deployment

279. The following principles of deployment are applicable to attack or defense. (295)

A premature deployment involves a long, disorganizing and fatiguing advance of the skirmish line, and should be avoided. A greater evil is to be caught by heavy fire when in dense column or other close order formation; hence advantage should be taken of cover in order to retain the battalion in close order formation until exposure to heavy hostile fire may reasonably be anticipated. (296)

The major regulates the depth of the deployment and the extent and density of the firing line, subject to such restrictions as a senior may have imposed.

Companies or designated subdivisions and detachments are conducted by their commanders in such manner as best to accomplish the mission assigned to them under the major's orders. Companies designated for the firing line march independently to the place of deployment, form skirmish line, and take up the advance. They conform, in general, to the base company. (297)

280. The commander of a battalion, whether it is operating alone or as part of a larger force, should hold a part of his command out of the firing line. By the judicious use of this force the major can exert an influence not otherwise possible over his firing line and can control, within reasonable limits, an action once begun. So if his battalion be assigned to the firing line the major will cause one, two, or three companies to be deployed on the firing line, retaining the remaining companies or company as a support for that firing line. The division of the battalion into firing line and support will depend upon the front to be covered and the nature and anticipated severity of the action. (298)

If the battalion be part of a larger command, the number of companies in the firing line will generally be determinable from the regimental commander's order; the remainder constitutes the support. If the battalion is acting alone, the support must be strong enough to maintain the original fire power of the firing line, to protect the flanks, and to perform the functions of a reserve, whatever be the issue of the action, See paragraph 346. (299)

If the battalion is operating alone, the support may, according to circumstances, be held in one or two bodies and placed behind the center, or one or both flanks of the firing line, or echeloned beyond a flank. If the battalion is part of a larger force, the support is generally held in one body. (300)

281. The distance between the firing line and the supporting group or groups will vary between wide limits; it should be as short as the necessity for protection from heavy losses will permit. When cover is available, the support should be as close as 50 to 100 yards; when such cover is not available, it should not be closer than 300 yards. It may be as far as 500 yards in rear if good cover is there obtainable and is not obtainable at a lesser distance. (301)

In exceptional cases, as in a meeting engagement, it may be necessary to place an entire battalion or regiment in the firing line at the initial deployment, the support being furnished by other troops. Such deployment causes the early mingling of the larger units, thus rendering leadership and control extremely difficult. The necessity for such deployment will increase with the inefficiency of the commander and of the service of information. (302)

Fire

282. Fire direction and fire control are functions of company and platoon commanders. The major makes the primary apportionment of the target—in defense, by assigning sectors of fire; in attack, by assigning the objective. In the latter case each company in the firing line takes as its target that part of the general objective which lies in its front. (303)

283. The major should indicate the point or time at which the fire fight is to open. He may do this in his order for deployment or he may follow the firing line close enough to do so at the proper time. If it be impracticable for him to do either, the senior officer with the firing line, in each battalion, selects the time for opening fire. (304)

Attack

284. The battalion is the attack unit, whether operating alone or as part of a larger unit. (305)

285. If his battalion be one of several in the firing line, the major, in executing his part of the attack, pushes his battalion forward as vigorously as possible within the front, or section, assigned to it. The great degree of independence allowed to him as to details demands, in turn, the exercise of good judgment on his part. Better leadership, better troops, and more favorable terrain enable one battalion to advance more rapidly in attack than another less fortunate, and such a battalion will insure the further advance of the others. The leading battalion should not, however, become isolated; isolation may lead to its destruction. (306)

The deployment having been made, the firing line advances without firing. The predominant idea must be to close with the enemy as soon as possible without ruinous losses. The limited supply of ammunition and the uncertainty of resupply, the necessity for securing fire superiority in order to advance within the shorter ranges, and the impossibility of accomplishing this at ineffective ranges, make it imperative that fire be not opened as long as the advance can be continued without demoralizing losses. The attack which halts to open fire at extreme range (over 1,200 yards) is not likely ever to reach its destination. Every effort should be made, by using cover or inconspicuous formations, or by advancing the firing line as a whole, to arrive within 800 yards of the enemy before opening fire. (307)

Except when the enemy's artillery is able to effect an unusual concentration of fire, its fire upon deployed infantry causes losses which are unimportant when compared with those inflicted by his infantry; hence the attacking infantry should proceed to a position as described above, and from which an effective fire can be directed against the hostile infantry with a view to obtaining fire superiority. The effectiveness of the enemy's fire must be reduced so as to permit further advance. The more effective the fire to which the enemy is subjected the less effective will be his fire. (308)

Occasionally the fire of adjacent battalions, or of infantry employing fire of position, or of supporting artillery, will permit the further advance of the entire firing line from this point, but it will generally be necessary to advance by rushes of fractions of the line.

The fraction making the rush should be as large as the hostile fire and the necessity for maintaining fire superiority will permit. Depending upon circumstances, the strength of the fraction may vary from a company to a few men.

The advance is made as rapidly as possible without losing fire superiority. The smaller the fraction which rushes, the greater the number of rifles which continue to fire upon the enemy. On the other hand, the smaller the fraction which rushes the slower will be the progress of the attack. (309)

286. Enough rifles must continue in action to insure the success of each rush. Frequently the successive advances of the firing line must be effected by rushes of fractions of decreased size; that is, advances by rushes may first be made by company, later by half company or platoon, and finally by squads or files; but no subsequent opportunity to increase the rate of advance, such as better cover or a decrease of the hostile fire, should be overlooked. (310)

287. Whenever possible, the rush is begun by a flank fraction of the firing line. In the absence of express directions from the major, each captain of a flank company determines when an advance by rushes shall be attempted. A flank company which inaugurates an advance by rushes becomes the base company, if not already the base. An advance by rushes having been inaugurated on one flank, the remainder of the firing line conforms; fractions rush successively from that flank and halt on the line established by the initial rush.

The fractions need not be uniform in size; each captain indicates how his company shall rush, having due regard to the ground and the state of the fire fight. (311)

A fraction about to rush is sent forward when the remainder of the line is firing vigorously; otherwise the chief advantage of this method of advancing is lost.

The length of the rush will vary from 30 to 80 yards, depending upon the existence of cover, positions for firing, and the hostile fire. (312)

When the entire firing line of the battalion has advanced to the new line, fresh opportunities to advance are sought as before. (313)

Two identical situations will never confront the battalion; hence at drill it is prohibited to arrange the details of an advance before the preceding one has been concluded, or to employ a fixed or prearranged method of advancing by rushes. (314)

288. The major posts himself so as best to direct the reënforcing of the firing line from the support. When all or nearly all of the support has been absorbed by the firing line, he joins, and takes full charge of, the latter. (315)

The reënforcing of the firing line by driblets of a squad or a few men has no appreciable effect. The firing line requires either no reënforcement or a strong one. Generally one or two platoons will be sent forward under cover of a heavy fire of the firing line. (316)

To facilitate control and to provide intervals in which reënforcements may be placed, the companies in the firing line should be kept closed in on their centers as they become depleted by casualties during the advance.

When this is impracticable reënforcements must mingle with and thicken the firing line. In battle the latter method will be the rule rather than the exception, and to familiarize the men with such conditions the combat exercises of the battalion should include both methods of reënforcing. Occasionally, to provide the necessary intervals for reënforcing by either of these methods, the firing line should be thinned by causing men to drop out and simulate losses during the various advances. Under ordinary conditions the depletion of the firing line for this purpose will be from one-fifth to one-half of its strength. (317)

289. The major or senior officer in the firing line determines when bayonets shall be fixed and gives the proper command or signal. It is repeated by all parts of the firing line. Each man who was in the front rank prior to deployment, as soon as he recognizes the command or signal, suspends firing, quickly fixes his bayonet, and immediately resumes firing; after which the other men suspend firing, fix bayonets, and immediately resume firing. The support also fixes bayonets. The concerted fixing of the bayonet by the firing line at drill does not simulate battle conditions and should not be required. It is essential that there be no marked pause in the firing. Bayonets will be fixed generally before or during the last, or second last, advance preceding the charge. (318)

290. Subject to orders from higher authority, the major determines the point from which the charge is to be made. The firing line having arrived at that point and being in readiness, the major causes the charge to be sounded. The signal is repeated by the musicians of all parts of the line. The company officers lead the charge. The skirmishers spring forward shouting, run with bayonets at charge, and close with the enemy.

The further conduct of the charging troops will depend upon circumstances; they may halt and engage in bayonet combat or in pursuing fire; they may advance a short distance to obtain a field of fire or to drive the enemy from the vicinity; they may assemble or reorganize, etc. If the enemy vacates his position every effort should be made to open fire at once on the retreating mass, reorganization of the attacking troops being of secondary importance to the infliction of further losses upon the enemy and to the increase of his confusion. In combat exercises the major will assume a situation and terminate the assault accordingly. (319)

Defense

291. In defense, as in attack, the battalion is the tactical unit best suited to independent assignment. Defensive positions are usually divided into sections and a battalion assigned to each. (320)

The major locates such fire, communicating, and cover trenches and obstacles as are to be constructed. He assigns companies to construct them and details the troops to occupy them. (321)

The major reënforces the firing line in accordance with the principles applicable to, and explained in connection with, the attack, maintaining no more rifles in the firing line than are necessary to prevent the enemy's advance. (322)

The supply of ammunition being usually ample, fire is opened as soon as it is possible to break up the enemy's formation, stop his advance, or inflict material loss, but this rule must be modified to suit the ammunition supply. (323)

The major causes the firing line and support to fix bayonets when an assault by the enemy is imminent. Captains direct this to be done if they are not in communication with the major and the measure is deemed advisable.

Fire alone will not stop a determined, skillfully conducted attack. The defender must have equal tenacity; if he can stay in his trench or position and cross bayonets, he will at least have neutralized the hostile first line, and the combat will be decided by reserves. (324)

If ordered or compelled to withdraw under hostile infantry fire or in the presence of hostile infantry, the support will be posted so as to cover the retirement of the firing line. (325)

When the battalion is operating alone, the support must be strong and must be fed sparingly into the firing line, especially if a counter-attack is planned. Opportunities for counter-attack should be sought at all times. (326)

THE REGIMENT

292. Normally, the regiment consists of three battalions, but these regulations are applicable to a regiment of two or more battalions

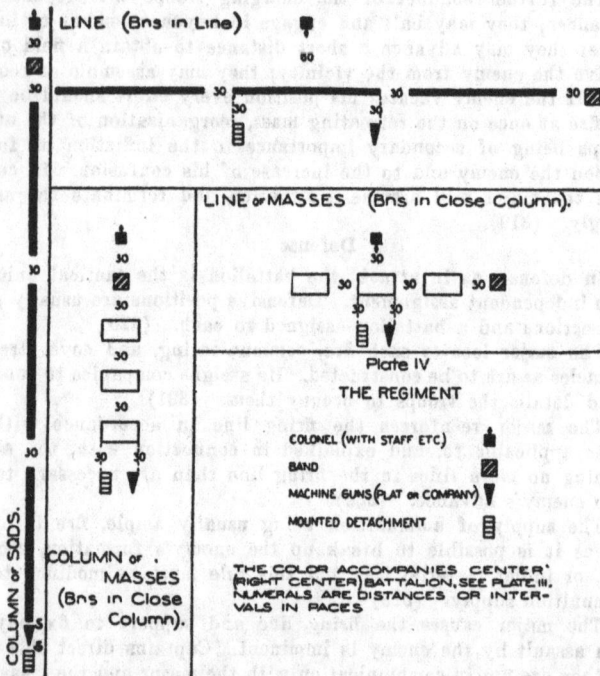

LINE (Bns in Line)

LINE or MASSES (Bns in Close Column).

Plate IV

THE REGIMENT

COLONEL (WITH STAFF ETC)

BAND

MACHINE GUNS (PLAT or COMPANY)

MOUNTED DETACHMENT

COLUMN or MASSES (Bns in Close Column).

COLUMN or SQDS.

THE COLOR ACCOMPANIES CENTER (RIGHT CENTER) BATTALION, SEE PLATE III. NUMERALS ARE DISTANCES OR INTERVALS IN PACES

Special units, such as band, machine-gun company, and mounted scouts, have special formations for their own use. Movements herein prescribed are for the battalions; special units conform thereto unless otherwise prescribed or directed. (327)

The colonel is responsible for the theoretical instruction and practical training of the regiment as a whole. Under his immediate supervision the training of the units of the regiment is conducted by their respective commanders. (328)

The colonel either gives his commands or orders orally, by bugle, or by signal, or communicates them by staff officers or orderlies.

Each major gives the appropriate commands or orders, and, in close-order movements, causes his battalion to execute the necessary movements at his command of execution. Each major ordinarily moves his battalion from one formation to another, in column of squads, in the most convenient manner, and, in the presence of the enemy, in the most direct manner consistent with cover.

Commanders of the special units observe the same principles as to commands and movements. They take places in the new formation as directed by the colonel; in the absence of such directions they conform as nearly as practicable to Plate IV, maintaining their relative positions with respect to the flank or end of the regiment on which they are originally posted. (329)

When the regiment is formed, and during ceremonies, the lieutenant colonel is posted 2 paces to the left of, and 1 pace less advanced than the colonel. In movements subsequent to the formation of the regiment and other than cermonies, the lieutenant colonel is on the left of the colonel. (330)

In whatever formation the regiment may be, the battalions retain their permanent administrative designations of first, second, third battalion. For convenience, they may be designated, when in line, as right, center, or left battalion; when in column, as leading, center, or rear battalion. These designations apply to the actual positions of the battalions in line or column. (331)

Except at cermonies, or when rendering honors, or when otherwise directed, after the regiment is formed, the battalions march and stand at ease during subsequent movements. (332)

CLOSE ORDER

To Form the Regiment

293. Unless otherwise directed, the battalions are posted from right to left, or from head to rear, according to the rank of the battalion commanders present, the senior on the right or at the head. A battalion whose major is in command of the regiment retains its place. (333)

For ordinary purposes, the regiment is formed in column of squads or in column of masses.

The adjutant informs the majors what the formation is to be. The battalions and special units having been formed, he posts himself and draws saber. **Adjutant's call** is sounded, or the adjutant signals **assemble.**

If forming in column of squads, the adjutant posts himself so as to be facing the column when formed, and 6 paces in front of the place to be occupied by the leading guide of the regiment; if forming in column of masses, he posts himself so as to be facing the right guides of the column when formed, and 6 paces in front of the place to be occupied by the right guide of the leading company. Later, he moves so as best to observe the formation.

The battalions are halted, at attention, in column of squads or close column, as the case may be, successively from the front in their proper order and places. The band takes its place when the leading battalion has halted. Other special units take their places in turn when the rear battalion has halted.

The majors and the commanders of the machine-gun company and mounted scouts (or detachment) each, when his command is in place, salutes the adjutant and commands: **At ease**; the adjutant returns the salutes. When all have saluted and the band is in place, the adjutant rides to the colonel, reports: **Sir, the regiment is formed,** and takes his post. The colonel draws saber.

The formation in column of squads may be modified to the extent demanded by circumstances. Prior to the formation the adjutant indicates the point where the head of the column is to rest and the direction in which it is to face; he then posts himself so as best to observe the formation. At **adjutant's call** or **assemble** the leading battalion marches to, and halts at, the indicated point. The other battalions take positions from which they may conveniently follow in their proper places. (334)

For ceremonies, or when directed, the regiment is formed in line or line of masses.

The adjutant posts himself so as to be 6 paces to the right of the right or leading company of the right battalion when the regiment is formed and faces in the direction in which the line is to extend. **Adjutant's call** is sounded; the band plays.

The adjutant indicates to the adjutant of the right battalion the point of rest and the direction in which the line is to extend, and then takes post facing the regiment midway between the post of the colonel and the center of the regiment. Each of the other battalion adjutants precedes his battalion to the line and marks its point of rest.

The battalions, arriving from the rear, each in line or close column, as the case may be, are halted on the line successively from right to left in their proper order and places. Upon halting, each major commands: 1. **Right**, 2. **DRESS**. The battalion adjutant assists in aligning the battalion and then takes his post.

The band, arriving from the rear, takes its place in line when the right battalion has halted; it ceases playing when the left battalion has halted. The machine-gun company and the mounted scouts (or detachment) take their places in line after the center battalion has halted.

The colonel and those who accompany him take post.

When all parts of the line have been dressed, and officers and all others have reached their posts, the adjutant commands: 1. **Present**

2. **ARMS.** He then turns about and reports to the colonel: **Sir, the regiment is formed**; the colonel directs the adjutant: **Take your post, Sir,** draws saber and brings the regiment to the order. The adjutant takes his post, passing to the right of the colonel. (335)

To Dismiss the Regiment

Being in any formation: **DISMISS YOUR BATTALIONS.** Each major marches his battalion off and dismisses it. (336)

Movements by the Regiment

The regiment executes the **halt, rests, facings, steps, and marchings, manual of arms,** resumes **attention, kneels, lies down, rises, stacks** and **takes arms,** as explained in the Schools of the Soldier and Squad, substituting in the commands, when necessary, **battalions for squad.**

The regiment executes **squads right (left), squads right (left) about, route step** and **at ease, obliques** and resumes the direct march as explained in the School of the Company.

The regiment in column of files, twos, squads, or platoons, changes direction, and in column of squads forms column of twos or files and reforms column of twos or squads, as explained in the School of the Company. In column of companies, it changes direction as explained in the School of the Battalion. (337)

When the formation admits of the simultaneous execution, by battalions, companies, or platoons, of movements prescribed in the School of the Company or Battalion, the colonel may cause such movements to be executed by prefixing, where necessary, **battalions (companies, platoons),** to the commands prescribed therein. (338)

The column of squads is the usual column of march; to shorten the column, if conditions permit, a double column of squads may be used, the companies of each battalion marching abreast in two columns. Preliminary to an engagement, the regiment or its units will be placed in the formation best suited to its subsequent tactical employment. (339)

To assume any formation, the colonel indicates to the majors the character of the formation desired, the order of the battalions, and the point of rest. Each battalion is conducted by its major, and is placed in its proper order in the formation, by the most convenient means and route.

Having halted in a formation, no movements for the purpose of correcting minor discrepancies in alignments, intervals, or distances are made unless specially directed by the colonel or necessitated by conditions of cover. (340)

To correct intervals, distances, and alignments, the colonel directs one or more of the majors to rectify their battalions. Each major so directed causes his battalion to correct its alignment, intervals, and distances, and places it in its proper position in the formation. (341)

CEREMONIES AND INSPECTIONS

CEREMONIES

General Rules for Ceremonies

294. The order in which the troops of the various arms are arranged for ceremonies is prescribed by *Army Regulations.*

When forming for ceremonies the companies of the battalion and the battalions of the regiment are posted from right to left in line and from head to rear in column, in the order of rank of their respective commanders present in the formation, the senior on the right or at the head.

The commander faces the command; subordinate commanders face to the front. (708)

295. At the command **present arms,** given by the colonel, the lieutenant colonel, and the colonel's staff salute; the major's staff salutes at the major's command. Each staff returns to the carry or order when the command **order arms** is given by its chief. (709)

At the **assembly** for a ceremony companies are formed on their own parades and informally inspected.

At **adjutant's call,** except for ceremonies involving a single battalion, each battalion is formed on its own parade, reports are received, and the battalion presented to the major. At the second sounding of **adjutant's call** the regiment is formed. (710)

REVIEWS

General Rules

296. The adjutant posts men or otherwise marks the points where the column changes direction in such manner that its flank in passing will be about 12 paces from the reviewing officer.

The post of the reviewing officer, usually opposite the center of the line, is indicated by a marker.

Officers of the same or higher grade, and distinguished personages invited to accompany the reviewing officer, place themselves on his left; their staffs and orderlies place themselves respectively on the left of the staff and orderlies of the reviewing officer; all others who accompany the reviewing officer place themselves on the left of his staff, their orderlies in rear. A staff officer is designated to escort distinguished personages and to indicate to them their proper positions. (711)

297. While riding around the troops, the reviewing officer may direct his staff, flag and orderlies to remain at the post of the reviewing officer, or that only his personal staff and flag shall accompany him; in either case the commanding officer alone accompanies the reviewing officer.

If the reviewing officer is accompanied by his entire staff, the staff officers of the commander place themselves on the right of the staff of the reviewing officer.

The reviewing officer and others at the reviewing stand salute the color as it passes; when passing around the troops, the reviewing officer and those accompanying him salute the color when passing in front of it.

The reviewing officer returns the salute of the commanding officer of the troops only. Those who accompany the reviewing officer do not salute. (712)

298. In passing in review, each staff salutes with its commander. (713)

After saluting the reviewing officer, the commanding officer of the troops turns out of the column and takes his post on the right of the reviewing officer, his staff on the right of the reviewing officer's staff. When the rear element of his command has passed, without changing his position, he salutes the reviewing officer and then rejoins his command. The commanding officer of the troops and his staff are the only ones who turn out of column. (714)

If the person reviewing the command is not mounted, the commanding officer and his staff on turning out of the column after passing the reviewing officer dismount preparatory to taking post. In such case, the salute of the commanding officer, prior to rejoining his command, is made with the hand before remounting. (715)

When the rank of the reviewing officer entitles him to the honor, each regimental color salutes at the command **present arms,** given or repeated by the major of the battalion with which it is posted; and again in passing in review. (716)

The band of an organization plays while the reviewing officer is passing in front of and in rear of the organization.

Each band, immediately after passing the reviewing officer, turns out of the column, takes post in front of and facing him, continues to play until its regiment has passed, then ceases playing and follows in rear of its regiment; the band of the following regiment commences to play as soon as the preceding band has ceased.

While marching in review but one band in each brigade plays at a time, and but one band at a time when within 100 paces of the reviewing officer. (717)

If the rank of the reviewing officer entitles him to the honor, the band plays the prescribed **national air** or the field music sounds **to the color, march, flourishes,** or **ruffles** when arms are presented. When passing in review at the moment the regimental color salutes, the musicians halted in front of the reviewing officer, sound **to the color, march, flourishes,** or **ruffles.** (718)

The formation for review may be modified to suit the ground, and the **present arms** and the ride around the line by the reviewing officer may be dispensed with. (719)

299. If the post of the reviewing officer is on the left of the column, the troops march in review with the guide left; the commanding officer and his staff turn out of the column to the left, taking post as prescribed above, but to the left of the reviewing officer; in saluting, the captains give the command: 1. **Eyes**, 2. **LEFT**. (720)

Except in the review of a single battalion, the troops pass in review in quick time only. (721)

In reviews of brigades or larger commands, each battalion, after the rear has passed the reviewing officer 50 paces, takes the double time for 100 yards in order not to interfere with the march of the column in rear; if necessary, it then turns out of the column and returns to camp by the most practicable route; the leading battalion of each regiment is followed by the other units of the regiment. (722)

In a brigade or larger review a regimental commander may cause his regiment to stand **at ease, rest,** or **stack arms** and **fall out** and **resume attention,** so as not to interfere with the ceremony. (723)

When an organization is to be reviewed before an inspector junior in rank to the commanding officer, the commanding officer receives the review and is accompanied by the inspector, who takes post on his left. (724)

Battalion Review

300. The battalion having been formed in line, the major faces to the front; the reviewing officer moves a few paces toward the major and halts; the major turns about and commands: 1. **Present**, 2. **ARMS**, and again turns about and salutes.

The reviewing officer returns the salute; the major turns about brings the battalion to order arms, and again turns to the front.

The reviewing officer approaches to about 6 paces from the major, the latter salutes, takes post on his right, and accompanies him around the battalion. The band plays. The reviewing officer proceeds to the right of the band, passes in front of the captains to the left of the line and returns to the right, passing in rear of the file closers and the band.

On arriving again at the right of the line, the major salutes, halts, and when the reviewing officer and staff have passed moves directly to his post in front of the battalion, faces it, and commands: 1. **Pass in review**, 2. **Squads right**, 3. **MARCH.**

At the first command the band changes direction if necessary, and halts.

At the third command, given when the band has changed direction, the battalion moves off, the band playing; without command from the major the column changes direction at the points indicated, and column of companies at full distance is formed successively to the left at the second change of direction; the major takes his post 30 paces in front of the band immediately after the second change; the band having passed the reviewing officer, turns to the left out of the column, takes post in front of and facing the reviewing officer, and remains there until the review terminates.

The major and staff salute, turn the head as in **eyes right**, and look toward the reviewing officer when the major is 6 paces from him; they return to the carry and turn the head and eyes to the front when the major has passed 6 paces beyond him.

Without facing about, each captain or special unit commander, except the drum major, commands: 1. **Eyes**, in time to add, 2. **RIGHT**, when at 6 paces from the reviewing officer, and commands **FRONT** when at 6 paces beyond him. At the command **eyes** the company officers and noncommissioned officers armed with the saber execute the first motion of present saber; at the command **right** all turn head and eyes to the right and the company officers complete **present saber**; at the command **front** all turn head and eyes to the front and resume the carry saber.

Noncommissioned staff officers, noncommissioned officers in command of subdivisions, and the drum major salute, turn the head and eyes, return to the front, resume the carry or drop the hand, at the points prescribed for the major. Officers and dismounted noncommissioned officers in command of subdivisions with arms in hand render the rifle or saber salute. Guides charged with the step, trace, and direction do not execute **eyes right**.

If the reviewing officer is entitled to a salute from the color the regimental color salutes when at 6 paces from him, and is raised when at 6 paces beyond him.

The major, having saluted, takes post on the right of the reviewing officer, remains there until the rear of the battalion has passed, then salutes and rejoins his battalion. The band ceases to play when the column has completed its second change of direction after passing the reviewing officer. (725)

When the battalion arrives at its original position in column, the major commands: 1. **Double time**, 2. **MARCH**.

The band plays in double time.

The battalion passes in review as before, except that in double time the command **eyes right** is omitted and there is no saluting except by the major when he leaves the reviewing officer.

The review terminates when the rear company has passed the reviewing officer; the band then ceases to play, and, unless otherwise directed by the major, returns to the position it occupied before marching in review, or is dismissed; the major rejoins the battalion and brings it to quick time. The battalion then executes such movements as the reviewing officer may have directed, or is marched to its parade ground and dismissed.

Marching past in double time may, in the discretion of the reviewing officer, be omitted; the review terminates when the major rejoins his battalion. (726)

At battalion review the major and his staff may be dismounted in the discretion of the commanding officer. (727)

Regimental Review

301. The regiment is formed in line or in line of masses.

In line the review proceeds as in the battalion, substituting "colonel" for "major" and "regiment" for "battalion."

To march the regiment in review, the colonel commands: **PASS IN REVIEW**. The band changes direction, if necessary, and halts. Each major then commands: 1. **Squads right**, 2. **MARCH**.

The band marches at the command of the major of the leading battalion.

At the second change of direction each major takes post 20 paces in front of his leading company.

The rear of the column having passed the reviewing officer, the battalions, unless otherwise directed, are marched to their parades and dismissed.

In line of masses, when the reviewing officer has passed around the regiment, the colonel commands: **PASS IN REVIEW**. The band changes direction, if necessary, and halts. The major of the right battalion then commands: 1. **Column of squads, first company, squads right**, 2. **MARCH**. At the command march the band and the leading company of the right battalion move off. Each company and battalion in rear moves off in time to follow at its proper distance. (728)

The review of a small body of troops composed of different arms is conducted on the principles laid down for the regiment. The troops of each arm are formed and marched according to the drill regulations for that arm. (729)

Review of Large Commands

302. A command consisting of one regiment, or less, and detachments of other arms is formed for review as ordered by the commanding officer. The principles of regimental review will be observed whenever practicable. (730)

In the review of a brigade or larger command the **present arms** and the ride around the line by the reviewing officer are omitted. The troops form and march in the order prescribed by the commanding officer. (731)

PARADES

General Rules

303. If dismounted, the officer receiving the parade, and his staff, stand at parade rest, with arms folded, while the band is sounding off; they resume attention with the adjutant. If mounted, they remain at attention. (732)

At the command **report**, given by a battalion adjutant, the captains in succession from the right salute and report; **A (or other) company, present or accounted for**; or, **A (or other) company, (so many) officers or enlisted men absent**, and resume the order saber; at the same command given by the regimental adjutant, the majors similarly report their battalions. (733)

Battalion Parade

304. At **adjutant's call** the battalion is formed in line but not presented. Lieutenants take their posts in front of the center of their respective platoons at the captain's command for dressing his company on the line. The major takes post at a convenient distance in front of the center and facing the battalion.

The adjutant, from his post in front of the center of the battalion, after commanding: 1. **Guides**, 2. **POSTS**, adds: 1. **Parade**, 2. **REST**; the battalion executes parade rest. The adjutant directs the band: **SOUND OFF.**

The band, playing in quick time, passes in front of the line of officers to the left of the line and back to its post on the right, when it ceases playing. At evening parade, when the band ceases playing, retreat is sounded by the field music and, following the last note and while the flag is being lowered, the band plays **The Star Spangled Banner.**

Just before the last note of retreat, the adjutant comes to attention and, as the last note ends, commands: 1. **Battalion**, 2. **ATTENTION**. When the band ceases playing he commands: 1. **Present**, 2. **ARMS**. He then turns about and reports: **Sir, the parade is formed.** The major directs the adjutant: **Take your post, Sir.** The adjutant moves at a trot (if dismounted, in quick time), passes by the major's right, and takes his post.

The major draws saber and commands: 1. **Order**, 2. **ARMS,** and adds such exercises in the manual of arms as he may desire. Officers, noncommissioned officers commanding companies or armed with the saber, and the color guard, having once executed order arms, remain in that position during the exercises in the manual.

The major then directs the adjutant: **Receive the reports, Sir.** The adjutant, passing by the major's right, advances at a trot (if dismounted, in quick time) toward the center of the line, halts midway between it and the major, and commands: **REPORT.**

304 (contd.)

The reports received, the adjutant turns about, and reports: **Sir, all are present or accounted for;** or **Sir, (so many) officers or enlisted men are absent,** including in the list of absentees those from the band and field music reported to him by the drum major prior to the parade.

The major directs: **Publish the orders, Sir.**

The adjutant turns about and commands: **Attention to orders;** he then reads the orders, and commands: 1. **Officers,** 2. **CENTER,** 0. **MARCH.**

At the command **center,** the company officers carry saber and face to the center. At the command **march,** they close to the center and face to the front; the adjutant turns about and takes his post.

The officers having closed and faced to the front, the senior commands: 1. **Forward,** 2. **MARCH.** The officers advance, the band playing; the left officer of the center or right center company is the guide, and marches on the major; the officers are halted at 6 paces from the major by the senior who commands: 1. **Officers,** 2. **HALT.** They halt and salute, returning to the carry saber with the major. The major then gives such instructions as he deems necessary, and commands: 1. **Officers,** 2. **POSTS,** 3. **MARCH.**

At the command **posts,** company officers face about.

At the command **march,** they step off with guide as before, and the senior commands: 1. **Officers,** 2. **HALT,** so as to halt 3 paces from the line; he then adds: 1. **Posts,** 2. **MARCH.**

At the command **posts,** officers face outward and, at the command **march,** step off in succession at 4 paces distance, resume their posts and order saber; the lieutenants march directly to their posts in rear of their companies.

The music ceases when all officers have resumed their posts.

The major then commands: 1. **Pass in review,** 2. **Squads right,** 3. **MARCH,** and returns saber.

The battalion marches according to the principles of review; when the last company has passed, the ceremony is concluded.

The band continues to play while the companies are in march upon the parade ground. Companies are formed in column of squads, without halting, and are marched to their respective parades by their captains.

When the company officers have saluted the major, he may direct them to form line with the staff, in which case they individually move to the front, passing to the right and left of the major and staff, halt on the line established by the staff, face about, and stand at attention. The music ceases when the officers join the staff. The major causes the companies to pass in review under the command of their first sergeants by the same commands as before. The company officers return saber with the major and remain at attention. (734)

Regimental Parade

305. The regiment is formed in line or in line of masses; the formation having proceeded up to, but not including the **present**, the parade proceeds as described for the battalion, with the following exceptions:

"Colonel" is substituted for "major", "regiment" for "battalion," in the description, and "battalions" for "battalion" in the commands.

Lieutenants remain in the line of file closers.

After publishing the orders, the adjutant commands: 1. **Officers, center**, 2. **MARCH.**

The company commanders remain at their posts with their companies.

The field and staff officers form one line, closing on the center. The senior commands: 1. **Forward**, 2. **MARCH.**

The second major is the guide and marches on the colonel.

After being dismissed by the colonel, each major moves individually to the front, turns outward, and followed by his staff resumes his post by the most direct line. The colonel directs the lieutenant colonel to march the regiment in review; the latter moves to a point midway between the colonel and the regiment and marches the regiment in review as prescribed. If the lieutenant colonel is not present the colonel gives the necessary commands for marching the regiment in review. (735)

ESCORTS

Escort of the Color

306. The regiment being being in line, the colonel details a company, other than the color company, to receive and escort the national color to its place in line. During the ceremony the regimental color remains with the color guard at its post with the regiment.

The band moves straight to its front until clear of the line of field officers, changes direction to the right, and is halted; the designated company forms column of platoons in rear of the band, the color bearer or bearers between the platoons.

The escort then marches without music to the colonel's office or quarters and is formed in line facing the entrance, the band on the right, the color bearer in the line of file closers.

The color bearer, preceded by the first lieutenant and followed by a sergeant of the escort, then goes to obtain the color.

When the color bearer comes out, followed by the lieutenant and sergeant, he halts before the entrance, facing the escort; the lieutenant places himself on the right, the sergeant on the left of the color bearer; the escort presents arms, and the field music sounds to the color; the first lieutenant and sergeant salute.

Arms are brought to the order; the lieutenant and sergeant return to their posts; the company is formed in column of platoons, the band taking post in front of the column; the color bearer places himself between the platoons; the escort marches in quick time, with guide left, back to the regiment, the band playing; the march is so conducted that when the escort arrives at 50 paces in front of the right of the regiment, the direction of the march shall be parallel to its front; when the color arrives opposite its place in line, the escort is formed in line to the left; the color bearer, passing between the platoons, advances and halts 12 paces in front of the colonel.

The color bearer having halted, the colonel, who has taken post 30 paces in front of the center of his regiment, faces about, commands: 1. **Present**, 2. **ARMS**, resumes his front, and salutes; the field music sounds to the color; and the regimental color bearer executes the color salute at the command **present arms**.

The colonel then faces about, brings the regiment to the order, at which the color bearer takes his post with the color company.

The escort presents arms and comes to the order with the regiment, at the command of the colonel, after which the captain forms it again in column of platoons, and, preceded by the band, marches it to its place in line, passing around the left flank of the regiment.

The band plays until the escort passes the left of the line, when it ceases playing and returns to its post on the right, passing in rear of the regiment.

The regiment may be brought to a rest when the escort passes the left of the line. (736)

Escort of the color is executed by a battalion according to the same principles. (737)

Escorts of Honor

307. Escorts of honor are detailed for the purpose of receiving and escorting personages of high rank, civil or military. The troops for this purpose are selected for their soldierly appearance and superior discipline.

The escort forms in line, opposite the place where the personage presents himself, the band on the flank of the escort toward which it will march. On the appearance of the personage, he is received with the honors due to his rank. The escort is formed into column of companies, platoons or squads, and takes up the march, the personage and his staff or retinue taking positions in rear of the column; when he leaves the escort, line is formed and the same honors are paid as before.

When the position of the escort is at a considerable distance from the point where the personage is to be received, as for instance, where a courtyard or wharf intervenes, a double line of sentinels is posted from that point to the escort, facing inward; the sentinels successively salute as he passes and are then relieved and join the escort.

An officer is appointed to attend him and bear such communication as he may have to make to the commander of the escort. (738)

Funeral Escort

308. The composition and strength of the escort are prescribed in *Army Regulations.*

The escort is formed opposite the quarters of the deceased; the band on that flank of the escort toward which it is to march.

Upon the appearance of the coffin, the commander commands: 1. **Present,** 2. **ARMS,** and the band plays an appropriate air; arms are then brought to the order.

The escort is next formed into column of companies, platoons, or squads. If the escort be small, it may be marched in line. The procession is formed in the following order: 1. **Music,** 2. **Escort,** 3. **Clergy,** 4. **Coffin and pallbearers,** 5. **Mourners,** 6. **Members of the former command of the deceased,** 7. **Other officers and enlisted men,** 8. **Distinguished persons,** 9. **Delegations,** 10. **Societies,** 11. **Civilians.** Officers and enlisted men (Nos. 6 and 7), with side arms, are in the order of rank, seniors in front.

The procession being formed, the commander of the escort puts it in march.

The escort marches slowly to solemn music; the column having arrived opposite the grave, line is formed facing it.

The coffin is then carried along the front of the escort to the grave; arms are presented, the music plays an appropriate air; the coffin having been placed over the grave, the music ceases and arms are brought to the order.

The commander next commands: 1. **Parade,** 2. **REST.** The escort executes **parade rest,** officers and men inclining the head.

When the funeral services are completed and the coffin lowered into the grave the commander causes the escort to resume attention and fire three rounds of blank cartridges, the muzzles of the pieces being elevated. When the escort is greater than a battalion, one battalion is designated to fire the volleys.

A musician then sounds **taps.**

The escort is then formed into column, marched in quick time to the point where it was assembled, and dismissed.

The band does not play until it has left the inclosure.

When the distance to the place of interment is considerable, the escort, after having left the camp or garrision, may march **at ease** in quick time until it approaches the burial ground, when it is brought to attention. The music does not play while marching **at ease.**

In marching at **attention,** the field music may alternate with the band in playing. (739)

When arms are presented at the funeral of a person entitled to any of the following honors, the band plays the prescribed **national air,** or the field music sounds to the color, march, flourishes, or ruffles, according to the rank of the deceased, after which the band plays an appropriate air. The commander of the escort, in forming column, gives the appropriate commands for the different arms. (740)

At the funeral of a mounted officer or enlisted man, his horse, in mourning caparison, follows the hearse. (741)

Should the entrance of the cemetery prevent the hearse accompanying the escort till the latter halts at the grave, the column is halted at the entrance long enough to take the coffin from the hearse, when the column is again put in march. The Cavalry and Artillery, when unable to enter the inclosure, turn out of the column, face the column, and salute the remains as they pass. (742)

When necessary to escort the remains from the quarters of the deceased to the church before the funeral service, arms are presented upon receiving the remains at the quarters and also as they are borne into the church. (743)

The commander of the escort, previous to the funeral, gives the clergyman and pallbearers all needful directions. (744)

INSPECTIONS

Company Inspection

309. Being in line at a halt: 1. **Open ranks**, 2. **MARCH**.

At the command march the front rank executes right dress; the rear rank and the file closers march backward 4 steps, halt, and execute right dress; the lieutenants pass around their respective flanks and take post, facing to the front, 3 paces in front of the center of their respective platoons. The captain aligns the front rank, rear rank, and file closers, takes post 3 paces in front of the right guide, facing to the left, and commands: 1. **Front**, 2. **PREPARE FOR INSPECTION**.

At the second command the lieutenants carry saber; the captain returns saber and inspects them, after which they face about, order saber, and stand at ease; upon the completion of the inspection they carry saber, face about, and order saber. The captain may direct the lieutenants to accompany or assist him, in which case they return saber and, at the close of the inspection, resume their posts in front of the company, draw and carry saber.

Having inspected the lieutenants, the captain proceeds to the right of the company. Each man, as the captain approaches him, executes **inspection arms**.

The captain takes the piece, grasping it with his right hand just above the rear sight, the man dropping his hands. The captain inspects the piece, and, with the hand and piece in the same position as in receiving it, hands it back to the man, who takes it with the left hand at the balance and executes **order arms**.

As the captain returns the piece the next man executes **inspection arms**, and so on through the company.

Should the piece be inspected without handling, each man executes **order arms** as soon as the captain passes to the next man.

The inspection is from right to left in front, and from left to right in rear, of each rank and of the line of file closers.

When approached by the captain the first sergeant executes **inspection saber.** Enlisted men armed with the pistol execute **inspection pistol** by drawing the pistol from the holster and holding it diagonally across the body, barrel up, and 6 inches in front of the neck, muzzle pointing up and to the left. The pistol is returned to the holster as soon as the captain passes.

Upon completion of the inspection the captain takes post facing to the left in front of the right guide and on line with the lieutenants and commands: 1. **Close ranks**, 2. **MARCH.**

At the command **march** the lieutenants resume their posts in line; the rear rank closes to 40 inches, each man covering his file leader; the file closers close to 2 paces from the rear rank. (745)

If the company is dismissed, rifles are put away. In quarters headdress and accoutrements are removed and the men stand near their respective bunks; in camp they stand covered, but without accoutrements, in front of their tents.

The captain, accompanied by the lieutenants, then inspects the quarters or camp. The first sergeant precedes the captain and calls the men to attention on entering each squad room or on approaching the tents; the men stand at attention, but do not salute. (746)

310. If the inspection is to include an examination of the equipment, the captain, after closing ranks, causes the company to stack arms, to march backward until 4 paces in rear of the stacks and to take intervals. He then commands: 1. **Unsling Equipment**, 2. **OPEN PACKS.**

At the first command each man unslings his equipment and places it on the ground at his feet, haversack to the front, end of the pack 1 foot in front of toes.

At the second command, pack carriers are unstrapped, packs removed and unrolled, the longer edges of the pack along the lower edge of the cartridge belt. Each man exposes shelter-tent pins; removes meat can, knife, fork, and spoon from the meat-can pouch, and places them on the right of the haversack, knife, fork, and spoon in the opened meat can; removes the canteen and cup from the cover and places them on the left side of the haversack; unstraps and spreads out haversack so as to expose its contents; folds up the carrier to uncover the cartridge pouches; opens same; unrolls toilet articles and places them on the outer flap of the haversack; opens first-aid pouch and exposes contents to view. Each man then resumes the attention. (Pl. VI.)

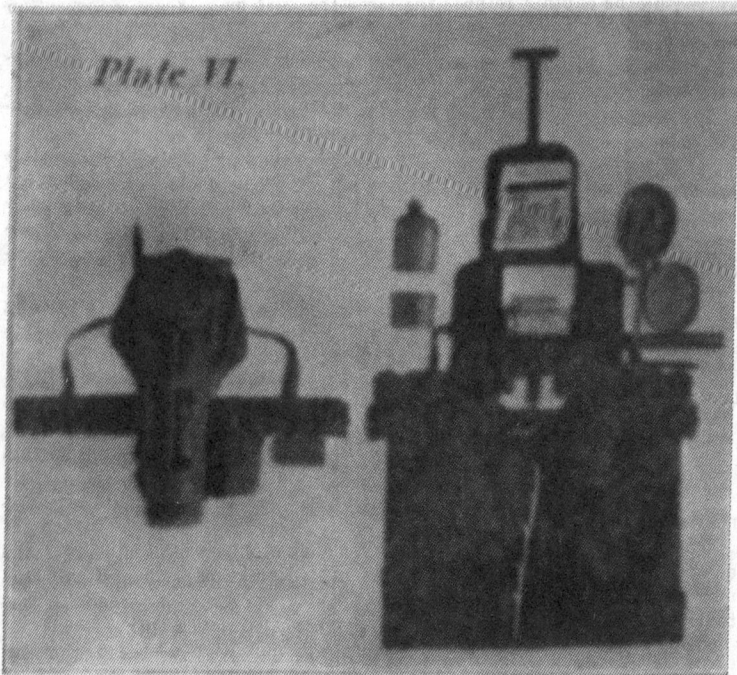

When the rations are not carried in the haversack the inspection proceeds as described, except that the toilet articles and bacon and condiment cans are displayed on the unrolled packs.

The captain then passes along the ranks and file closers as before, inspects the equipments, returns to the right, and commands: **CLOSE PACKS.**

Each man rolls up his toilet articles, straps up his haversack and its contents, replaces the meat can, knife, fork, and spoon, and the canteen and cup; closes cartridge pockets and first-aid pouch; rolls up and replaces pack in carrier, and, leaving the equipment in its position on the ground, resumes the position of attention.

All equipments being packed, the captain commands: **SLING EQUIPMENT.**

The equipments are slung and belts fastened.

The captain then causes the company to assemble and take

arms. The inspection is completed as already explained. (747) (See Par. 346.)

Should the inspector be other than the captain, the latter, after commanding **front**, adds **REST**, and faces to the front. When the inspector approaches, the captain faces to the left, brings the company to attention, faces to the front, and salutes. The salute acknowledged, the captain carries saber, faces to the left, commands: **Prepare for inspection**, and again faces to the front.

The inspection proceeds as before; the captain returns saber and accompanies the inspector as soon as the latter passes him. (748)

Battalion Inspection

311. If there be both inspection and review, the inspection may either precede or follow the review.

The battalion being in column of companies at full distance, all officers dismounted, the major commands: 1. **Prepare for inspection**, 2. **MARCH.**

At the first command each captain commands: **Open ranks.**

At the command **march** the ranks are opened in each company, as in the inspection of the company.

The field musicians join their companies.

The drum major conducts the band to a position 30 paces in rear of the column, if not already there, and opens ranks.

The major takes post facing to the front and 20 paces in front of the center of the leading company. The staff takes post as if mounted. The color takes post 5 paces in rear of the staff.

Field and staff officers senior in rank to the inspector do not take post in front of the column but accompany him.

The inspector inspects the major, and, accompanied by the latter, inspects the staff officers.

The major then commands: **REST**, returns saber, and, with his staff, accompanies the inspector.

If the major is the inspector he commands: **REST**, returns saber, and inspects his staff, which then accompanies him.

The inspector, commencing at the head of the column, then makes a minute inspection of the color guard, the noncommissioned staff, and the arms, accoutrements, dress, and ammunition of each soldier of the several companies in succession, and inspects the band.

The adjutant gives the necessary commands for the inspection of the color guard, noncommissioned staff, and band.

The color guard and noncommissioned staff may be dismissed as soon as inspected. (749)

As the inspector approaches each company its captain commands: 1. **Company**, 2. **ATTENTION**, 3. **PREPARE FOR INSPECTION**, and faces to the front; as soon as inspected he returns saber and accompanies the inspector. The inspection proceeds as in company inspection. At its completion the captain closes ranks and commands: **REST**. Unless otherwise directed by the inspector, the major directs that the company be marched to its parade and dismissed. (750)

If the inspection will probably last a long time the rear companies may be permitted to stack arms and fall out, before the inspector approaches they fall in and take arms. (751)

The band plays during the inspection of the companies.

When the inspector approaches the band the adjutant commands: **PREPARE FOR INSPECTION**.

As the inspector approaches him each man raises his instrument in front of the body, reverses it so as to show both sides, and then returns it.

Company musicians execute inspection similarly. (752)

At the inspection of quarters or camp the inspector is accompanied by the captain, followed by the other officers or by such of them as he may designate. The inspection is conducted as described in the company inspection. (753)

Regimental Inspection

312. The commands, means, and principles are the same as described for a battalion.

The colonel takes post facing to the front and 20 paces in front of the major of the leading battalion. His staff takes post as if mounted. The color takes post 5 paces in rear of the staff.

The inspector inspects the colonel and the lieutenant colonel, and accompanied by the colonel, inspects the staff officers.

The colonel then commands: **REST**, returns saber, and, with the lieutenant colonel and staff, accompanies the inspector.

If the colonel is the inspector he commands: **REST**, returns saber, and inspects the lieutenant colonel and staff, all of whom then accompany him.

The inspector, commencing at the head of the column, makes a minute inspection of the color guard, noncommissioned staff, each battalion in succession, and the band.

On the approach of the inspector each major brings his battalion to attention. Battalion inspection follows. (754)

MUSTER
Regimental, Battalion, or Company Muster

313. Muster is preceded by an inspection, and, when practicable, by a review.

The adjutant is provided with the muster roll of the field, staff, and band, the surgeon with the hospital roll; each captain with the roll of his company. A list of absentees, alphabetically arranged, showing cause and place of absence, accompanies each roll. (755)

Being in column of companies at open ranks, each captain, as the mustering officer approaches, brings his company to right shoulder arms, and commands: **ATTENTION TO MUSTER.**

The mustering officer or captain then calls the names on the roll; each man, as his name is called, answers **Here** and brings his piece to order arms.

After muster, the mustering officer, accompanied by the company commanders and such other officers as he may designate, verifies the presence of the men reported in hospital, on guard, etc. (756)

A company may be mustered in the same manner on its own parade ground, the muster to follow the company inspection. (757)

THE COLOR

316. The word "color" implies the national color; it includes the regimental color when both are present.

The rules prescribing the colors to be carried by regiments and battalions on all occasions are contained in *Army Regulations.* (766)

In garrison the colors, when not in use, are kept in the office or quarters of the colonel, and are escorted thereto and therefrom by the color guard. In camp the colors, when not in use, are in front of the colonel's tent. From reveille to retreat, when the weather permits, they are displayed uncased; from retreat to reveille and during inclement weather they are cased.

Colors are said to be cased when furled and protected by the oil-cloth covering. (767)

The regimental color salutes in the ceremony of escort of the color, and when saluting an officer entitled to the honor, but in no other case.

If marching, the salute is executed when at 6 paces from the officer entitled to the salute; the carry is resumed when 6 paces beyond him.

The national color renders no salute. (768)

The Color Guard

317. The color guard consists of two color sergeants, who are the color bearers, and two experienced privates selected by the colonel. The senior color sergeant carries the national color; the junior color sergeant carries the regimental color. The regimental color, when carried, is always on the left of the national color, in whatever direction they may face. (769)

The color guard is formed and marched in one rank, the color bearers in the center. It is marched in the same manner and by the same commands as a squad, substituting, when necessary, **guard** for **squad.** (770)

The color company is the center or right center company of the center or right center battalion. The color guard remains with that company unless otherwise directed. (771)

In line the color guard is in the interval between the inner guides of the right and left center companies.

In line of columns or in close line, the color guard is midway between the right and left center companies and on line with the captains.

In column of companies or platoons the color guard is midway between the color company and the company in rear of the color company and equidistant from the flanks of the column.

In close column the color guard is on the flank of the color company.

In column of squads the color guard is in the column between the color company and the company originally on its left.

When the regiment is formed in line of masses for ceremonies, the color guard forms on the left of the leading company of the center (right center) battalion. It rejoins the color company when the regiment changes from line of masses. (772)

The color guard when with a battalion that takes the battle formation, joins the regimental reserve, whose commander directs the color guard to join a certain company of the reserve. (773)

The color guard executes neither loadings nor firings; in rendering honors, it executes all movements in the manual; in drill, all movements unless specially excused. (774)

To Receive the Color

318. The color guard, by command of the senior color sergeant, presents arms on receiving and parting with the color. After parting with the color, the color guard is brought to order arms by command of the senior member who is placed as the right man of the guard. (775)

At drills and ceremonies, excepting escort of the color, the color, if present, is received by the color company after its formation.

The formation of the color company completed, the captain faces to the front; the color guard, conducted by the senior sergeant, approaches from the front and halts at a distance of 10 paces from the captain, who then faces about, brings the company to the present, faces to the front, salutes, again faces about and brings the company to the order. The color guard comes to the present and order at the command of the captain, and is then marched by the color sergeant directly to its post on the left of the color company. (776)

When the battalion is dismissed the color guard escorts the color to the office or quarters of the colonel. (777)

Manual of the Color

319. At the **carry** the heel of the pike rests in the socket of the sling; the right hand grasps the pike at the height of the shoulder.

At the **order** the heel of the pike rests on the ground near the right toe, the right hand holding the pike in a vertical position.

At **parade rest** the heel of the pike is on the ground, as at the **order**; the pike is held with both hands in front of the center of the body, left hand uppermost.

The **order** is resumed at the command **attention**.

The left hand assists the right when necessary.

The **carry** is the habitual position when the troops are at a shoulder, port, or trail.

The **order** and **parade rest** are executed with the troops.

The color salute: Being at a carry, slip the right hand up the pike to the height of the eye then lower the pike by straightening the arm to the front. (778)

THE BAND

320. The band is formed in two or more ranks, with sufficient intervals between the men and distances between the ranks to permit of a free use of the instruments.

The field music, when united, forms with and in rear of the band; when the band is not present the posts, movements, and duties of the field music are the same as prescribed for the band; when a musician is in charge his position is on the right of the front rank. When the battalion or regiment turns about by squads, the band executes the countermarch; when the battalion or regiment executes **right, left,** or **about face,** the band faces in the same manner.

In marching, each rank dresses to the right.

In executing **open ranks** each rank of the band takes the distance of 3 paces from the rank next in front; the drum major verifies the alignment.

The field music sounds the **march, flourishes, or ruffles,** and **to the** color at the signal of the drum major. (779)

The drum major is 3 paces in front of the center of the front rank, and gives the signals or commands for the movements of the band as for a squad, substituting in the commands **band for squad.** (780)

Signals of the Drum Major

321. Preparatory to a signal the staff is held with the right hand near the head of the staff, hand below the chin, back to the front, ferrule pointed upward and to the right.

Prepare to play: Face toward the band and extend the right arm to its full length in the direction of the staff: **Play:** Bring the arm back to its original position in front of the body.

Prepare to cease playing: Extend the right arm to its full length in the direction of the staff. **Cease playing:** Bring the arm back to its original position in front of the body.

To march: Turn the wrist and bring the staff to the front, the ferrule pointing upward and to the front; extend the arm to its full length in the direction of the staff.

To halt: Lower the staff into the raised left hand and raise the staff horizontally above the head with both hands, the arms extended; lower the staff with both hands to a horizontal position at the height of the hips.

To countermarch: Face toward the band and give the signal to march. The countermarch is executed by each front-rank man to the right of the drum major turning to the right about, each to the left, turning to the left about, each followed by the men covering him. The drum major passes through the center.

To oblique: Bring the staff to a horizontal position, the head of the staff opposite the neck, the ferrule pointing in the direction the oblique is to be made; extend the arm to its full length in the direction of the staff.

To march by the right flank: Extend the arm to the right, the staff vertical, ferrule upward, back of the hand to the rear.

To march by the left flank: Extend the arm to the left, the staff vertical, ferrule upward, back of the hand to the front.

To diminish front: Let the ferrule fall into the left hand at the height of the eyes, right hand at the height of the hip.

To increase front: Let the ferrule fall into the left hand at the height of the hip, right hand at the height of the neck.

The march, flourishes, or ruffles: Bring the staff to a vertical position, hand opposite the neck, back of the hand to the front, ferrule pointing down.

To the color: Bring the staff to a horizontal position at the height of the neck, back of the hand to the rear, ferrule pointing to the left.

When the band is playing, in marching, the drum major beats the time with his staff and supports the left hand at the hip, fingers in front, thumb to the rear.

The drum major, with staff in hand, salutes by bringing his staff to a vertical position, head of the staff up and opposite the left shoulder.

The drum major, with staff in hand, salutes by bringing his staff to a vertical position, head of the staff up and opposite the left shoulder.

At a halt, and the band not playing, the drum major holds his staff with the ferrule touching the ground about 1 inch from toe of right foot, at an angle of about 60°, ball pointing upward to the right,

right hand grasping staff near the ball, back of the hand to the front; left hand at the hip, fingers in front, thumb to the rear. (781)

MANUAL OF THE SABER

322. 1. Draw, 2. SABER.

At the command **draw** unhook the saber with the thumb and first two fingers of the left hand, thumb on the end of the hook, fingers lifting the upper ring; grasp the scabbard with the left hand at the upper band, bring the hilt a little forward, seize the grip with the right hand, and draw the blade 6 inches out of the scabbard, pressing the scabbard against the thigh with the left hand.

At the command **saber** draw the saber quickly, raising the arm to its full extent to the right front, at an angle of about 45° with the horizontal, the saber, edge down, in a straight line with the arm; make a slight pause and bring the back of the blade against the shoulder, edge to the front, arm nearly extended, hand by the side, elbow back, third and fourth fingers back of the grip; at the same time hook up the scabbard with the thumb and first two fingers of the left hand, thumb through the upper ring, fingers supporting it; drop the left hand by the side.

This is the position of carry saber dismounted.

Par. 782.

Par 782

Par. 782.

Par. 784.

Officers and noncommissioned officers armed with the saber un-hook the scabbard before mounting; when mounted, in the first motion of **draw saber** they reach with the right hand over the bridle hand and without the aid of the bridle hand draw the saber as before; the right hand at the **carry** rests on the right thigh.

On foot the scabbard is carried hooked up.   (782)

When publishing orders, calling the roll, etc., the saber is held suspended from the right wrist by the saber knot; when the saber knot is used it is placed on the wrist before drawing saber and taken off after returning saber. (783)

322a. Being at the order or carry: 1. Present, 2. **SABER** (or **ARMS**).

At the command **present** raise and carry the saber to the front, base of the hilt as high as the chin and 6 inches in front of the neck, edge to the left, point 6 inches farther to the front than the hilt, thumb extended on the left of the grip, all fingers grasping the grip.

At the command **saber**, or **arms**, lower the saber, point in prolongation of the right foot and near the ground, edge to the left, hand by the side, thumb on left of grip, arm extended. If mounted, the hand is held behind the thigh, point a little to the right and front of the stirrup.

In rendering honors with troops officers execute the first motion of the salute at the command **present**, the second motion at the command **arms**; enlisted men with the saber execute the first motion at the command **arms** and omit the second motion. (784)

Being at a carry: 1. Order, 2. **SABER** (or **ARMS**).

Drop the point of the saber directly to the front, point on or near the ground, edge down, thumb on back of grip.

Being at the **present saber**, should the next command be **order arms**, officers and noncommissioned officers armed with the saber **order saber**; or the command be other than **order arms**, they execute **carry saber**.

When arms are brought to the order the officers or enlisted men with the saber drawn **order saber**. (785)

The saber is held at the carry while giving commands, marching at attention, or changing position in quick time.

When at the order sabers are brought to the carry when arms are brought to any position except the **present** or **parade rest**. (786)

Being at the order: 1. Parade, 2. **REST**.

Take the position of parade rest except that the left hand is uppermost and rests on the right hand, point of saber on or near the ground in front of the center of the body, edge to the right.

At the command **attention** resume the order saber and the position of the soldier. (787)

In marching in double time the saber is carried diagonally across the breast, edge to the front; the left hand steadies the scabbard. (788)

Par. 787.

Par. 785.

Par. 790.

Par. 788.

Officers and noncommissioned officers armed with the saber, on all duties under arms draw and return saber without waiting for command. All commands to soldiers under arms are given with the saber drawn. (789)

Being at a carry: 1. Return, 2. SABER.

At the command **return** carry the right hand opposite to and 6 inches from the left shoulder, saber vertical, edge to the left; at the same time unhook and lower the scabbard with the left hand and grasp it at the upper band.

At the command **saber** drop the point to the rear and pass the blade across and along the left arm; turn the head slightly to the left, fixing the eyes on the opening of the scabbard, raise the right hand. insert and return the blade; free the wrist from the saber knot (if inserted in it), turn the head to the front, drop the right hand by the side; hook up the scabbard with the left hand, drop the left hand by the side.

Officers and noncommissioned officers armed with the saber, when mounted, return saber without using the left hand; the scabbard is hooked up on dismounting. (790)

At inspection enlisted men with the saber drawn execute the first motion of **present saber** and turn the wrist to show both sides of the blade, resuming the carry when the inspector has passed. (791)

MANUAL OF TENT PITCHING

Shelter Tents

323. Being in line or in column of platoons, the captain commands: **FORM FOR SHELTER TENTS.**

The officers, first sergeant, and guides fall out; the cooks form a file on the flank of the company nearest the kitchen, the first sergeant and right guide fall in, forming the right file of the company; blank files are filled by the file closers or by men taken from the front rank; the remaining guide, or guides, and file closers form on a convenient flank. Before forming column of platoons, preparatory to pitching tents, the company may be redivided into two or more platoons, regardless of the size of each. (792) (See Par. 347.)

The captain then causes the company to take intervals as described in the School of the Squad, and commands: **PITCH TENTS.**

At the command **pitch tents**, each man steps off obliquely to the right with the right foot and lays his rifle on the ground, the butt of the rifle near the toe of the right foot, muzzle to the front, barrel to the left, and steps back into his place; each front-rank man then draws his bayonet and sticks it in the ground by the outside of the right heel.

Equipments are unslung, packs opened, shelter half and pins removed; each man then spreads his shelter half, small triangle to the rear, flat upon the ground the tent is to occupy, the rear-rank man's half on the right. The halves are then buttoned together; the guy loops at both ends of the lower half are passed through the buttonholes provided in the lower and upper halves; the whipped end of the guy rope is then passed through both guy loops and secured, this at both

ends of the tent. Each front-rank man inserts the muzzle of his rifle
under the front end of the ridge and holds the rifle upright, sling to
the front, heel of butt on the ground beside the bayonet. His rear-
rank man pins down the front corners of the tent on the line of bayonets,
stretching the tent taut, he then inserts a pin in the eye of the front
guy rope and drives the pin at such a distance in front of the rifle as
to hold the rope taut; both men go to the rear of the tent, each pins
down a corner, stretching the sides and rear of the tent before securing;
the rear-rank man then inserts an intrenching tool, or a bayonet in its
scabbard, under the rear end of the ridge inside the tent, the front-rank
man pegging down the end of the rear guy ropes; the rest of the pins
are then driven by both men, the rear-rank man working on the right.

The front flaps of the tent are not fastened down, but thrown
back on the tent.

As soon as the tent is pitched each man arranges his equipment
and the contents of his pack in the tent and stands at attention in front
of his own half on line with the front guyrope pin.

To have a uniform slope when the tents are pitched, the guy ropes
should all be of the same length. In shelter tents camps, in localities where
suitable material is procurable, tent poles may be improvised and used
in lieu of the rifle and bayonet or intrenching tool as supports for the
shelter tent. (793) (See Par. 348.)

When the pack is not carried the company is formed for shelter
tents, intervals are taken, arms are laid aside or on the ground, the
men are dismissed and proceed to the wagon, secure their packs, return
to their places, and pitch tents as heretofore described. (794)

Double shelter tents may be pitched by first pitching one tent
as heretofore described, then pitching a second tent against the opening
of the first, using one rifle to support both tents, and passing the front
guy ropes over and down the sides of the opposite tents. The front
corner of one tent is not pegged down, but is thrown back to permit
an opening into the tent. (795)

Single Sleeping Bag

324. Spread the poncho on the ground, buttoned end at the feet,
buttoned side to the left; fold the blanket once across its short dimen-
sion and lay it on the poncho, folded side along the right side of the
poncho; tie the blanket together along the left side by means of the
tapes provided; fold the left half of the poncho over the blanket and
button it together along the side and bottom. (796) (See Par. 351.)

Double Sleeping Bag

Spread one poncho on the ground, buttoned end at the feet,
buttoned side to the left; spread the blankets on top of the poncho;

tie the edges of the blankets together with the tapes provided; spread a second poncho on top of the blankets, buttoned end at the feet, buttoned side to the right; button the two ponchos together along both sides and across the end. (797) (See Par. 352.)

To Strike Shelter Tents

325. The men standing in front of their tents: **STRIKE TENTS**.

Equipments and rifles are removed from the tent; the tents are lowered, packs made up, and equipments slung, and the men stand at attention in the places originally occupied after taking intervals. (798) (See Par. 353.)

Common and Wall Tents

326. Four men pitch each tent.

Drive a pin to mark the center of the door; spread the tent on the ground to be occupied; place door loops over door pin; draw front corners taut, align, and peg them down; lace rear door, if necessary; draw rear corners taut in both directions and peg them down; the four corner guy pins are then driven in prolongation of the diagonals of the tent and about 2 paces beyond the corner pins; temporarily loosen the front door and the lee corner loops from the pins; insert uprights and ridge pole, inserting the pole pins in ridge pole and in eyelets of tent and fly; raise the tent; hold it in position; replace lee corner loops, and secure corner and fly guy ropes; tighten same to hold poles vertical; drive wall pins through the loops as they hang; drive intermediate guy pins, aligning them on corner pins already driven. (799)

The Pyramidal Tent

327. One squad pitches each tent.

The corporal drives a pin to mark the center of the door. The others of the squad unfold the tent and spread it out on the ground to be occupied, pole and tripod underneath. The corporal places the door loops over the door pin; one man goes to each corner of the tent; the two men in front draw the front corners taut, align the front of the tent with the company line of tents, and peg the corners down; the two men in rear draw rear corners taut in both directions and peg them down. The same four men drive the four corner guy pins in prolongation of the diagonals of the tent, about 2 paces beyond the corner pins. Meantime, the other men of the squad, having crept under the tent, insert the tent pole spindle in top plate, the corporal placing the hood in position; the pole is raised and the lower end inserted in the tripod socket; the tripod is raised to its proper height. Under the supervision of the corporal the men inside the tent shift the tripod and the men outside the tent handle the corner guy lines in such manner as to erect the tent with the corner eaves directly above the corner pins. Each outside man, moving to the left, drives pins for the wall loops along one side of the tent and, returning, drives the intermediate guy pins; in both cases the pins are aligned on the corner pins. The inside men assist. (800)

Conical Wall Tent

328. Drive the door pin and center pin 8 feet 3 inches apart. Using the hood lines with center pin as center, describe two concentric circles with radii 8 feet 3 inches and 11 feet 3 inches. In the outer circle drive two door guy pins 3 feet apart. At intervals of about 3 feet drive the other guy pin.

In other respects conical tents are erected practically as in the case of pyramidal tents. (801)

To Strike Common, Wall, Pyramidal, and Conical Wall Tents

329. STRIKE TENTS.

The men first remove all pins except those of the four corner guy ropes, or the four quadrant guy ropes in the case of the conical wall tent. The pins are neatly piled or placed in their receptacle.

One man holds each guy, and when the ground is clear the tent is lowered, folded, or rolled and tied, the poles or tripod and pole fastened together, and the remaining pins collected. (802)

To Fold Tents

330. For folding common, wall, hospital, and storage tents: Spread the tent flat on the ground, folded at the ridge so that bottoms of side walls are even, ends of tent forming triangles to the right and left; fold the triangular ends of the tent in toward the middle, making it rectangular in shape; fold the top over about 9 inches; fold the tent in two by carrying the top fold over clear to the foot; fold again in two from the top to the foot; throw all guys on tent except the second from each end; fold the ends in so as to cover about two-thirds of the second cloths; fold the left end over to meet the turned-in edge of the right end, then fold the right end over the top, completing the bundle; tie with the two exposed guys.

Method of Folding Pyramidal Tent

331. The tent is thrown toward the rear and the back wall and roof canvas pulled out smooth. This may be most easily accomplished by leaving the rear-corner wall pins in the ground with the wall loops attached, one man at each rear-corner guy, and one holding the square iron in a perpendicular position and pulling the canvas to its limit away from the former front of the tent. This leaves the three remaining sides of the tent on top of the rear side, with the door side in the middle.

Now carry the right-front corner over and lay it on the left-rear corner. Pull all canvas smooth, throw guys toward square iron, and pull bottom edges even. Then take the right-front corner and return to the right, covering the right-rear corner. This folds the right side of the tent on itself, with the crease in the middle and under the front side of tent.

Next carry the left-front corner to the right and back as described above; this when completed will leave the front and rear sides of the tent lying smooth and flat and the two side walls folded inward, each on itself.

Place the hood in the square iron which has been folded downward toward the bottom of tent, and continue to fold around the square iron as a core, pressing all folds down flat and smooth, and parallel with the bottom of the tent. If each fold is compactly made and the canvas kept smooth, the last fold will exactly cover the lower edge of the canvas. Lay all exposed guys along the folded canvas except the two on the center width, which should be pulled out and away from bottom edge to their extreme length for tying. Now, beginning at one end, fold toward the center on the first seam (that joining the first and second widths) and fold again toward the center so that the already folded canvas will come to within about 3 inches of the middle width. Then fold over to the opposite edge of middle width of canvas. Then begin folding from opposite end, folding the first width in half, then making a second fold to come within about 4 or 5 inches of that already folded; turn this fold entirely over that already folded. Take the exposed guys and draw them taut across each other, turn bundle over on the under guy, cross guys on top of bundle drawing tight. Turn bundle over on the crossed guys and tie lengthwise.

When properly tied and pressed together this will make a package 11 by 23 by 34 inches, requiring about 8,855 cubic inches to store or pack.

Stencil the organization designation on the lower half of the middle width of canvas in the back wall. (803)

MANUAL OF THE BUGLE

Morning Calls

332. First call, guard mounting, full dress, overcoats, drill, stable, water, and boots and saddles precede the assembly by such interval as may be prescribed by the commanding officer.

Mess, church, and fatigue, classed as service calls, may also be used as warning calls.

First call is the first signal for formation for roll call and for all ceremonies except guard mounting.

Guard mounting is the first signal for guard mounting.

The field music assembles at first call and guard mounting.

In a mixed command, boots and saddles is the signal to mounted troops that their formation is to be mounted; for mounted guard mounting or mounted drill, it immediately follows the signal guard mounting or drill.

When full dress or overcoats are to be worn, the full dress or overcoat call immediately follows first call, guard mounting, or boots and saddles. (804)

Formation Calls

333. **Assembly:** The signal for companies or details to fall in.

Adjutant's call: The signal for companies to form battalion; also for the guard details to form for guard mounting on the camp or garrison parade ground; it follows the assembly at such interval as may be prescribed by the commanding officer.

It is also used as a signal for the battalions to form regiment, following the first adjutant's call at such interval as the commanding officer may prescribe.

To the color: Is sounded when the color salutes. (805)

Alarm Calls

334. **Fire call:** The signal for the men to fall in, without arms, to extinguish fire.

To arms: The signal for the men to fall in, under arms, on their company parade grounds as quickly as possible.

To horse: The signal for mounted men to proceed under arms to their horses, saddle, mount and assemble at a designated place as quickly as possible. In extended order this signal is used to remount troops. (806)

Service Calls

335. **Tattoo, taps, mess, sick, church, recall, issue, officers', captains', first sergeants', fatigue, school, and the general.**

The general is the signal for striking tents and loading wagons preparatory to marching.

Reveille precedes the assembly for roll call; retreat follows the assembly, the interval between being only that required for formation and roll call, except when there is parade.

Taps is the signal for extinguishing lights; it is usually preceded by call to quarters by such interval as prescribed by *Army Regulations*.

Assembly, reveille, retreat, adjutant's call, to the color, the flourishes, ruffles, and the marches are sounded by all the field music united; the other calls, as a rule, are sounded by the musician of the guard or orderly musician; he may also sound the assembly when the musicians are not united.

The morning gun is fired at the first note of reveille, or, if marches be played before reveille, it is fired at the commencement of the first march.

The evening gun is fired at the last not of retreat. (807)

APPENDIX A

WAR DEPARTMENT,
OFFICE OF THE CHIEF OF STAFF,
Washington, December 2, 1911.

The Infantry Drill Regulations, 1911, have been prepared for the use of troops armed with the United States magazine rifle, model 1903. For the guidance of organizations armed with the United States magazine rifle, model 1898, the following alternative paragraphs are published and will be considered as substitute paragraphs for the corresponding paragraphs in the text: 75 (in part), 96, 98, 99, 134, 139, 141, 142, 148, and 150.

By order of the Secretary of War:

LEONARD WOOD,
Major General, Chief of Staff.

336 * * *

Third. The cut-off is kept turned down, except when using the magazine. (75)

* * * * * * *

337. Being at order arms: 1. **Unfix, BAYONET.**

If the bayonet scabbard is carried on the belt: Take the position of parade rest, grasp the handle of the bayonet firmly with the right hand, press the spring with the forefinger of the left hand, raise the bayonet until the handle is about 6 inches above the muzzle of the piece, drop the point to the left, back of hand toward the body, and, glancing at the scabbard, return the bayonet, the blade passing between the left arm and body; regrasp the piece with the right hand and resume the order.

If the bayonet scabbard is carried on the haversack: Take the bayonet from the rifle with the left hand and return it to the scabbard in the most convenient manner.

If marching or lying down, the bayonet is fixed and unfixed in the most expeditious and convenient manner and the piece returned to the original position.

Fix and unfix bayonet are executed with promptness and regularity, but not in cadence. (96)

338. Being at order arms: 1. **Inspection, 2. ARMS.**

At the second command, take the position of port arms. **(TWO)** With the right hand open the magazine gate, turn the bolt handle up, draw the bolt back and glance at the magazine and chamber. Having found them empty, or having emptied them, raise the head and eyes to the front. (98)

339. Being at inspection arms: 1. **Order (Right shoulder, port), 2. ARMS.**

At the preparatory command, push the bolt forward, turn the handle down, close the magazine gate, pull the trigger, and resume port arms. At the command **arms**, complete the movement ordered. (99)

340. Pieces being loaded and in the position of load, to execute other movements with the pieces loaded. 1, **Lock**, 2 **PIECES.**

At the command **pieces** turn the safety lock fully to the right.

The safety lock is said to be at the "ready" when turned to the left, and at the "safe" when turned to the right.

The cut-off is said to be "on" when turned up and "off" when turned down. (134)

341. Being in line or skirmish line at halt: 1. **With dummy (blank or ball) cartridges,** 2. **LOAD.**

At the command **load** each front-rank man or skirmisher faces half right and carries the right foot to the right, about one foot, to such position as will insure the greatest firmness and steadiness of the body; raises or lowers the piece and drops it into the left hand at the balance, left thumb extended along the stock, muzzle at the height of the breast. With the right hand he turns and draws the bolt back, takes a cartridge between the thumb and first two fingers and places it in the receiver; places palm of the hand against the back of the bolt handle; thrusts the bolt home with a quick motion, turning down the handle, and carries the hand to the small of the stock. Each rear-rank man moves to the right front, takes a similar position opposite the interval to the right of his front-rank man, muzzle of the piece extending beyond the front rank, and loads.

A skirmish line may load while moving, the pieces being held as nearly as practicable in the position of load.

If kneeling or sitting the position of the piece is similar; if kneeling the left forearm rests on the left thigh; if sitting the elbows are supported by the knees. If lying down the left hand steadies and supports the piece at the balance, the toe of the butt resting on the ground, the muzzle off the ground.

For reference, these positions (standing, kneeling, and lying down) are designated as that of **load.** (139)

342. FILL MAGAZINE.

Take the position of load, if not already there, open the gate of the magazine with the right thumb, take five cartridges from the box or belt, and place them, with the bullets to the front, in the magazine, turning the barrel slightly to the left to facilitate the insertion of the cartridges; close the gate and carry the right hand to the small of the stock.

To load from the magazine the command **from magazine** will be given preceding that of **load;** the **cut-off** will be turned up on coming to the position of **load.**

[144]

To resume loading from the belt the command **from belt** will be given preceding the command **load**; the cut-off will be turned down on coming to the position of **load**.

The commands **from magazine** and **from belt**, indicating the change in the manner of loading, will not be repeated in subsequent commands.

The words **from belt** apply to cartridge box as well as belt.

In loading from the magazine care should be taken to push the bolt fully forward and turn the handle down before drawing the bolt back, as otherwise the extractor will not catch the cartridge in the chamber, and jamming will occur with the cartridge following.

To fire from the magazine, the command **magazine fire** may be given at any time. The cut-off is turned up and an increased rate of fire is executed. After the magazine is exhausted the cut-off is turned down and the firing continued, loading from the belt.

Magazine fire is employed only when, in the opinion of the platoon leader or company commander, the maximum rate of fire becomes necessary. (141)

343. UNLOAD.

All take the position of load, turn the cut-off up, if not already there, turn the safety lock to the left, and alternately open and close the chamber until all the cartridges are ejected. After the last cartridge is ejected the chamber is closed and the trigger pulled. The cartridges are then picked up, cleaned, and returned to the box or belt, and the piece brought to the order. (142)

344. CLIP FIRE.

Turn the cut-off up; **fire at will** (reloading from the magazine) until the cartridges in the piece are exhausted; turn the cut-off down; fill magazine; reload and take the position of **suspend firing**. (148)

345. CEASE FIRING.

Firing stops; pieces not already there are brought to the position of load, the cut-off turned down if firing from magazine, the cartridge is drawn or the empty shell is ejected, the trigger is pulled, sights are laid down, and the piece is brought to the order.

Cease firing is used for long pauses to prepare for changes of position or to steady the men. (150)

APPENDIX B

WAR DEPARTMENT,
OFFICE OF THE CHIEF OF STAFF,
Washington, December 2, 1911.

Paragraphs 747, 792, 793, 794, 795, 796, 797, and 798, Infantry Drill Regulations, 1911, apply only to troops equipped with the Infantry Equipment, model 1910. For troops equipped under General Orders, No. 23,

War Department, 1906, and orders amendatory thereof, the alternative paragraphs published herewith will govern.

By order of the Secretary of War:

LEONARD WOOD,

Major General, Chief of Staff.

346. If the inspection is to include an examination of the blanket rolls the captain, before dismissing the company and after inspecting the file closers, directs the lieutenants to remain in place, closes ranks, stacks arms, dresses the company back to four paces from the stacks, takes intervals, and commands: 1. **Unsling,** 2. **PACKS,** 3. **Open,** 4. **PACKS.**

At the second command each man unslings his roll and places it on the ground at his feet, rounded end to the front, square end of shelter half to the right.

At the fourth command the rolls are untied, laid perpendicular to the front with the triangular end of the shelter half to the front, opened, and unrolled to the left; each man prepares the contents of his roll for inspection and resumes the attention.

The captain then returns saber, passes along the ranks and file closers as before, inspects the rolls, returns to the right, draws saber and commands: 1. **Close,** 2. **PACKS.**

At the second command each man, with his shelter half smoothly spread on the ground with buttons up and triangular end to the front, folds his blanket once across its length and places it upon the shelter half, fold toward the bottom edge one-half inch from the square end, the same amount of canvas uncovered at the top and bottom. He then places the parts of the pole on the side of the blanket next the square end of shelter half, near and parallel to the fold, end of pole about 6 inches from the edge of the blanket; nests the pins similarly near the opposite edge of the blanket and distributes the other articles carried in the roll; folds the triangular end and then the exposed portion of the bottom of the shelter half over the blanket.

The two men in each file roll and fasten first the roll of the front and then of the rear rank man. The file closers work similarly two and two, or with the front rank man of a blank file. Each pair stands on the folded side, rolls the blanket roll closely and buckles the straps, passing the end of the strap through both keeper and buckle, back over the buckle and under the keeper. With the roll so lying on the ground that the edge of the shelter half can just be seen when looking vertically downward one end is bent upward and over to meet the other, a clove hitch is taken with the guy rope first around the end to which it is attached and then around the other end, adjusting the length of rope between hitches to suit the wearer.

As soon as a file completes its two rolls each man places his roll in the position it was in after being unslung and stands at attention.

All the rolls being completed, the captain commands: 1. **Sling,** 2. **PACKS.**

At the second command the rolls are slung, the end containing the pole to the rear.

The company is assembled, takes arms, and the captain completes the inspection as before. (747)

347. Being in line or in column of platoons, the captain commands: **FORM FOR SHELTER TENTS.**

The officers, first sergeant, and guides fall out; the cooks form a file on the flank of the company nearest the kitchen, the first sergeant and right guide fall in, forming the right file of the company; blank files are filled by the file closers or by men taken from the front rank; the remaining guide or guides, and file closers form on a convenient flank. Before forming column of platoons, preparatory to pitching tents, the company may be redivided into two or more platoons regardless of the size of each. (792)

348. The captain then causes the company to take intervals as described in the School of the Squad, and commands: **PITCH TENTS.**

At the command **pitch tents,** each man steps off obliquely to the right with the right foot and lays his rifle on the ground, the butt of the rifle near the toe of the right foot, muzzle to the front, barrel to the left, and steps back into his place; each front-rank man then draws his bayonet and sticks it in the ground by the outside of the right heel. All unsling and open the blanket rolls and take out the shelter half, poles, and pins. Each then spreads his shelter half, triangle to the rear, flat upon the ground the tent is to occupy, rear-rank man's half on the right. The halves are then buttoned together. Each front-rank man joins his pole, inserts the top in the eyes of the halves, and holds the pole upright beside the bayonet placed in the ground; his rear rank man, using the pins in front, pins down the front corners of the tent on the line of bayonets, stretching the canvas taut; he then inserts a pin in the eye of the rope and drives the pin at such distance in front of the pole as to hold the rope taut. Both then go to the rear of the tent; the rear-rank man adjusts the pole and the front-rank man drives the pins. The rest of the pins are then driven by both men, the rear-rank man working on the right.

As soon as the tent is pitched each man arranges the contents of the blanket roll in the tent and stands at attention in front of his own half on line with the front guy rope pin.

The guy ropes, to have a uniform slope when the shelter tents are pitched, should all be of the same length. (793)

349. When the blanket roll is not carried, intervals are taken as described above; the position of the front pole is marked with a bayonet

and equipments are laid aside. The men then proceed to the wagon, secure their rolls, return to their places, and pitch tents as heretofore described. (794)

350. To pitch double shelter tent, the captain gives the same commands as before, except take half interval is given instead of take interval. In taking interval each man follows the preceding man at 2 paces. The captain then commands: **PITCH DOUBLE TENTS.**

The first sergeant places himself on the right of the right guide and with him pitches a single shelter tent.

Only the odd numbers of the front rank mark the line with the bayonet.

The tent is formed by buttoning together the square ends of two single tents. Two complete tents, except one pole, are used. Two guy ropes are used at each end, the guy pins being placed in front of the corner pins.

The tents are pitched by numbers 1 and 2, front and rear rank; and by numbers 3 and 4, front and rear rank; the men falling in on the left are numbered, counting off if necessary.

All the men spread their shelter halves on the ground the tent is to occupy. Those of the front rank are placed with the triangular ends to the front. All four halves are then buttoned together, first the ridges and then the square ends. The front corners of the tent are pinned by the front-rank men, the odd number holding the poles, the even number driving the pins. The rear-rank men similarly pin the rear corners.

While the odd numbers steady the poles, each even number of the front rank takes his pole and enters the tent, where, assisted by the even number of the rear rank, he adjusts the pole to the center eyes of the shelter halves in the following order: (1) The lower half of the front tent; (2) the lower half of the rear tent; (3) the upper half of the front tent; (4) the upper half of the rear tent. The guy ropes are then adjusted.

The tents having been pitched, the triangular ends are turned back, contents of the rolls arranged, and the men stand at attention, each opposite his own shelter half and facing out from the tent. (795)

351. Omitted. (796)
352. Omitted. (797)

CHAPTER II

MANUAL OF THE BAYONET

(The numbers following the paragraphs are those of the Manual of the Bayonet, and references in the text to certain paragraph numbers refer to these numbers and not to the numbers preceding the paragraphs.)

354. The infantry soldier relies mainly on fire action to disable the enemy, but he should know that personal combat is often necessary to obtain success. Therefore, he must be instructed in the use of the rifle and bayonet in hand-to-hand encounters. (1)

The object of this instruction is to teach the soldier how to make effective use of the rifle and bayonet in personal combat; to make him quick and proficient in handling his rifle; to give him an accurate eye and a steady hand; and to give him confidence in the bayonet in offense and defense. When skill in these exercises has been acquired, the rifle will still remain a most formidable weapon at close quarters should the bayonet be lost or disabled. (2)

Efficiency of organizations in bayonet fighting will be judged by the skill shown by individuals in personal combat. For this purpose pairs or groups of opponents, selected at random from among recruits and trained soldiers, should engage in assaults, using the fencing equipment provided for the purpose. (3)

Officers and specially selected and thoroughly instructed non-commissioned officers will act as instructors. (4)

Instruction in bayonet combat should begin as soon as the soldier is familiar with the handling of his rifle and will progress, as far as practicable, in the order followed in the text. (5)

Instruction is ordinarily given on even ground, but practice should also be had on uneven ground, especially in the attack and defense of intrenchments. (6)

These exercises will not be used as a calesthenic drill. (7)

The principles of the commands are the same as those given in paragraphs 9, 15, and 38, Infantry Drill Regulations. Intervals and distances will be taken as in paragraphs 109 and 111, Infantry Drill Regulations, except that, in formations for bayonet exercises, the men should be at least four paces apart in every direction. (8)

Before requiring soldiers to take a position or execute a movement for the first time, the instructor executes the same for the purpose of illustration, after which he requires the soldiers to execute the movement individually. Movements prescribed in this manual will not be executed in cadence as the attempt to do so results in incomplete execution and lack of vigor. Each movement will be executed correctly as quickly as possible by every man. As soon as the movements are executed accurately, the commands are given rapidly, as expertness with the bayonet depends chiefly upon quickness of motion. (9)

The exercises will be interrupted at first by short and frequent rests. The rests will be less frequent as proficiency is attained. Fatigue and exhaustion will be specially guarded against as they prevent proper interest being taken in the exercises and delay the progress of the instruction. Rests will be given from the position of order arms in the manner prescribed in *Infantry Drill Regulations.* (10)

THE BAYONET

NOMENCLATURE AND DESCRIPTION

354a. The bayonet is a cutting and thrusting weapon consisting of three principal parts, viz, the *blade, guard,* and *grip.* (11)

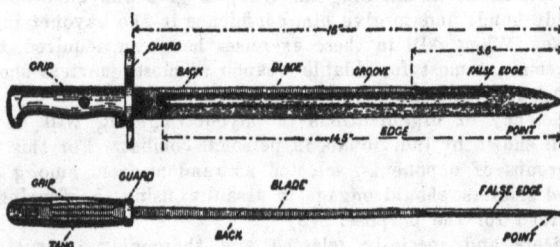

The blade has the following parts: Edge, false edge, back, grooves, point, and tang. The length of the blade from guard to point is 16 inches, the edge 14.5 inches, and the false edge 5.6 inches. Length of the rifle, bayonet fixed, is 59.4 inches. The weight of the bayonet is 1 pound; weight of rifle without bayonet is 8.69 pounds. The center of gravity of the rifle, with bayonet fixed, is just in front of the rear sight. (12)

I. INSTRUCTION WITHOUT THE RIFLE

The instructor explains the importance of good footwork and impresses on the men the fact that quickness of foot and suppleness of body are as important for attack and defense as is the ability to parry and deliver a strong point or cut. (13)

All foot movements should be made from the position of *guard.* As far as practicable, they will be made on the balls of the feet to insure quickness and agility. No hard and fast rule can be laid down as to the length of the various foot movements; this depends entirely on the situations occurring in combat. (14)

The men having taken intervals or distances, the instructor commands:

1. **Bayonet exercise,** 2. **GUARD.**

At the command **guard,** half face to the right, carry back and place the right foot about once and a half its length to the rear and about 3 inches to the right, the feet forming with each other an angle of about 60°, weight of the body balanced equally on the balls of the feet, knees slightly bent, palms of hands on hips, fingers to the front, thumbs to the rear, head erect, head and eyes straight to the front. (15)

To resume the attention, 1. **Squad** 2. **ATTENTION.** The men take the position of the soldier and fix their attention. (16)

ADVANCE. Advance the left foot quickly about once its length follow immediately with the right foot the same distance. (17)

RETIRE. Move the right foot quickly to the rear about once its length, follow immediately with the left foot the same distance. (18)

1. **Front,** 2. **PASS.** Place the right foot quickly about once its length in front of the left, advance the left foot to its proper position in front of the right. (19)

1. **Rear,** 2. **PASS.** Place the left foot quickly about once its length in rear of the right, retire the right foot to its proper position in rear of the left.

The passes are used to get quickly within striking distance or to withdraw quickly therefrom. (20)

1. **Right,** 2. **STEP.** Step to the right with the right foot about once its length and place the left foot in its proper relative position. (21)

1. **Left,** 2. **STEP.** Step to the left with the left foot about once its length and place the right foot in its proper relative position.

These steps are used to circle around an enemy, to secure a more favorable line of attack, or to avoid the opponent's attack. Better ground or more favorable light may be gained in this way. In bayonet fencing and in actual combat the foot first moved in stepping to the right or left is the one which at the moment bears the least weight. (22)

II. INSTRUCTION WITH THE RIFLE

The commands for and the execution of the foot movements are the same as already given for movements without the rifle. (23)

The men having taken intervals or distances, the instructor commands:

1. **Bayonet exercise,** 2. **GUARD.**

At the second command take the position of guard (see par. 15); at the same time throw the rifle smartly to the front, grasp the rifle with the left hand just below the lower band, fingers between the stock and gun sling, barrel turned slightly to the left, the right hand grasping the small of the stock about 6 inches in front of the right hip, elbows free from the body, bayonet point at the height of the chin. (24)

354a (contd.)

1. Order, 2. ARMS.

Bring the right foot up to the left and the rifle to the position of order arms, at the same time resuming the position of attention. (25)

During the preliminary instruction, attacks and defenses will be executed from guard until proficiency is attained, after which they may be executed from any position in which the rifle is held. (26)

ATTACKS

1. THRUST.

Thrust the rifle quickly forward to the full length of the left arm, turning the barrel to the left, and direct the point of the bayonet at the point to be attacked, butt covering the right forearm. At the same time straighten the right leg vigorously and throw the weight of the body forward and on the left leg, the ball of the right foot always on the ground. Guard is resumed immediately without command.

Par. 27.

Par. 24

The force of the thrust is delivered principally with the right arm, the left being used to direct the bayonet. The points at which the attack should be directed are, in order of their importance, stomach, chest, head, neck, and limbs. (27)

1. LUNGE.

Executed in the same manner as the thrust, except that the left foot is carried forward about twice its length. The left heel must always be in rear of the left knee. Guard is resumed immediately without command. Guard may also be resumed by advancing the right foot if, for any reason, it is desired to hold the ground gained in lunging. In the latter case, the preparatory command forward will be given. Each method should be practiced. (28)

1. Butt, 2. STRIKE.

Straighten right arm and right leg vigorously and swing butt of rifle against point of attack, pivoting the rifle in the left hand at about the height of the left shoulder, allowing the bayonet to pass to the rear on the left side of the head. Guard is resumed without command.

The points of attack in their order of importance are, head, neck, stomach, and crotch. (29)

Par. 28.

Par. 29.

1. Cut, 2. DOWN.

Execute a quick downward stroke, edge of bayonet directed at point of attack. Guard is resumed without command. (30)

1. Cut, 2. RIGHT (LEFT).

With a quick extension of the arms execute a cut to the right (left), directing the edge toward the point attacked. Guard is resumed without command.

The cuts are especially useful against the head, neck, and hands of an enemy. In executing left cut it should be remembered that the false, or back edge, is only 5.6 inches long. The cuts can be executed in continuation of strokes, thrusts, lunges, and parries. (31)

To direct an attack to the right, left, or rear the soldier will change front as quickly as possible in the most convenient manner, for example: 1. To the right rear, 2. Cut, 3. DOWN; 1. To the right, 2. LUNGE; 1. To the left, 2. THRUST, etc.

Whenever possible the impetus gained by the turning movement of the body should be thrown into the attack. In general this will be best accomplished by turning on the ball of the right foot.

These movements constitute a change of front in which the position of guard is resumed at the completion of the movement. (32)

Good judgment of distance is essential. Accuracy in thrusting and lunging is best attained by practicing these attacks against rings or other convenient openings, about 3 inches in diameter, suitably suspended at desired heights. (33)

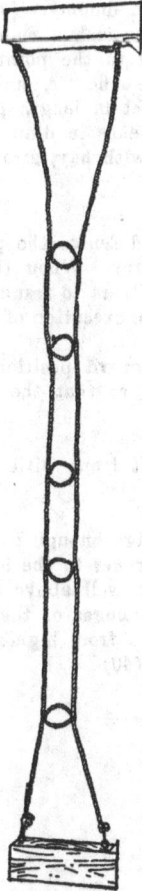

Par. 33.

Par. 36.

354a (contd.)

The thrust and lunges at rings should first be practiced by endeavoring to hit the opening looked at. This should be followed by directing the attack against one opening while looking at another. (34)

The soldier should also experience the effect of actual resistance offered to the bayonet and the butt of the rifle in attacks. This will be taught by practicing attacks against a dummy. (35)

Dummies should be constructed in such a manner as to permit the execution of attacks without injury to the point or edge of the bayonet or to the barrel or stock of the rifle. A suitable dummy can be made from pieces of rope about 5 feet in length plaited closely together into a cable between 6 and 12 inches in diameter. Old rope is preferable. Bags weighted and stuffed with hay, straw, shavings, etc., are also suitable. (36)

DEFENSES.

In the preliminary drills in the defenses the position of guard is resumed, by command, after each parry. When the men have become proficient, the instructor will cause them to resume the position of guard instantly without command after the execution of each parry. (37)

1. Parry, 2. **RIGHT**.

Keeping the right hand in the guard position, move the rifle sharply to the right with the left arm, so that the bayonet point is about 6 inches to the right. (38)

1. Parry, 2. **LEFT**.

Move the rifle sharply to the left front with both hands so as to cover the point attacked. (39)

1. Parry, 2. **HIGH**.

Raise the rifle with both hands high enough to clear the line of vision, barrel downward, point of the bayonet to the left front.

When necessary to raise the rifle well above the head, it may be supported between the thumb and forefinger of the left hand. This position will be necessary against attacks from higher elevations, such as men mounted or on top of parapets. (40)

Par. 40.   Par. 41.

1. Low parry, 2. RIGHT (LEFT).

Carry the point of the bayonet down until it is at the height of the knee, moving the point of the bayonet sufficiently to the right (left) to keep the opponent's attacks clear of the point threatened.

These parries are rarely used, as an attack below the waist leaves the head and body exposed. (41)

Par. 41.   Par. 44

Parries must not be too wide or sweeping, but sharp, short motions, finished with a jerk or quick catch. The hands should, as far as possible, be kept in the line of attack. Parries against **butt strike** are made by quickly moving the guard so as to cover the point attacked. (42)

To provide against attack from the right, left, or rear the soldier will change front as quickly as possible in the most convenient manner; for example: 1. **To the left rear**, 2. **Parry**, 3. **HIGH**; 1. **To the right**, 2. **Parry**, 3. **RIGHT**, etc.

These movements constitute a change of front in which the position of guard is resumed at the completion of the movement.

In changing front for the purpose of attack or defense, if there is danger of wounding a comrade, the rifle should first be brought to a vertical position. (43)

III. INSTRUCTION WITHOUT THE BAYONET

1. **Club rifle**, 2. **SWING**.

Being at order arms, at the preparatory command quickly raise and turn the rifle, regrasping it with both hands between the rear sight and muzzle, barrel down, thumbs around the stock and toward the butt; at the same time raise the rifle above the shoulder farthest from the opponent, butt elevated and to the rear, elbows slightly bent and knees straight. Each individual takes such position of the feet, shoulders, and hands as best accords with his natural dexterity. **SWING.** Tighten the grasp of the hands and swing the rifle to the front and downward, directing it at the head of the opponent and immediately return to the position of **club rifle** by completing the swing of the rifle downward and to the rear. Repeat by the command, **SWING.**

The rifle should be swung with sufficient force to break through any guard or parry that may be interposed.

Being at **club rifle**, order arms is resumed by command.

The use of this attack against dummies or in fencing is prohibited. (44)

Par. 44.

The position of **club rifle** may be taken from any position of the rifle prescribed in the Manual of Arms. It will not be taken in personal combat unless the emergency is such as to preclude the use of the bayonet. (45)

IV. COMBINED MOVEMENTS

The purpose of combined movements is to develop more vigorous attacks and more effective defenses than are obtained by the single movements; to develop skill in passing from attack to defense and the reverse. Every movement to the front should be accompanied by an attack, which is increased in effectiveness by the forward movement of the body. Every movement to the rear should ordinarily be accompained by a parry and should always be followed by an attack. Move ments to the right or left may be accompanied by attacks or defenses. (46)

Not more than three movements will be used in any combination. The instructor should first indicate the number of movements that are to be combined as two movements or three movements. The execution is determined by one command of execution, and the position of guard is taken upon the completion of the last movement only.

EXAMPLES

Front pass and LUNGE.
Right step and THRUST.
Left step and low parry RIGHT.
Rear pass, parry left and LUNGE.
Lunge and cut RIGHT.
Parry right and parry HIGH.
Butt strike and cut DOWN.
Thrust and parry HIGH.
Parry high and LUNGE.
Advance, thrust and cut RIGHT.
Right step, parry left and cut DOWN.
To the left, butt strike and cut DOWN.
To the right rear, cut down and butt STRIKE. (47)

Attacks against dummies will be practiced. The approach will be made against the dummies both in quick time and double time. (48)

V. PRACTICAL BAYONET COMBAT

The principles of practical bayonet combat should be taught as far as possible during the progress of instruction in bayonet exercises. (49)

The soldier must be continually impressed with the extreme importance of the offensive due to its moral effect. Should an attack fail, it should be followed immediately by another attack before the opponent has an opportunity to assume the offensive. Keep the opponent on the defensive. If, due to circumstances, it is necessary to take the defensive, constantly watch for an opportunity to assume the offensive and take immediate advantage of it. (50)

Observe the ground with a view to obtaining the best footing. Time for this will generally be too limited to permit more than a single hasty glance. (51)

In personal combat watch the opponent's eyes if they can be plainly seen, and do not fix the eyes on his weapon nor upon the point of your attack. If his eyes can not be plainly seen, as in night attacks, watch the movements of his weapon and of his body. (52)

Keep the body well covered and deliver attacks vigorously. The point of the bayonet should always be kept as nearly as possible in the line of attack. The less the rifle is moved upward, downward, to the right, or to the left, the better prepared the soldier is for attack or defense. (53)

Constantly watch for a chance to attack the opponent's left hand. His position of **guard** will not differ materially from that described in paragraph 24. If his bayonet is without a cutting edge, he will be at a great disadvantage. (54)

The butt is used for close and sudden attacks. It is particularly useful in riot duty. From the position of port arms a sentry can strike a severe blow with the butt of the rifle. (55)

Against a man on foot, armed with a sword, be careful that the muzzle of the rifle is not grasped. All the swordsman's energies will be directed toward getting past the bayonet. Attact him with short stabbing thrusts, and keep him beyond striking distance of his weapon. (56)

The adversary may attempt a greater extension in the thrust and lunge by quitting the grasp of his piece with the left hand and advancing the right as far as possible. When this is done, a sharp parry may cause him to lose control of his rifle, leaving him exposed to a counter-attack, which should follow promptly. (57)

Against odds a small number of men can fight to best advantage by grouping themselves so as to prevent their being attacked from behind. (58)

In fighting a mounted man armed with a saber every effort must be made to get on his near or left side, because here his reach is much shorter and his parries much weaker. If not possible to disable such an enemy, attack his horse and then renew the attack on the horseman (59)

In receiving night attacks the assailant's movements can be best observed from the kneeling or prone position, as his approach generally brings him against the sky line. When he arrives within attacking distance rise quickly and lunge well forward at the middle of his body. (60)

VI. FENCING EXERCISES

Fencing exercises in two lines consist of combinations of thrusts, parries, and foot movements executed at command or at will, the opponent replying with suitable parries and returns. (61)

354a (contd.)

The instructor will inspect the entire fencing equipment before the exercise begins and assure himself that everything is in such condition as will prevent accidents. (62)

The men equip themselves and form in two lines at the order, facing each other, with intervals of about 4 paces between files and a distance of about 2 paces between lines. One line is designated as number 1; the other, number 2. Also as attack and defense. (63)

The opponents being at the order facing each other, the instructor commands: **SALUTE.**

Each man, with eyes on his opponent, carries the left hand smartly to the right side, palm of the hand down, thumb and fingers extended and joined, forearm horizontal, forefinger touching the bayonet. (Two) Drop the arm smartly by the side.

This salute is the fencing salute.

All fencing exercises and all fencing at will between individuals will begin and terminate with the formal courtesy of the fencing salute. (64)

After the fencing salute has been rendered the instructor commands: 1. **Fencing exercise,** 2. **GUARD.**

At the command **guard** each man comes to the position of **guard,** heretofore defined, bayonets crossed, each man's bayonet bearing lightly to the right against the corresponding portion of the opponent's bayonet. This position is known as the **engage** or **engage right.** (65)

Being at the **engage right: ENGAGE LEFT.**

The attack drops the point of his bayonet quickly until clear of his opponent's rifle and describes a semicircle with it upward and to the right; bayonets are crossed similarly as in the engaged position, each man's bayonet bearing lightly to the left against the corresponding portion of the opponent's bayonet. (66)

Being at **engage left: ENGAGE RIGHT.**

The attack quickly drops the point of his bayonet until clear of his opponent's rifle and describes a semicircle with it upward and to the left and **engages.** (67)

Being engaged: **ENGAGE LEFT AND RIGHT.**

The attack **engages** left and then immediately **engages right.** (68)

Being engaged left: **ENGAGE RIGHT AND LEFT.**

The attack **engages** right and then immediately **engages left.** (69)

1. **Number one, ENGAGE RIGHT (LEFT);** 2. **Number two, COUNTER.**

Number one executes the movement ordered, as above; number two quickly drops the point of his bayonet and circles it upward to the original position. (70)

In all fencing while maintaining the pressure in the engage, a certain freedom of motion of the rifle is allowable, consisting of the play, or up-and-down motion, of one bayonet against the other. This is necessary to prevent the opponent from divining the intended attack. It also prevents his using the point of contact as a pivot for his assaults. In changing from one engage to the other the movement is controlled by the left hand, the right remaining stationary. (71)

After some exercise in **engage, engage left,** and **counter,** exercises will be given in the **assaults.** (72)

Assaults

The part of the body to be attacked will be designated by name as head, neck, chest, stomach, legs. No attacks will be made below the knees. The commands are given and the movements for each line are first explained thoroughly by the instructor; the execution begins at the command **assault.** Number one executes the attack, and number two parries; conversely, at command, number two attacks and number one parries. (73)

For convenience in instruction assaults are divided into **simple attacks, counter-attacks, attack on the rifle,** and **feints.** (74)

Simple Attacks

Success in these attacks depends on quickness of movement There are three simple attacks—the **straight,** the **disengagement,** and the **counter disengagement.** They are not preceded by a feint. (75)

In the **straight** the bayonet is directed straight at an opening from the engaged position. Contact with the opponent's rifle may, or may not, be abandoned while making it. If the opening be high or low, contact with the rifle will usually be abandoned on commencing the attack. If the opening be near his guard, the light pressure used in the engage may be continued in the attack.

Example: Being at the **engage right,** 1. **Number one, at neck** (head, chest, right leg, etc.), **thrust;** 2. **Number two, parry right;** 3. **ASSAULT.** (76)

In the **disengagement** contact with the opponent's rifle is abandoned and the point of the bayonet is circled **under** or **over** his bayonet or rifle and directed into the opening attacked. This attack is delivered by one continuous spiral movement of the bayonet from the moment contact is abandoned.

Example: Being at the **engage right,** 1. **Number one, at stomach** (left chest, left leg, etc.), **thrust** 2. **Number two, parry left** (etc.); 3. **ASSAULT.** (77)

In the **counter disengagement** a swift attack is made into the opening disclosed while the opponent is attempting to change the engagement of his rifle. It is delivered by one continuous spiral movement of the bayonet into the opening.

354a (contd.)

Example: Being at the engage right, 1. Number two, engage left; 2. Number one, at chest, thrust; 3. Number two, parry left; 4. ASSAULT.

Number two initiates the movement, number one thrusts as soon as the opening is made, and number two then attempts to parry. (78)

A counter-attack or return is one made instantly after or in continuation of a parry. The parry should be as narrow as possible. This makes it more difficult for the opponent to recover and counter parry. The counter-attack should also be made at, or just before, the full extension of the opponent's attack, as when it is so made, a simple extension of the arms will generally be sufficient to reach the opponent's body.

Example: Being at engage, 1. Number two, at chest, lunge; 2. Number one, parry right, and at stomach (chest, head, etc.), thrust; 3. ASSAULT. (79)

ATTACKS ON THE RIFLE

These movements are made for the purpose of forcing or disclosing an opening into which an attack can be made. They are the press, the beat, and the twist. (80)

In the press the attack quickly presses against the opponent's bayonet or rifle with his own and continues the pressure as the attack is delivered.

Example: Being at the engage, 1. Number one, press, and at chest, thrust; 2. Number two, parry right; 3. ASSAULT. (81)

The attack by disengagement is particularly effective following the press.

Example: Being at the engage, 1. Number one, press, and at stomach, thrust; 2. Number two, low parry left; 3. ASSAULT. (82)

The beat is an attack in which a sharp blow struck against the opponent's rifle for the purpose of forcing him to expose an opening into which an attack immediately follows. It is used when there is but slight opposition or no contact of rifles.

Example: Being at the engage, 1. Number one, beat, and at stomach (chest, etc.), thrust; 2. Number two, parry left; 3. ASSAULT. (83)

In the twist the rifle is crossed over the opponent's rifle or bayonet and his bayonet forced downward with a circular motion and a straight attack made into the opening. It requires superior strength on the part of the attack.

Example: Being at the engage, 1. Number one, twist, and at stomach, thrust; 2. Number two, low parry, left; 3. ASSAULT. (84)

FEINTS

Feints are movements which threaten or simulate attacks and are made with a view to inducing an opening or parry that exposes

[164]

the desired point of attack. They are either single or double, according to the number of such movements made by the attack. (85)

In order that the attack may be changed quickly, as little force as possible is put into a feint.

Example: Being at the **engage, Number one, feint** head **thrust** at stomach, lunge; 2. **Number two, parry right and low parry right;** 3. **ASSAULT.**

Number one executes the feint and then the attack. Number two executes both parries. (86)

In double feints first one part of the body and then another is threatened and a third attacked.

Example: Being at the **engage,** 1. **Number one, feint straight thrust** at chest; **disengagement** at chest; at stomach, **lunge;** 2. **Number two, parry right, parry left, and low parry left;** 3. **ASSAULT.** (87)

An opening may be offered or procured by opposition, as in the **press** or **beat.** (88)

In fencing exercises every feint should at first be parried. When the defense is able to judge or divine the character of the attack the feint is not necessarily parried, but may be nullified by a counter feint. (89)

A **counter feint** is a feint following the opponent's feint or following a parry of his attack and generally occurs in combined movements. (90)

Combined Movements

When the men have become thoroughly familiar with the various foot movements, parries, guards, attacks, feints, etc., the instructor combines several of them and gives the commands in quick succession, increasing the rapidity and number of movements as the men become more skillful. Opponents will be changed frequently.

1. Example: Being at the **engage,** 1. **Number one, by disengagement** at chest, **thrust;** 2. **Number two, parry left, right step** (left foot first), and **lunge;** 3. **ASSAULT.**

2. Example: Being at **engage left,** 1. **Number one, press and lunge;** 2. **Number two, parry right, left step, and thrust;** 3. **ASSAULT.**

3. Example: Being at the **engage,** 1. **Number one, by disengagement** at chest, **thrust;** 2. **Number two, parry left, front pass, and at head butt strike;** 3. **Number one, right step;** 4. **ASSAULT.** (91)

Examples 1 and 2 are typical of movements known as **cross counters,** and example No. 3 of movements known as **close counters.** (92)

A **chancery** is an attack by means of which the opponent is disarmed, which causes him to lose control of his rifle, or which disables his weapon. (93)

When the different combinations are executed with sufficient skill the instructor will devise series of movements to be memorized

354a (contd.)

and executed at the command **assault.** The accuracy and celerity of the movements will be carefully watched by the instructor, with a view to the correction of faulty execution. (94)

It is not intended to restrict the number of movements, but to leave to the discretion of company commanders and the ingenuity of instructors the selection of such other exercises as accord with the object of the drill. (95)

VII. FENCING AT WILL

As satisfactory progress is made the instructor will proceed to the exercises at will, by which is meant assaults between two men, each endeavoring to hit the other and to avoid being hit himself. Fencing at will should not be allowed to degenerate into random attacks and defenses. (96)

The instructor can supervise but one pair of combatants at a time. Frequent changes should be made so that the men may learn different methods of attack and defense from each other. (97)

The contest should begin with simple, careful movements, with a view to forming a correct opinion of the adversary; afterwards everything will depend on coolness, rapid and correct execution of the movements and quick perception of the adversary's intentions. (98)

Continual retreat from the adversary's attack and frequent dodging to escape attacks should be avoided. The offensive should be continually encouraged. (99)

In fencing at will, when no commands are given, opponents facing each other at the position of order arms, salute. They then immediately and simultaneously assume the position of guard, rifles engaged. Neither man may take the position of guard before his opponent has completed his salute. The choice of position is decided before the salute. (100)

The opponents being about two paces apart and the fencing salute having been rendered, the instructor commands, **at will, 2. ASSAULT,** after which either party has the right to attack. To interrupt the contest the instructor will command **HALT,** at which the combatants will immediately come to the order. To terminate the contest the instructor will command, 1. **Halt,** 2. **SALUTE,** at which the combatants will immediately come to the order, salute, and remove their masks. (101)

When men have acquired confidence in fencing at will, one opponent should be required to advance upon the other in quick time at **charge bayonet,** from a distance not to exceed 10 yards, and deliver an attack. As soon as a hit is made by either opponent the instructor commands, **HALT,** and the assault terminates. Opponents alternate in assaulting. The assailant is likewise required to advance at double time from a distance not exceeding 20 yards and at a run from a distance not exceeding 30 yards. (102)

The instructor will closely observe the contest and decide doubt-ful points. He will at once stop the contest upon the slightest indication of temper. After conclusion of the combat he will comment on the action of both parties, point out errors and deficiencies and explain how they may be avoided in the future. (103)

As additional instruction, the men may be permitted to wield the rifle left handed, that is on the left side of the body, left hand at the small of the stock. Many men will be able to use this method to ad-vantage. It is also of value in case the left hand is wounded. (104)

Par. 104.

After men have fenced in pairs, practice should be given in fencing between groups, equally and unequally divided. When practic-able, intrenchments will be used in fencing of this character.

In group fencing it will be necessary to have a sufficient number of umpires to decide hits. An individual receiving a hit is withdrawn at once from the bout, which is decided in favor of the group having the numerical superiority at the end. The fencing salute is not required in group fencing. (105)

RULES FOR FENCING AT WILL

1. Hits on the legs below the knees will not be counted. No hit counts unless, in the opinion of the instructor, it has sufficient force to disable.

354a (contd.)

2. Upon receiving a hit, call out "hit."

3. After receiving a fair hit a counter-attack is not permitted. A position of engage is taken.

4. A second or third hit in a combined attack will be counted only when the first hit was not called.

5. When it is necessary to stop the contest—for example, because of breaking of weapons or displacement of means of protection—take the position of the order.

6. When it is necessary to suspend the assault for any cause, it will not be resumed until the adversary is ready and in condition to defend himself.

7. Attacks directed at the crotch are prohibited in fencing.

8. Stepping out of bounds, when established, counts as a hit. (106)

Suggestions for Fencing at Will

When engaging in an assault, first study the adversary's position and proceed by false attacks, executed with speed, to discover, if possible, his instinctive parries. In order to draw the adversary out and induce him to expose that part of the body at which the attack is to be made, it is advisable to simulate an attack by a feint and then make the real attack. (107)

Return attacks should be frequently practiced, as they are difficult to parry, and the opponent is within easier reach and more exposed. The return can be made a continuation of the parry, as there is no previous warning of its delivery, although it should always be expected. Returns are made without lunging if the adversary can be reached by thrusts or cuts. (108)

Endeavor to overcome the tendency to make a return without knowing where it will hit. Making returns blindly is a bad habit and leads to instinctive returns—that is, habitual returns with certain attacks from certain parries—a fault which the skilled opponent will soon discover. (109)

Do not draw the rifle back preparatory to thrusting and lunging (110)

The purpose of fencing at will is to teach the soldier as many forms of simple, effective attacks and defenses as possible. Complicated and intricate movements should not be attempted. (111)

Hints for Instructors

The influence of the instructor is great. He must be master of his weapon, not only to show the various movements, but also to lead in the exercises at will. He should stimulate the zeal of the men and arouse pleasure in the work. Officers should qualify themselves as instructors by fencing with each other. (112)

The character of each man, his bodily conformation, and his degree of skill must always be taken into account. When the instructor is demonstrating the combinations, feints, returns, and parries the rapidity of his attack should be regulated by the skill of the pupil and no more force than is necessary should be used. If the pupil exposes himself too much in the feints and parries the instructor will, by an attack convince him of his error; but if these returns be too swiftly or too strongly made the pupil will become overcautious and the precision of his attack will be impaired. The object is to teach the pupil, not to give exhibitions of superior skill. (113)

Occasionally the instructor should leave himself uncovered and fail to parry, in order to teach the pupil to take quick advantages of such opportunities. (114)

VIII. COMPETITIONS

In competitions between different organizations none but skillful fencers will be allowed to participate. (115)

In contests between two men judges may assign values to hits as follows:

	Thrusts and lunges.	Cuts.	Butt of rifle.
Stomach	4	....	1
Chest	3	....	....
Head	3	2	3
Neck	2	2	2
Legs	1	1	....
Arms and hands	1	1	....

Stepping out of bounds, 4 points. (116)

When superiority between two men is decided by bouts, each bout will be decided by itself, i. e., points won in one bout can not be carried over to another. (117)

Details other than those mentioned above will be arranged by the officials of the competition. (118)

CHAPTER III

MANUAL OF PHYSICAL TRAINING

(Extracts)

METHODS

355. In the employment of the various forms of physical training it is necessary that well-defined methods should be introduced in order that the object of this training may be attained in the most thorough and systematic manner. Whenever it is possible this work should be conducted out of doors. In planning these methods the following factors must be considered:

(a) The condition and physical aptitude of the men.
(b) The facilities.
(c) The time.
(d) Instruction material.

The question of the *physical aptitude* and *general condition*, etc., of the men is a very important one, and it should always determine the nature and extent of the task expected of them; never should the work be made the determining factor. In general, it is advisable to divide the men into three classes, viz, the recruit class, the intermediate class, and the advanced class. The work for each class should fit the capabilities of the members of that class and in every class it should be arranged progressively.

Facilities are necessarily to be considered in any plan of instruction, but as most posts are now equipped with better than average facilities the plan laid down in this Manual will answer all purposes.

Time is a decidedly important factor, and no plan can be made unless those in charge of this work know exactly how much time they have at their disposal. During the suspension of drills five periods a week, each of 45 minutes duration, should be devoted to physical training; during the drill period a 15-minute drill in setting-up exercises should be ordered on drill days. The time of day, too, is important. When possible, these drills should be held in the morning about two hours after breakfast, and at no time should they be held immediately before or after a meal.

The proper use of the *instruction material* is undoubtedly the most important part of an instructor's duty, for it not only means the selection of the proper material but its application. Every exercise has a function peculiarly it own; in other words, it has a certain affect upon a certain part of the body and plays a rôle in the development of the men. It is, therefore, the sum of these various exercises properly grouped that constitutes the method. So far as possible, every lesson should be planned to embrace setting-up exercises that call into action all parts of the body, applied gymnastics, apparatus work, and exercises that develop coördination and skill, such as jumping and vaulting.

The best results are obtained when these exercises which affect the extensor muscles chiefly are followed by those affecting the flexors; i. e., flexion should always be followed by extension, or vice versa. It is also advisable that a movement requiring a considerable amount of muscular exertion should be followed by one in which this exertion is reduced to a minimum. As a rule, especially in the setting-up exercises, one portion of the body should not be exercised successively; thus, arm exercises should be followed by a trunk exercise, and that in turn by a leg, shoulder, and neck exercise.

The following program of a week's work illustrates the application of the instruction material as described above: each drill is of 45 minutes duration:

FIRST DAY'S PROGRAM

1. Marching in quick and double time (5 minutes).
2. Setting-up exercises (15 minutes).
3. Applied gymnastics, flexor work, horizontal bar (15 minutes).
4. Jumping exercises (8 minutes).
5. Trunk and arm stretching exercises in conjunction with breath ing exercises (2 minutes).

SECOND DAY'S PROGRAM

1. Exercises in marching, combined with arm and leg exercises (10 minutes).
2. Setting-up exercises, chiefly trunk exercises (5 minutes).
3. Applied gymnastics, extensor work, parallel bars (15 minutes).
4. Vaulting, low vaulting bars (13 minutes).
5. Stretching and breathing exercises (2 minutes).

THIRD DAY'S PROGRAM

1. Marching in double and quick time (5 minutes).
2. Setting-up exercises, general work (15 minutes).
3. Applied gymnastics, flexor work, rings (15 minutes).
4. Jumping exercises (8 minutes).
5. Stretching and breathing exercises (2 minutes).

FOURTH DAY'S PROGRAM

1. Running and walking (5 minutes).
2. Setting-up exercises, general work (10 minutes).
3. Applied gymnastics, extensor work, side horse (15 minutes).

355 (contd.)

4. Climbing (13 minutes).
5. Stretching and breathing exercises (2 minutes).

FIFTH DAY'S PROGRAM

1. Marching quick time, running, and exercises while marching in quick time (10 minutes).
2. Setting-up exercises, trunk movements (5 minutes).
3. Applied gymnastics, flexor work, horizontal bar (15 minutes).
4. Vaulting, side horse vaults (13 minutes).
5. Stretching and breathing exercises (2 minutes).

Clubs, dumb-bells, bar bells, wands, or rifles may be substituted for the setting-up exercises occasionally, and the gymnastic contests may also be used in place of the jumping and vaulting exercises.

Large numbers may be employed in a body in the setting-up exercises and also in the exercises with the clubs, etc. In the applied or apparatus work, unless the facilities afford a sufficient number of the same kind of apparatus, it is advisable to divide the men into small squads.

Officers who have been placed in charge of this work must not for an instant lose sight of the fact that to them has been intrusted a part of the soldier's training which is of great importance, and that success or failure is dependent entirely upon themselves. Work as important as this is worthy of the best efforts, and it should never be intrusted to those who are not enthusiastic about it.

Whenever possible the officer in charge should conduct the work personally, as in no profession does the individuality and personal influence of a leader carry such weight as it does in the military.

A well-defined program should be mapped out before the drill begins, and this should be carried out faithfully. Every day's work should dovetail into the next and be progressive.

Instructors should not fail to do as much as possible themselves, as an example is always more impressive than a precept; it will also serve to keep the officer in fit condition.

Where commands are large, the athletic officer should be given officer assistants, whom he should train so that they may be able to carry out his program intelligently. If officers are not available, he should select likely enlisted men and train them to be leaders capable of taking charge of a squad.

The work laid down in this manual should not be followed blindly; every instructor should select such portions, and if necessary vary them, as in his opinion are productive of the best results under the conditions under which he is laboring.

The work should be so conducted that the men are developed harmoniously; that is, any tendency to develop one side or one portion of the body at the expense of the other should be avoided.

Insist upon accurate and precise execution of every movement. By doing so those other essential qualities, besides strength and endurance—activity, agility, gracefulness, and accuracy—will also be developed.

Exercises which require activity and agility, rather than those that require strength only, should be selected.

It should be constantly borne in mind that these exercises are the means and not the end; and if there be a doubt in the mind of the instructor as to the effect of an exercise, it is always well to err upon the side of safety. *Underdoing is rectifiable; overdoing is often not.* The object of this work is not the development of expert gymnasts, but the development of physically sound men by means of a system in which the chances of bodily injury are reduced to a minimum. When individuals show a special aptitude for gymnastics they may be encouraged, within limits, to improve this ability, but never at the expense of their fellows.

The drill should be made as attractive as possible, and this can best be accomplished by employing the mind as well as the body. The movements should be as varied as possible, thus constantly offering the men something new to make them keep their minds on their work. A movement many times repeated presents no attraction and is executed in a purely mechanical manner, which should always be discountenanced.

Short and frequent drills should be given in preference to long ones, which are liable to exhaust all concerned, and exhaustion means lack of interest and benefit. All movements should be carefully explained, and, if necessary, illustrated by the instructor.

The lesson should begin with the less violent exercises, gradually working up to those that are more so, then gradually working back to the simpler ones, so that the men at the close of the drill will be in as nearly a normal condition as possible.

When one portion of the body is being exercised, care should be taken that the other parts remain quiet as far as the conformation of the body will allow. The men must learn to exercise any one part of the body independent of the other part.

Everything in connection with physical training should be such that the men look forward to it with pleasure, not with dread, for the mind exerts more influence over the human body than all the gymnastic paraphernalia that was ever invented.

Exercise should be carried on as much as possible in the open air; at all times in pure, dry air.

[173]

Only those men whom the post surgeon declares to be physically qualified and sure to be benefited thereby should be put through a course of physical training.

Never exercise the men to the point of exhaustion. If there is evidence of panting, faintness, fatigue, or pain, the exercise should be stopped at once, for it is nature's way of saying "too much."

By constant practice the men should learn to breathe slowly through the nostrils during all exercises, especially running.

A fundamental condition of exercise is unimpeded respiration. Proper breathing should always be insisted upon; "holding the breath" and breathing only when it can no longer be held is injurious. Every exercise should be accompanied by an unimpeded and, if possible, by an uninterrupted act of respiration, the inspiration and respiration of which depends to a great extent upon the nature of the exercise. Inhalation should always accompany that part of an exercise which tends to elevate and distend the thorax—as raising arms over head laterally, for instance; while that part of an exercise which exerts a pressure against the walls of the chest should be accompanied by exhalation, as for example, lowering arms laterally from shoulders or overhead.

If after exercising, the breathing becomes labored and distressed, it is an unmistakable sign that the work has been excessive. Such excessiveness is not infrequently the cause of serious injury to the heart and lungs, or to both. In cases where exercise produces palpitation, labored respiration, etc., it is advisable to recommend absolute rest, or to order the execution of such exercises as will relieve the oppressed and overtaxed organ. Leg exercises slowly executed will afford great relief. By drawing the blood from the upper to the lower extremities they equalize the circulation, thereby lessening the heart's action and quieting the respiration.

Never exercise immediately after a meal; digestion is more important at this time than extraneous exercise.

Never eat or drink immediately after exercise; allow the body to recover its normal condition first, and the most beneficial results will follow. If necessary, pure water, not too cold, may be taken in small quantities, but the exercise should be continued, especially if in a state of perspiration.

Never, if at all possible, allow the underclothing to dry on the body. Muscular action produces an unusual amount of bodily heat; this should be lost gradually, otherwise the body will be chilled; hence, after exercise, never remove clothing to cool off, but, on the contrary, wear some wrap in addition. In like manner, be well wrapped on leaving the gymnasium.

Cold baths, especially when the body is heated, as in the case after exercising violently, should be discouraged. In individual instances such baths may appear apparently beneficial, or at least not injurious; in a majority of cases, however, they can not be used with impunity. Tepid baths are recommended. When impossible to bathe, the flannels worn while exercising should be stripped off; the body sponged with tepid water, and then rubbed thoroughly with coarse towels. After such a sponge the body should be clothed in clean, warm clothing.

Flannel is the best material to wear next to the body during physical drill, as it absorbs the perspiration, protects the body against drafts and, in a mild manner, excites the skin. When the conditions permit it the men may be exercised in the ordinary athletic costume, sleeveless shirt, flappers, socks, and gymnasium shoes.

COMMANDS—SETTING-UP EXERCISES

COMMANDS

There are two kinds of commands:

The preparatory indicates the movement to be executed.

The command of execution causes the execution.

In the command: 1. **Arms forward**, 2. **RAISE**, the words **Arms forward** constitute the preparatory command, and Raise, the command of execution. Preparatory commands are printed in **bold face**, and those of execution in **CAPITALS**.

The tone of command is animated, distinct, and of a loudness proportioned to the number of men for whom it is intended.

The various movements comprising an exercise are executed by commands and, unless otherwise indicated, the continuation of an exercise is carried out by repeating the command, which usually takes the form of numerals, the numbers depending upon the number of movements, that an exercise comprises. Thus, if an exercise consists of two movements, the counts will be one, two; or if it consists of eight movements, the counts will be correspondingly increased; thus every movement is designated by a separate command.

Occasionally, especially in exercises that are to be executed slowly, words rather than numerals are used, and these must be indicative of the nature of the various movements.

In the continuation of an exercise the preparatory command is explanatory, the command of execution causes the execution and the continuation is caused by a repetition of numerals denoting the number of movements required, or of words describing the movements if words are used. The numerals or words preceding the command **halt** should always be given with a rising inflection on the first numeral or word of command of the last repetition of the exercise in order to prepare the men for the command **halt**.

[175]

355 (contd.)

For example:

1. Arms to thrust, 2. RAISE, 3. Thrust arms upward, 4. EXER-CISE, ONE, TWO, ONE, TWO, ONE, HALT; the rising inflection preparatory to the command halt being placed on the "one" preceding the "halt."

Each command must indicate, by its tone, how that particular movement is to be executed; thus, if an exercise consists of two movements, one of which is to be energized, the command corresponding to that movement must be emphasized.

Judgment must be used in giving commands, for rarely is the cadence of two movements alike; and a command should not only indicate the cadence of an exercise, but also the nature if its execution.

Thus, many of the arm exercises are short and snappy; hence the command should be given in a smart tone of voice, and the interval between the commands should be short.

The leg exercises can not be executed as quickly as those of the arms; therefore, the commands should be slightly drawn out and follow one another in slow succession.

The trunk exercises, owing to the deliberateness of execution, should be considerably drawn out and follow one another in slow succession.

The antagonistic exercises, where one group of muscles is made to antagonize another, tensing exercises, the commands are drawn still more. In these exercises words are preferable to numerals. In fact it should be the object of the instructor to convey to the men, by the manner of his command, exactly the nature of the exercise.

All commands should be given in a clear and distinct tone of voice, articulation should be distinct, and an effort should be made to cultivate a voice which will inspire the men with enthusiasm and tend to make them execute the exercises with willingness, snap, and precision. It is not the volume, but the quality, of the voice which is necessary to successful instruction.

THE POSITION OF ATTENTION

This is the position an unarmed dismounted soldier assumes when in ranks. During the setting-up exercises, it is assumed whenever the command attention is given by the instructor.

Having allowed his men to rest, the instructor commands: 1. Squad, 2. ATTENTION. Figs. A and B.

The words class, section, or company may be substituted for the word "squad."

At the command attention, the men will quickly assume and retain the following position:

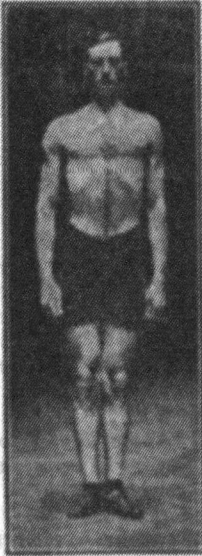 

Fig. A.   Fig. B.

Heels on same line and as near each other as the conformation of the man permits.

Feet turned out equally and forming an angle of about 45 degrees.

Knees straight without stiffness.

The body erect on the hips, the spine extended throughout its entire length.

The shoulders falling naturally, are forced back until they are square.

Chest arched and slightly raised.

The arms hang naturally; thumbs along seams of trousers; back of hands out and elbows turned back.

Head erect, chin drawn in so that the axis of the head and neck is vertical; eyes straight to the front and, when the nature of the terrain permits it, fixed on an object at their own height.

Too much attention can not be given to this position, and instructors are cautioned to insist that the men accustom themselves to it. As a rule, it is so exaggerated that it not only becomes ridiculous, but positively harmful. The men must be taught to assume a natural and graceful position, one from which all rigidity is eliminated and from which action is possible without first relaxing muscles that have been constrained in an effort to maintain the position of attention. In other words, coördination rather than strength should be depended upon.

In the position described the weight rests principally upon the balls of the feet, the heels resting lightly upon the ground.

The knees are extended easily, but never locked.

The body is now inclined forward until the front of the thighs is directly over the point of the toes; the hips are square and the waist is extended by the erection of the entire spine, but never to such a degree that mobility of the waist is lost.

In extending the spine, the chest is naturally arched and the abdomen is drawn in, but never to the extent where it interferes with respiration.

In extending the spinal column, the shoulders must not be raised, but held loosely in normal position and forced back until the points of the shoulders are at right angles with an anterior-posterior plane running through the shoulders.

The chin should be square; i. e., horizontal and forced back enough to bring the neck in a vertical plane; the eyes fixed to the front and the object on which they are fixed must be at their own height whenever the nature of the terrain permits it.

When properly assumed, a vertical line drawn from the top of the head should pass in front of the ear, just in front of the shoulder and of the thigh, and find its base at the balls of the feet.

All muscles should be contracted only enough to maintain this position, which at all times should be a lithesome one, that can be maintained for a long period without fatigue—one that makes for activity and that is based upon a correct anatomical and physicological basis.

Instructors will correct the position of attention of every man individually and they will ascertain, when the position has been properly assumed, whether the men are "on their toes," i. e., carrying the weight on the balls of the feet, whether they are able to respire properly, and whether they find a strain across the small of the back, which should be as flat as possible. This should be repeated until the men are able to assume the position correctly without restraint or rigidity.

At the command **rest** or **at ease** the men, while carrying out the provisions of the drill regulations, should be cautioned to avoid assuming any position that has a tendency to nullify the object of the position of attention; standing on leg for instance; allowing the shoulders to slope forward; drooping the head; folding arms across chest, etc. The weight should always be distributed equally upon both legs; the head, trunk, and shoulders remain erect and the arms held in a position that does not restrict the chest or derange the shoulders. The positions illustrated here have been found most efficacious. Figs. C. and D.

Fig. C.

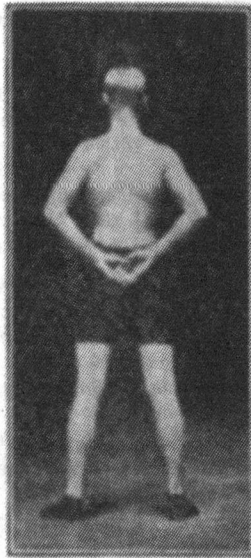

Fig. D.

FORMATIONS

The men form in a single or double rank, the tallest men on the right.

The instructor commands: 1. **Count off.**

At this command, all except the right file execute "eyes right" and, beginning on the right, the men in each rank count 1, 2, 3, 4; each man turns his head and eyes to the front as he counts.

The instructor then commands: 1. **Take distance,** 2. **MARCH,** 3. **Squad,** 4. **HALT.**

At the command **march,** No. 1 of the front rank moves straight to the front; Nos. 2, 3, and 4 of the front and Nos. 1, 2, 3, and 4 of the rear rank in the order named move straight to the front, each stepping off, so as to follow the preceding man at four paces; the command halt is given when all have their distances.

If it is desired that a less distance than four paces be taken, the distance desired should be indicated in the preparatory command. The men of the squad may be caused to cover No. 1 front rank by command **cover.**

The instructor then commands: 1. **Right (left),** 2. **FACE,** 3. **COVER.**

355 (contd.)

At these commands the men face in the direction indicated and cover in file.

To assemble the squad the instructor commands: 1. **Right (left)**, 2. **FACE**, 3. Assemble, 4. **MARCH**.

After facing and at command march, No. 1 of the front rank stands fast, the other members of both ranks resuming their original positions, or for convenience in the gymnasium they may be assembled to the rear, in which case the assemblage is made on No. 4 of the rear rank.

Unless otherwise indicated, the guide is always right.

SPECIAL TRAINING

In addition to the regular squad or class work instructors should, when they notice a physical defect in any man, recommend some exercise which will tend to correct it.

The most common physical defects and corresponding corrective exercises are noted here.

DROOPING HEAD

Exercise the muscles of the neck by bending, turning, and circling the head, muscles tense.

ROUND AND STOOPED SHOULDERS

Stretch arms sideward from front horizontal, turning palms upward, muscles tense.

Swing arms forward and backward, muscles relaxed.

Circle arms forward and backward slowly, energize backward motion, muscles tense; forward motion with muscles relaxed.

Circle shoulders backward, move them forward first, then raise them, then move them backward as far as possible in the raised position, muscles tense, and then lower to normal position, muscles relaxed.

WEAK BACK

Bend trunk forward as far as possible and erect it slowly.

Bend trunk forward, back arched and head thrown back.

Bend trunk sideward, without moving hips out of normal position, right and left.

Lie on floor, face down, and raise head and shoulders.

WEAK ABDOMEN

Circle trunk right or left.

Bend trunk backward or obliquely backward.

Bend head and trunk backward without moving hips out of normal plane.

Lie on floor, face up, and raise head and shoulders slightly; or to sitting position or raise legs slightly; or to a vertical position.

To increase depth and width of chest

Arm stretchings, sideward and upward, muscles tense.

Same, with deep inhalations.

Arm swings and arm circles outward, away from the body.

Raise extended arms over head laterally and cross them behind the head.

Breathing exercises in connection with arm and shoulder exercises.

STARTING POSITIONS

In nearly all the arm exercises it is necessary to hold the arms in some fixed position from which the exercises can be most advantageously executed, and to which position the arms are again returned upon completing the exercise. These positions are termed **starting positions**; and though it may not be absolutely necessary to assume one of them before or during the employment of any other portion of the body, it is advisable to do so, since they give to the exercise a finished, uniform, and graceful appearance.

In the following positions, at the command **down**, resume the **attention**. Practice in assuming the starting position may be had by repeating the commands of execution, such as **raise, down**.

Intervals having been taken and attention assumed, the instructor commands:

1. 1. **Arms forward**, 2. **RAISE**, 3. **Arms**, 4. **DOWN**. Fig. 1.

At the command **raise**, raise the arms to the front smartly, extended to their full length, till the hands are in front of and at the height of the shoulders, palms down, fingers extended and joined, thumbs under forefingers. At **Arms, DOWN**, resume position of attention.

FIG. 1.   [181]

FIG. 2.

2. 1. **Arms sideward**, 2. **RAISE**, 3. **Arms**, 4. **DOWN**. Fig. 2.

At the command **raise**, raise the arms laterally until horizontal, palms down, fingers as in 1.

The arms are brought down smartly without allowing them to touch the body.

FIG. 3.

3. 1. **Arms upward**, 2. **RAISE**, 3. **Arms**, 4. **DOWN**. Fig. 3.

At the command **raise**, raise the arms from the sides, extended to their full length, with the forward movement, until they are vertically overhead, backs of hands turned outward, fingers as in 1.

This position may also be assumed by raising the arms laterally until vertical. The instructor cautions which way he desires it done.

4. 1. **Arms backward**, 2. **CROSS**, 3. **Arms**, 4. **DOWN**, Fig. 4.

FIG. 4.

FIG. 5.

FIG. 6.

At the command **cross**, the arms are folded across the back; hands grasping forearms.

5. 1. **Arms to thrust**, 2. **RAISE**, 3. **Arms**, 4. **DOWN**. Fig. 5.

At the command **raise**, raise the forearms to the front until horizontal, elbow forced back, upper arms against the chest, hands tightly closed, knuckles down.

6. 1. **Hands on hips**, 2. **PLACE**, 3. **Arms**, 4. **DOWN**. Fig. 6.

At the command **place**, place the hands on the hips, the finger tips in line with trouser seams; fingers extended and joined, thumbs to the rear, elbows pressed back.

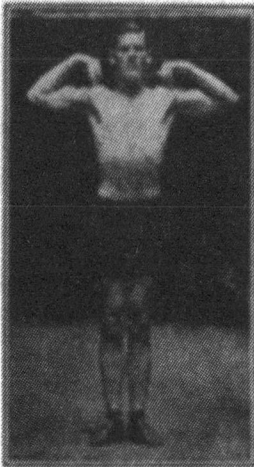

FIG. 7.

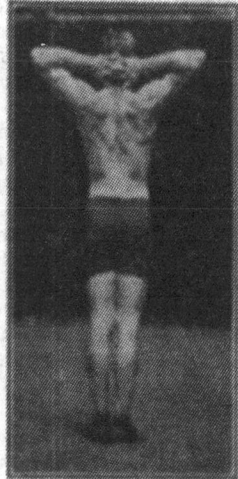

FIG. 8.

7. 1. **Hands on shoulders**, 2. **PLACE**, 3. **Arms**, 4. **DOWN**. Fig. 7.

At the command **place**, raise the forearms to the vertical position, palms inward, without moving the upper arms; then raise the elbows upward and outward until the upper arms are horizontal; at the same time bending the wrist and allowing the finger tips to rest lightly on the shoulders.

[183]

8.   1. **Fingers in rear of head,** 2. **LACE,** 3. Arms, 4. **DOWN.**   Fig. 8.

At the command lace, raise the arms and forearms as described in 7, and lace the fingers behind the lower portion of the head, elbows well up and pressed well back.

These positions should be practiced frequently, and instead of recovering the position of attention after each position, the instructor may change directly from one to another by giving the proper commands instead of commanding **arms, DOWN.**

For instance: To change from the position described in paragraph 8 to that described in paragraph 9 (having commanded: 1. **Hands on shoulders,** 2. **PLACE)** he commands: 1. **Hands on hips,** 2. **PLACE.**

These changes should, however, be made only after the positions are thoroughly understood and correctly assumed.

SETTING-UP EXERCISES

356. As has been stated previously (see par. 2), the exercises form the basis upon which the entire system of physical training in the service is founded.   Therefore too much importance can not be attached to them.   Through the number and variety of movements they offer it is possible to develop the body harmoniously with little if any danger of injurious results.   They develop the muscles and impart vigor and tone to the vital organs and assist them in their functions; they develop endurance and are important factors in the development of smartness, grace, and precision.   They should be assiduously practiced.   The fact that they require no apparatus of any description makes it possible to do this out of doors or even in the most restricted room, proper sanitary conditions being the only adjunct upon which their success is dependent.   No physical training drill is complete without them.   They should always precede the more strenuous forms of training, as they prepare the body for the greater exertion these forms demand.

The following series prescribed for the recruit and trained soldier's instruction is indicated here to illustrate the nature and amount and arrangement of work that should be required of each class.   At the discretion of instructors these exercises may be substituted by others of a similar character.   Instructors are cautioned, however, to employ all the parts of the body in every lesson and to suit the exercise as far as practicable to the natural function of the particular part of the body which they employ.

In these lessons only the preparatory command is given here; the command of execution, which is invariably **Exercise,** and the commands of continuance, as well as the command to discontinue, having been explained in paragraph 6, are omitted.

Every preparatory command should convey a definite description of the exercise required; by doing so long explanations are avoided and the men will not be compelled to memorize the various movements.

RECRUIT INSTRUCTION

First Series

1. Position of attention, from at ease and rest.
2. Starting position, Par. 10, Figs. 1 to 8.
3. 1. Raise and lower arms to side horizontal.
   Two counts; repeat 8 to 10 times, Fig. 2.
   The arms rigidly extended are brought to the sides smartly without coming in contact with the thighs. Inhale on first and exhale on second count.
4. 1. **Hands on hips, 2. PLACE, 3. QUARTER BEND TRUNK FORWARD.**
   Two counts; repeat 8 to 10 times, Fig. 9.
   The trunk is inclined forward at the waist about 45° and then extended again; the hips are as perpendicular as possible; execute slowly; exhale on first and inhale and raise chest on second count.

FIG. 9.

5. 1. **Arms to thrust, 2. RAISE, 3. RAISE SHOULDERS.**
   Two counts; repeat 8 to 10 times, Fig. 10.
   The shoulders are raised as high as possible without deranging the position of the body or head and lowered back to position; execute briskly; inhale on first and exhale on second count.

[185]

FIG. 10. FIG. 11.

6. 1. **Hands on hips,** 2. **PLACE,** 3. **QUARTER BEND KNEES.**

Two counts; repeat 8 to 10 times, Fig 11.

The knees are flexed until the point of the knee is directly over the toes; whole foot remains on ground; heels closed; head and body erect; execute moderately fast, emphasizing the extension; breathe naturally.

7. 1. **Arms backward,** 2. **CROSS,** 3. **RISE ON TOES.**

Two counts; repeat 8 to 10 times, Fig. 12.

The body is raised smartly until the toes and ankles are extended as much as possible; heels closed; head and trunk erect; in recovering position heels are lowered gently; breathe naturally.

8. 1. **Breathing exercise,** 2. **INHALE,** 3. **EXHALE.**

At inhale the arms are stretched forward overhead and the lungs are inflated; at exhale the arms are lowered laterally and the lungs deflated; execute slowly; repeat four times.

FIG. 12.

Second Series

FIG. 13.

1. Position of attention, as in first series.
2. Repeat first lesson.
3. 1. Hands on shoulders, 2. **PLACE**, 3. **EXTEND ARMS FORWARD.**

Two counts; repeat 8 to 10 times.

The arms are extended forward forcibly, palms down, and brought back to position smartly, elbows being forced back; exhale on first and inhale on second count.

4. 1. Hands on hips, 2. **PLACE**, 3. **BEND TRUNK BACKWARD.**

Two counts; repeat 6 to 8 times, Fig. 13.

The trunk is bent backward as far as possible; head and shoulders fixed; knees extended; feet firmly on the ground; hips as nearly perpendicular as possible; in recovering care should be taken not to sway forward; execute slowly; inhale on first and exhale on second count.

5. 1. Arms to thrust, 2. **RAISE**, 3. **MOVE SHOULDERS FORWARD.**

FIG. 14.

Two counts; repeat 8 to 10 times, Fig. 14.

The shoulders are relaxed and moved forward and in as far as possible and then moved backward without jerking; head and trunk erect; execute slowly; exhale on first and inhale on second count.

6. 1. Arms backward, 2. **CROSS**, 3. **HALF BEND KNEES.**

Two counts; repeat 8 to 10 times, Fig. 15.

The knees are separated and bent halfway to the ground, point of knee being forced downward; head and trunk erect; execute smartly and emphasize the extension; breathe naturally.

FIG. 15.

[187]

FIG. 16.

7. 1. **Hands on hips, 2. PLACE, 3. HALF BEND TRUNK FORWARD.**

Two counts; repeat 8 to 10 times, Fig. 16.

The trunk is inclined forward until it is at right angles to the legs, hips perpendicular; knees extended; head and shoulders fixed; execute moderately slow; exhale on first and inhale and raise chest on second count.

8. 1. **Hands on shoulders, 2. PLACE, 3. STRIKE ARMS SIDEWARD.**

The arms, knuckles down, hands closed, are flung outward forcibly and brought back to shoulders smartly; execute fast; breathe naturally.

9. Breathing exercise, as in first lesson.

Third Series

1. Position of attention, as in first series.

2. Repeat second lesson.

3. 1. Raise arms overhead laterally.

Two counts; repeat 8 to 10 times, Fig. 3.

The arms, rigidly extended at the elbows, are raised overhead, palms inward, smartly and brought down the same way; execute moderately fast; inhale on the first and exhale on the second count.

4. 1. **Hands on hips, 2. PLACE, 3. BEND TRUNK SIDEWARD, RIGHT OR LEFT.**

Two counts; repeat 6 to 8 times, Fig. 17.

The trunk, stretched at the waist, is inclined sideward as far as possible; head and shoulders fixed; knees extended and feet firmly on the ground; execute slowly; inhale on first and exhale on second count.

FIG. 17.

5. 1. **Arms to thrust**, 2. **RAISE**, 3. **BEND HEAD FORWARD AND BACKWARD.**

Four counts; repeat 6 to 8 times, Fig. 18.

The chin is drawn in and the head bent forward, back muscles of neck being stretched upward; shoulders remain fixed; in recovering the muscles are relaxed; execute slowly; inhale and raise chest on first and exhale on second count. In bending the head backward the muscles of the neck are stretched upward; breathe as before.

6. 1. Curl shoulders forward.

Two counts; repeat 6 to 8 times, Fig. 19.

FIG. 18.

FIG. 19.

The shoulders relaxed are rolled forward as far as possible; arms being rotated forward; they are then rolled backward and the arms are rotated backward; execute slowly; exhale on first and inhale on second count.

7. 1. **Hands on hips**, 2. **PLACE**, 3. **FULL BEND KNEES.**

Two counts; repeat 6 to 8 times, Fig. 20.

The knees are separated and bent as much as possible; point of knees forced forward and downward; heels together; trunk and head erect; execute slowly; breathe naturally.

FIG. 20.

356 (contd.)

8. 1. **Hands in rear of head,** 2. **LACE,** 3. **On toes,** 4. **RISE,** 5. **ROCK.**
Two counts; repeat 6 to 8 times.

The body is raised on toes and then by short and quick extensions and flexions of the toes it is lowered and raised, knees extended; heels together and free from the ground; breathe naturally.

9. Breathing exercise as in first lesson.

FIG. 21.

FIG. 22.

Fourth Series

1. Repeat third series.
2. 1. **Arms to thrust,** 2. **RAISE,** 3. **THRUST ARMS FORWARD.**
Two counts; repeat 8 to 10 times, Fig. 21.

The arms, knuckles up, are thrust forward forcibly; in recovering the elbows are forced back; execute moderately fast; exhale on first and inhale on the second count.

3. 1. **Hands on shoulders,** 2. **PLACE,** 3. **TWIST TRUNK SIDEWARD, RIGHT OR LEFT.**

Two counts; repeat 6 to 8 times, Fig. 22.

The trunk is turned to the right or left as far possible; hips as nearly perpendicular as possible; shoulders square and head erect; knees extended and feet firm; execute slowly; inhale on first and exhale on second count.

4. 1. **Arms to thrust,** 2. **RAISE,** 3. **TURN HEAD RIGHT, OR LEFT.**

Two counts; repeat 6 to 10 times, Fig. 23.

The head, chin square, is turned to the right, or left as far as possible, muscles of the neck being stretched; shoulders remain square; execute slowly; breathe naturally.

5. 1. **Hands on hips,** 2. **PLACE,** 3. **RAISE KNEE.**

Two counts; repeat 10 to 12 times, Fig. 24.

FIG. 23.

FIG. 24.

The thigh and knee are flexed until they are at right angles, thigh horizontal; toes depressed; the right knee is raised at one and the left at two; trunk and head erect; execute in cadence of quick time; breathe naturally.

6. 1. **Fingers in rear of head,** 2. **LACE,** 3. **FULL BEND TRUNK FORWARD.**

Two counts; repeat 6 to 8 times, Fig. 25.

The trunk is bent forward as far as possible; knees extended; feet firm; head and shoulders fixed; execute slowly; exhale on first and inhale on second count.

7. 1. **Hands on hips,** 2. **PLACE,** 3. **On toes,** 4. **RISE,** 5. **HOP.**

Two counts; repeat 12 to 16 times.

The body is raised on toes and the hopping is performed with knees extended; execute fast; breathe naturally.

FIG. 25.

8. Breathing exercise, as in first lesson.

356 (contd.)

Fifth Series

1. Repeat fourth series.
2. 1. Arms forward, 2. RAISE, 3. STRETCH ARMS SIDEWARD.
   Two counts; repeat 6 to 8 times, Fig. 26.

   From the front horizontal the arms are extended to their fullest extent and then stretched sideward, the arms rotating till the palms are up; the sideward movement is performed slowly; the recovery relaxed and quick; inhale on first and exhale on the second count.

FIG. 26.

3. 1. **Hands on hips, 2. PLACE, 3. BEND TRUNK OBLIQUELY FORWARD, RIGHT OR LEFT.**

Two counts; repeat 4 to 8 times, Fig. 27.

The trunk is turned to the right and bent forward to the half-bend position; shoulders remain square, in the plane of the ground; head fixed; knees straight; feet firm; hips as nearly perpendicular as possible; execute slowly; exhale on the first and inhale and raise chest on second count.

FIG. 27.

FIG. 28.

4. 1. **Arms to thrust, 2. RAISE, 3. EXTEND LEG FORWARD.**

Two counts; repeat 8 to 10 times, Fig. 28.

The knee and ankle are extended forward with a snap, the toes just escaping the ground; all extensor muscles contracted; in recovering relax; trunk and head erect; execute briskly; breathe naturally.

FIG. 29.

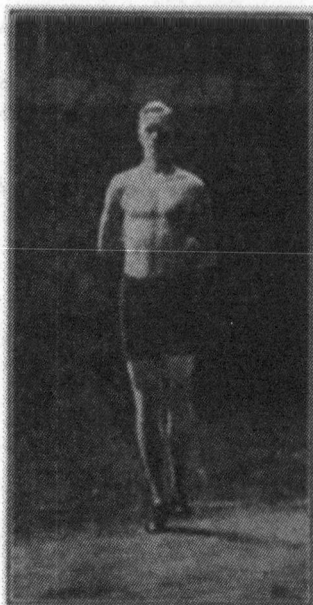

FIG. 30.

5. 1. Hands on shoulders, 2. **PLACE**, 3. **MOVE ELBOWS FOR-WARD.**

Two counts; repeat 8 to 10 times, Fig. 29.

The elbows are brought together horizontally in front and then forced back as far as possible; the forward movement relaxed, the backward a stretch not a jerk; execute moderately fast; exhale on the first and inhale on the second count.

6. 1. Hands on hips, 2. **PLACE**, 3. **BEND TRUNK FORWARD AND BACKWARD.**

Two counts; repeat 6 to 8 times.

Bend trunk forward to the half-bend position, Fig. 16, and then backward, Fig. 13; execute slowly; exhale on first and inhale on second count.

7. 1. Arms backward, 2. **CROSS**, 3. **RISE ON TOES, RIGHT AND LEFT ALTERNATELY.**

Four counts; repeat 10 to 12 times, Fig. 30.

The body is extended on the toes of the right foot and then on those of the left; heels closed; trunk and head erect; execute mod-erately fast; breathe naturally.

8. Breathing exercise, as the first lesson.

Fig. 31.

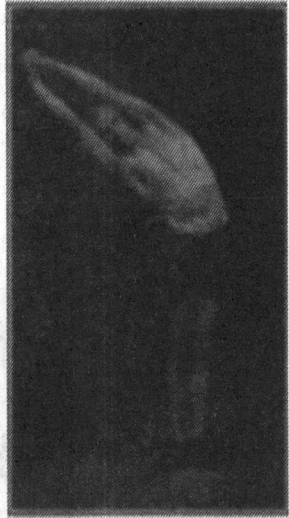

Fig. 32.

Sixth Series

1. Repeat fifth series.

2. 1. **Arms forward overhead,** 2. **RAISE,** 3. **SWING ARMS DOWN-WARD AND UPWARD.**

Two counts; repeat 8 to 10 times, Fig. 31.

3. 1. **Arms sideward overhead,** 2. **RAISE,** 3. Fingers, 4. **LACE,** 5. **BEND TRUNK SIDEWARD, RIGHT AND LEFT.**

Two counts; repeat 6 to 8 times, Fig. 32.

The arms are fully extended and the body, stretched at the waist, is bent sideward to the right and left; knees straight; feet firm; head erect; execute slowly; breathe naturally.

4. 1. **Knees to squatting position, hands on hips,** 2. **BEND,** 3. **ROCK.**

Two counts; repeat 6 to 8 times.

The knees are bent as in Fig. 20; extend and bend the knees in quick succession; trunk and head erect; heels closed; execute moderately fast; breathe naturally.

5. 1. **Arms to thrust,** 2. **RAISE,** 3. **MOVE SHOULDERS FORWARD, UP, BACK, AND DOWN.**

356 (contd.)

Four counts; repeat 8 to 10 times.

The shoulders are relaxed and brought forward; in that position they are raised; then they are forced back without lowering them; and then they are dropped back to position; execute slowly; exhale on the first; inhale on the second and third and exhale on the last count.

6. 1. **Arms to thrust, 2. RAISE, 3. THRUST ARMS FORWARD; SWING THEM SIDEWARD, FORWARD, AND BACK TO POSITION.**

Four counts; repeat 8 to 10 times.

The arms are thrust forward, then relaxed and swung sideward, then forward and finally brought back to position, pressing elbows well to the rear; execute moderately fast; exhale on the first and third and inhale on the second and fourth counts.

FIG. 33.

7. 1. **HOP TO SIDE STRADDLE AND SWING ARMS OVER HEAD LATERALLY AND RECOVER POSITION OF ATTENTION.**

Two counts; repeat 8 to 10 times, Fig. 33.

The distance between the legs is about 30 inches; in alighting the toes come in contact with the ground first and knees are bent

[196]

slightly; trunk and head erect; arms extended; execute moderately fast; breathe naturally.

8. Breathing exercises, as in first lesson.

Trained Soldier's Instruction

First Series

1. 1. **Stretch arms FORWARD, SIDEWARD, FORWARD AND DOWN.**

Four counts; repeat 6 to 8 times.

The arms, stretched to their utmost, are raised forward horizontally, then moved sideward, knuckles down; in returning and lowering the arms the muscles are relaxed; trunk and head erect; execute first two motions slowly; second two moderately fast; inhale on first and second, and exhale on third and fourth counts.

Fig. 34.

2. 1. Hands on shoulders, 2. **PLACE**, 3. **HALF BEND TRUNK FORWARD AND EXTEND ARMS SIDEWARD.**

Two counts; repeat 6 to 8 times, Fig. 34.

The trunk is bent as in Fig. 16. and arms are extended forcibly; in the recovery the elbows are forced back and the chest raised; execute slowly; inhale on first, exhale on second count.

FIG. 35.

3. 1. Hands on hips, 2. **PLACE**, 3. **FULL BEND KNEES AND EXTEND ARMS SIDEWARD.**

Two counts; repeat 6 to 8 times, Fig. 35.

The knees are bent as in Fig. 20, and arms are extended sideward forcibly; execute moderately slow; breathe naturally.

4. 1. Arms sideward, 2. **RAISE**, 3. **ROLL SHOULDERS AND ARMS FORWARD AND BACK.**

Two counts; repeat 6 to 10 times, Fig. 36.

The arms are rotated and the shoulders rolled forward and backward as far as possible; execute slowly; exhale on first and inhale and raise chest on second count.

FIG. 36.

[198]

5. 1. **Hands on shoulders, 2. PLACE, 3. TWIST TRUNK SIDE-WARD RIGHT, OR LEFT, AND EXTEND ARMS SIDEWARD.**

Two counts; repeat 6 to 8 times, Fig. 37.

The trunk is twisted as in Fig. 22; execute moderately fast; inhale on the first and exhale on the second count.

FIG. 37.   FIG. 38.

6. 1. **RAISE ARMS AND RIGHT OR LEFT LEG FORWARD, MOVE ARMS SIDEWARD AND LEG BACKWARD; MOVE ARMS AND LEG FORWARD AND RECOVER.**

Four counts; repeat 8 to 10 times, Fig. 38.

On the first count, the arms and legs are raised forward, arms horizontal, leg extended; toes depressed; foot at height of knee; on the second count the arms are moved smartly to side horizontal and the leg is moved backward, knee and toes extended; at **three** the first position is assumed and at **four** the position of attention; execute moderately fast; inhale on first two and exhale on last two counts.

7. 1. Forearms vertical, 2. RAISE, 3.
EXTEND UPWARD AND RAISE ON
TOES; RESUME VERTICAL POSITION;
AND RECOVER POSITION OF ATTEN-
TION.

Four counts; repeat 8 to 10 times,
Fig. 39.

The forearms are raised vertically at
one; at two they are extended upward and
the body is raised on toes; at three the first
position is assumed, and at four the position
of attention; execute briskly; inhale on first
two and exhale on last two counts.

8. Breathing exercise.

Second Series

1. Repeat first series.

2. 1. Arms to thrust, 2. RAISE, 3.
THRUST ARMS UPWARD, SWING DOWN-
WARD AND BACKWARD, SWING UP-
WARD AND RECOVER.

Four counts; repeat 6 to 10 times,
Fig. 40.

The arms
are thrust up-
ward forcibly at
one; at two the
arms, relaxed,
are swung down-
ward to the
front and back
as far as pos-
sible; at three
they are swung
upward, and at
four the position
of attention is
resumed; trunk
and head erect;
knees extended;
execute moder-
ately fast; inhale
on first three and
exhale on last
count.

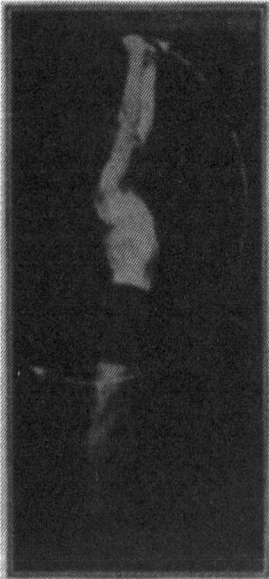

FIG. 39.

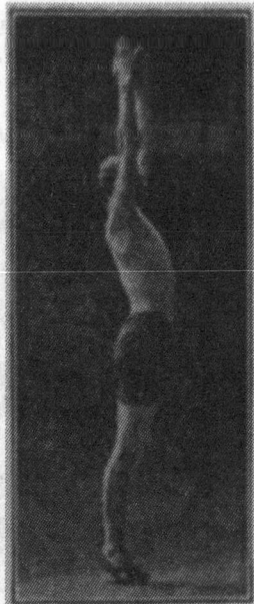

FIG. 41.

FIG 40.

[200]

3. 1. Hands on shoulders, 2. **PLACE**, 3. **BEND TRUNK BACK-
WARD AND EXTEND ARMS SIDEWARD, KNUCKLES UP.**

Two counts; repeat 6 to 8 times, Fig. 41.

The trunk is bent backward as in Fig. 13, and the arms, knuckles
down, are extended to the side horizontal; head fixed; knees extended;
feet firm; execute slowly; inhale on first and exhale on second count.

4. 1. **Full bend
knees and raise arms
knuckles down, to
side horizontal,** 2.
BEND, 3. **ROCK,
AND CIRCLE ARMS
BACKWARD.**

Two counts; re-
peat 6 to 10 times,
Fig. 42.

The knees, bent
to the squatting posi-
tion, are slightly ex-
tended and flexed as
in Exercise 4, Sixth
Lesson, Recruit In-
struction, and the
arms are circled back-
ward in circles of

FIG. 42.

about 12 inches; head and trunk erect; arms extended; execute moderately fast; breathe naturally.

5. 1. Hands on hips, 2. **PLACE,** 3. **CIRCLE
TRUNK RIGHT, OR LEFT.**

Six counts; repeat 4 to 6 times,
Fig. 43.

The trunk is half forward at one;
at two it is moved to the right side bend
position; at three to the back bend; at four
to the left bend; at five to the front bend
position and raised at six; knees extended;
feet firm; head fixed; execute slowly; exhale on first; inhale on second; hold breath
on third and fourth; exhale on fifth and
inhale on sixth count.

6. 1. Hands on hips, 2. **PLACE,** 3.
**SWING RIGHT AND LEFT LEG FOR-
WARD, BREAST HIGH, AND EXTEND
RIGHT AND LEFT ARM FORWARD
HORIZONTALLY, ALTERNATING RIGHT AND LEFT.**

FIG. 43.

Four counts; repeat 6 to 10 times, Fig. 44.

The right leg, knee extended, is swung forward high enough to come in contact with the hand; supporting leg extended; body inclined as little as possible; execute moderately fast; breathe naturally.

FIG. 44.

FIG. 45 a.   FIG. 45 b.

7. 1. LEANING REST ON FOUR COUNTS.
Repeat 6 to 8 times, Fig. 45 a and b.

At **one** knees are bent to squatting position, hands on the ground between knees; at **two** the legs are extended backward to the leaning rest; at **three** the first position is resumed, and at **four** the position of attention; hands should be directly under shoulders; back arched; knees straight; head fixed; execute moderately fast; breathe naturally.

8. Breathing exercise.

Third Series

1. Repeat second series.

2. 1. **STRETCH ARMS FORWARD, SIDEWARD, UPWARD, SIDE-WARD, FORWARD, AND DOWN.**

Six counts; repeat 6 to 10 times.

First five counts arms are extended as much as possible; in the last they are relaxed; execute slowly; inhale on first five counts and exhale on last.

FIG. 46.

3. 1. **HALF BEND TRUNK FORWARD AND ROTATE ARMS IN-WARD; RAISE AND BEND TRUNK BACKWARD, RAISING AND ROTATING ARMS BACKWARD, PALMS UP; RESUME FIRST POSITION AND RECOVER.**

Four counts; repeat 4 to 8 times, Fig. 46.

In the first position the body and arms are relaxed; in the second the body and arms are tense, Fig. 41; the third position is the same as the first, and at **four** the position of attention is resumed; execute slowly; exhale on first and third and inhale on second and fourth counts.

4. 1. **Hands on hips,** 2. **PLACE,** 3. **RAISE ON TOES, BEND KNEES TO SQUATTING POSITION; EXTEND KNEES AND RECOVER.**

Four counts; repeat 6 to 8 times.

The body is raised on toes slowly at **one**; at **two** the knees are bent slowly to squatting position; at **three** they are extended slowly and at **four** the starting position is resumed; execute slowly; breathe naturally.

356 (contd.)

FIG. 47.

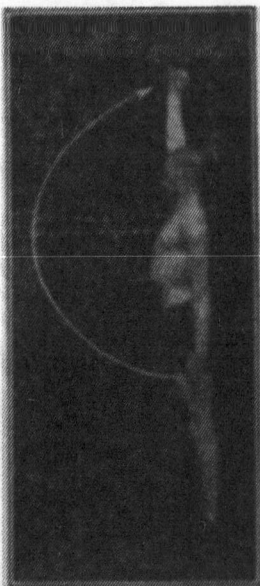

FIG. 48.

5. 1. HOP TO SIDE STRADDLE POSITION, HANDS ON HIPS, BEND TRUNK FORWARD AND EXTEND ARMS DOWNWARD, FINGERS TOUCHING GROUND; RESUME STRADDLE WITH HANDS ON HIPS AND HOP TO ATTENTION.

Four counts; repeat 6 to 8 times, Fig. 47.

Execute moderately fast; breathe naturally.

6. 1. Arms to thrust, 2. RAISE, 3. THRUST ARMS FORWARD; SWING RIGHT, (LEFT) ARM UP, LEFT, (RIGHT) DOWN; SWING TO FRONT HORIZONTAL AND RECOVER.

Four counts, or alternating in eight counts; repeat 8 to 10 times, Fig. 48.

The thrust and recovery are forcible, the swings brisk but relaxed; execute moderately fast; exhale on first and third count and inhale on second and fourth.

7. 1. STEP POSITION FORWARD RIGHT. OR LEFT, AND RAISE ARMS TO FRONT HORIZONTAL; LUNGE FORWARD AND SWING ARMS TO SIDE HORIZONTAL; RESUME FIRST POSITION AND RECOVER POSITION OF ATTENTION.

Fig. 49 a.

Fig. 49 b.

Four counts; repeat 6 to 10 times, Fig. 49 a and b.

The right foot, knee extended and toes depressed, is moved forward once its length, the toes resting on the ground lightly, the weight resting on the left leg, and the arms are raised to the front horizontal, at one; at two the right foot is advanced and planted smartly, the distance between heels being about 3-foot lengths, and the arms are swung to side horizontal; right knee is well bent, left extended; trunk and head erect; at three the first position, and at four the position of attention are resumed; execute moderately fast; exhale first and third and inhale on second and fourth counts.

8. Breathing exercise.

Fourth Series

1. Repeat third series.

2. 1. Hands on shoulders, 2. PLACE, 3. EXTEND ARMS UPWARD; SWING ARMS DOWNWARD LATERALLY, UPWARD LATERALLY, AND RECOVER STARTING POSITION.

Four counts; repeat 6 to 10 times.

The first and fourth motions are energetic; the second and third relaxed; execute moderately fast; inhale on first and third counts and exhale on second and fourth.

3. 1. To side straddle with arms overhead, 2. Hop, 3. BEND TRUNK FORWARD AND BACK AND SWING ARMS DOWNWARD AND UPWARD.

356 (contd.)

FIG. 50.

Two counts; repeat 6 to 8 times, Fig. 50.

Being in the straddle position, the body is bent forward as far as possible and the arms are swung between the legs; the arms are then swung upward and the body bent backward; knees extended; execute moderately fast; exhale on first and inhale on second count.

4. 1. Arms to thrust, 2. **RAISE**, 3. **THRUST ARMS SIDEWARD AND LUNGE SIDEWARD RIGHT AND LEFT ALTERNATELY.**

Four counts; repeat 8 to 10 times, Fig. 51.

FIG. 51.

FIG. 52.

The starting position is resumed at two and four; the distance of the lunge is three times the length of the feet; supporting leg extended; head and trunk erect; execute moderately fast; inhale on first and third and exhale on second and fourth counts.

5. 1. **Hands on shoulders, 2. PLACE, 3. BEND TRUNK SIDEWARD RIGHT AND EXTEND LEFT ARM OBLIQUELY UPWARD AND RIGHT OBLIQUELY DOWNWARD; SWING TRUNK SIDEWARD LEFT AND RIGHT AND RECOVER.**

Four counts; repeat 6 to 8 times, Fig. 52.

The trunk is bent to the right, the left arm, palm down, is extended obliquely upward and the right arm obliquely downward, at one; at two the body is bent to the left; at three to the right and at four the starting position is resumed; arms extended; knees straight; head fixed; execute moderately fast; breathe naturally.

FIG. 53.

6. 1. **To squatting position, hands on ground, 2. BEND, 3. EXTEND RIGHT AND LEFT LEGS BACKWARD, ALTERNATELY.**

Four counts; repeat 6 to 10 times, Fig. 53.

The squatting position is the starting position, from there the right and left legs are extended backward and brought back to the squatting position again; execute moderately fast; breathe naturally.

7. 1. **Hands on shoulders, 2. PLACE, 3. EXTEND ARMS SIDEWARD AND SWING RIGHT AND LEFT LEGS SIDEWARD ALTERNATELY.**

Four counts; repeat 8 to 10 times, Fig. 54.

The legs are extended and swung loosely to the side as high as possible, arms being extended with each leg movement; execute moderately fast; inhale on one and three and exhale on two and four.

8. Breathing exercise.

Fig. 54.

Fifth Series

This series is composed of three groups, each group containing four exercises, and together they form a combination which can be adapted to music.

Each exercise is composed of four movements and should be repeated four times, twice to the right and twice to the left, alternately, except the last, which is repeated in the same direction. The third position always corresponds to the first, and the fourth to the position of attention.

When performed to music it is advisable to employ "two-four" time, allowing two beats to every movement, or four measures to an exercise, the action occurring on the first beat and a pause in position during the second beat. If this is done, and the tempo is made to suit the movements, it will be possible to execute the exercises with precision and vigor, and slurring a movement for the sake of keeping time will be eliminated. Every group should be preceded by an introduction of four measures.

FIRST GROUP
First Exercise

FIG. 55.

FIG. 56.

COUNTS.

1—2. Raise arms overhead laterally and step position forward right. Fig. 55.

3—4. Lunge forward right and swing the arms downward and backward laterally. Fig. 56.

5—6. Resume first position.

7—8. Resume position of attention.
Repeat left, right, left.

Second Exercise.

FIG. 57.

FIG. 58.

FIG. 59.

1—2. Raise right arm obliquely upward to the right, and left arm obliquely backward to the left, and step position forward to the right with the right foot. Fig. 57.

3—4. Lunge obliquely forward to the right and swing right arm downward to the rear, and left arm obliquely upward. Fig. 58.

5—6. Resume first position.

7—8. Resume position of attention.
Repeat left, right, left.

1—2. Flex arms over shoulders with lateral motion, knuckles to the rear, hands closed, and step position sideward right. Fig. 59.

Third Exercise

FIG. 60.

3—4. Lunge sideward right and extend arms to side horizontal, knuckles to the rear. Fig. 60.

5—6. Resume first position.

7—8. Resume position of attention.

Repeat left, right, left.

Fourth Exercise

1—2. Raise arms to side horizontal and step position backward right. Fig. 61.

3—4. Lunge backward right and raise arms overhead, knuckles out. Fig. 62.

5—6. Resume first position.

7—8. Resume position of attention.

Repeat left, right, left.

FIG. 61.

FIG. 62.

SECOND GROUP
First Exercise

FIG. 63.

1—2. Lunge forward right and raise arms to side horizontal. Fig. 63
3—4. Bend trunk forward and move arms downward. Fig. 64.
5—6. Resume first position.

FIG. 64.
[212]

7—8. Resume position of attention.
Repeat left, right, left.

Second Exercise

Fig. 65.

1—2. Lunge sideward right and raise right arm obliquely upward, and left arm obliquely downward. Fig. 65.

3—4. Bend trunk sideward right and swing left arm upward, knuckles out, and right arm downward in rear of body, knuckles out. Fig. 66.

5—6. Resume first position.

7—8. Resume position of attention.
Repeat left, right, left.

FIG. 66.

Third Exercise

1—2. Lunge obliquely forward to the right, and raise arms overhead laterally.   Fig. 67.

3—4. Bend trunk forward and swing arms downward and upward. Fig. 68.

5—6. Resume first position.

7—8. Resume position of attention.

Repeat left, right, left.

FIG. 67.

Fourth Exercise

1—2. Lunge backward right and raise arms to side horizontal, knuckles up. Fig. 69.

3—4. Bend trunk and head backward and raise arms overhead, palms in. Fig. 70.

5—6. Resume first position.

7—8. Resume position of attention.

Repeat left, right, left.

FIG. 68.

FIG. 69.

THIRD GROUP
First Exercise

Fig. 70.

Fig. 71.

Fig. 72.

1—2. Stride forward right and flex arms over shoulders laterally, hands closed, knuckles up. Fig. 71.

3—4. Face to the left on both heels, bending knees and striking arms sideward, knuckles down. Fig. 72.

5—6. Resume first position.

7—8. Resume position of attention.

Repeat left, right, left.

Second Exercise

FIG. 73.

1—2. Stride sideward right and raise and circle arms inward, crossing them above. Fig. 73.

3—4. Bend right knee and trunk obliquely forward, clasp thigh with arms. Fig. 74.

5—6. Resume first position.

7—8. Resume position of attention.
   Repeat left, right, left.

Third Exercise

Fig. 74.

Fig. 75.

Fig. 76.

1—2. Stride backward right, and raise arms overhead laterally, palms in. Fig. 75.

3—4. Turn about on both heels, bend left knee and trunk backward and lower arms to side horizontal, palms up. Fig. 76.

5—6. Resume first position.

7—8. Resume position of attention.
Repeat left, right, left.

Fourth Exercise

FIG. 77.

1—2. Bend to the squatting position, hands on the ground. Same as Fig. 45a.

3—4. Extend to the leaning-rest with legs straddled. Fig. 77.

5—6. Resume first position.

7—8. Resume position of attention.

Repeat left, right, left.

Note.—Length of stride in these exercises should be 28 inches between heels.

HOPPING EXERCISES

Hopping is executed by raising the body on the balls of the feet and forcing the body from the ground by a series of quick extensions of the toe and ankle joints; knees remain easily extended, heels together and free from the floor.

Having assumed a position for the arms, the instructor commands: 1. **On toes**, 2. **RISE**, 3. **HOP**.

At the command hop, execute one spring, alighting on the balls of the feet. Continue by repeating one, two.

1. Hop and turn to the right or left at every second, fourth or sixth hop.

2. Hop and turn about at every second, fourth, or sixth hop.

3. Hop to side straddle in four and return to attention in four hops.

4. Hop to side straddle and continue to hop in that direction.

5. Hop to side straddle in one hop and return to attention in next hop.

6. Hop to cross straddle and return to attention in next hop.

7. Hop on right leg and hold left forward, sideward, or backward.

8. Hop on left leg and hold right forward, sideward, or backward.

9. Hop four times on right leg, and then change and hop four times on left leg, holding the unemployed leg forward, sideward, or backward.

356 (contd.)

10. Same as in 9, hopping twice on each leg.
11. Same as in 9, hopping once on each leg.
12. Hop forward, sideward, or backward.

LEAPING.

Leaping or jumping as a setting-up exercise has for its object the raising of the body from 8 to 12 inches from the ground, and there is, however, no gaining of ground as in gymnastic or athletic jumping.

At the first command, the arms are raised to the front horizontal and the body is elevated on the toes. (See Fig. 1, Jumping.)

At the command leap, the arms are swung downward and backward and the knees are slightly bent (See Fig. 2, Jumping); without pausing the arms are swung forward again and as they pass through the vertical plane the knees are extended and the body is forced from the floor.

The moment the feet leave the floor the knees are extended; feet are closed and toes depressed; the arms are in the front horizontal; the back is arched and the head is erect. (See Fig. 3, Jumping.)

In alighting, the balls of the feet touch the floor first, knees slightly bent; the latter are quickly extended, however, and the arms brought down by the sides and the position of attention is assumed.

Continue by repeating **leap**.

LEAPING EXERCISES

1. Leap and execute a quarter of a turn to the right or left.
2. Leap and execute a half turn to the right or left.
3. Leap and straddle legs sideward (legs are closed) before alighting.
4. Leap and cross straddle, right or left leg forward.
5. Leap and cross legs, right over left or left over right.
6. Leap and raise heels.
7. Leap and raise knees.
8. Leap and strike feet together.
9. Leap and strike feet together twice.
10. Leap and strike feet together three times.
11. Leap and cross and recross legs.
12. Leap and raise heels and touch them with hands.
13. Leap and swing arms sideward.
14. Leap and swing arms upward.
15. Leap and circle arms forward.
16. Leap and circle arms backward.
17. Leap and circle arms inward.
18. Leap and circle arms outward.
19. Leap and swing arms upward and execute a whole body turn.

WALKING AND MARCHING

The length of the full step in quick time is 30 inches, measured from heel to heel, and the cadence is at the rate of 120 steps per minute.

Proper posture and carriage have ever been considered very important in the training of soldiers. In marching, the head and trunk should remain immobile, but without stiffness; as the left foot is carried forward the right forearm is swung forward and inward obliquely across the body until the thumb, knuckles being turned out, reaches a point about the height of the belt plate. The upper arm does not move beyond the perpendicular plane while the forearm is swung forward, though the arm hangs loosely from the shoulder joint. The forearm swing ends precisely at the moment the left heel strikes the ground; the arm is then relaxed and allowed to swing down and backward by its own weight until it reaches a point where the thumb is about the breadth of a hand to the rear of the buttocks. As the right arm swings back, the left arm is swung forward with the right leg. The forward motion of the arm assists the body in marching by throwing the weight forward and inward upon the opposite foot as it is planted. The head is held erect; body well stretched from the waist; chest arched; and there should be no rotary motion of the body about the spine.

As the leg is thrown forward the knee is smartly extended, the heel striking the ground first.

The instructor having explained the principles and illustrated the step and arm swing, commands: 1. **Forward,** 2. **MARCH**—and to halt the squad he commands: 1. **Squad,** 2. **HALT.**

In executing the setting-up exercises on the march the cadence should at first be given slowly and gradually increased as the men become more expert; some exercises require a slow and others a faster pace; it is best in these cases to allow the cadence of the exercise to determine the cadence of the step.

The men should march in a single file at proved intervals. The command that causes and discontinues the execution should be given as the left foot strikes the ground.

On the march, to discontinue the exercise, command: 1. **Quick time,** 2. **MARCH,** instead of **HALT,** as when at rest.

All of the arm, wrist, finger, and shoulder exercises, and some of the trunk and neck, may be executed on the march by the same commands and means as when at rest.

The following leg and foot exercises are executed at the command march; the execution always beginning with the left leg or foot,

1. 1. **On toes,** 2. **MARCH.**
2. 1. **On heels,** 2. **MARCH.**
3. 1. **On right heel and left toe,** 2. **MARCH.**

4. 1. On left heel and right toe, 2. **MARCH.**
5. 1. On toes with knees stiff, 2. **MARCH.**
6. 1. Swing extended leg forward, ankle high, 2. **MARCH.**
7. 1. Swing extended leg forward, knee high, 2. **MARCH.**
8. 1. Swing extended leg forward, waist high, 2. **MARCH.**
9. 1. Swing extended leg forward, shoulder high, 2. **MARCH.**
10. 1. Raise heels, 2. **MARCH.**
11. 1. Raise knees, thigh horizontal, 2. **MARCH.**
12. 1. Raise knees, chest high, 2. **MARCH.**
13. 1. Circle extended leg forward, ankle high, 2. **MARCH.**
14. 1. Circle extended leg forward, knee high, 2. **MARCH.**
15. 1. Circle extended leg forward, waist high, 2. **MARCH.**
16. 1. Swing extended leg backward, 2. **MARCH.**
17. 1. Swing extended leg sideward, 2. **MARCH.**
18. 1. Raise knee and extend leg forward, 2. **MARCH.**
19. 1. Raise heels and extend leg forward, 2. **MARCH.**

Steps

In the steps, the rules given above apply, viz, the command march given as the left foot strikes the ground, determines the execution, which always begins with the left foot, and is continued until the command: 1. Quick time, 2. **MARCH,** is given, when the direct step is resumed.

The different steps are executed at the following commands:

1. Cross step, 2. **MARCH.**

As the legs move forward they are crossed. The body does not turn.

1. Halting step, 2. **MARCH.**

The left foot is advanced and planted; the right foot is brought directly in rear of the left, resting on the ball only; the right is then advanced and planted and the left brought up, and so on.

1. Foot-balancing step, 2. **MARCH.**

The left foot is advanced and planted; the right foot is brought up beside it, heels touching; the body is then raised on the toes and lowered. The right foot is then advanced and planted and the left brought up, and so on.

1. Continuous change step, 2. **MARCH.**

The left foot is advanced and planted; the toes of the right are then advanced near the heel of the left in the halting step; the left foot is then advanced about half a step (15 inches) and the right foot is advanced with the full step and planted; the toes of the left foot are then brought up to the heel of the right foot, which advances a half step, when the left foot is advanced a full step, and so on.

1. **Knee-rocking step, 2. MARCH.**

As each foot is planted it is accompained by a slight bending and extension in the corresponding knee; the other leg remaining fully extended, heel raised.

1. **Lunging step, 2. MARCH.**

The length of the step is 45 inches, the knee in advance being well bent; the other leg remaining fully extended, heel raised; trunk erect.

1. **Leg-balance step, 2. MARCH.**

The left foot is advanced, ankle high; it is then swung backward and forward and planted, the body during the swinging balanced on the right leg. The right foot is then advanced, swung backward and forward and planted, and so on.

1. **Body-balance step, 2. MARCH.**

The left foot is advanced, ankle high, body being bent slightly to the rear; the left foot is then swung backward, body being bent slightly to the front; the same foot is then swung forward again and planted, the body in the meantime becoming erect. This is repeated with the right foot, and so on.

1. **Heel-and-toe step, 2. MARCH.**

The left foot is advanced and allowed to rest on the heel; it is then swung backward and allowed to rest on the toes; it is once more advanced and planted. This is repeated with the right foot, and so on.

1. **Cross step, raising knees, 2. MARCH.**

Execute the cross step and raise the knees. The cross step may also be executed in combination with the swings of the extended leg.

The change step may be combined with the following: Cross step, halting step, raising knees, foot-rocking step, on toes, raising heels, swinging and circling legs, heel and toe step. These may also be combined with the change step hop.

1. **Continuous change step hop, 2. MARCH.**

Execute the ordinary change step, hopping with the change.

1. **Forward gallop hop, 2. MARCH.**

The left foot is advanced and planted, the right is brought up in rear as in the halting step; this is done four times in succession. The same is done four times with the right foot in advance, and so on.

1. **Sideboard gallop hop, 2. MARCH.**

The left foot is advanced, body turning on the right; four hops are then executed sideward on the left foot followed by the right; at the fourth hop the body is turned to the left about and four hops executed sideward on the right foot followed by the left, and so on.

DOUBLE TIMING

The length of the step in double time is 36 inches; the cadence is at the rate of 180 steps per minute. To march in double time the instructor commands: 1. **Double time. 2. MARCH.**

356 (contd.)

If at a halt, at the first command shift the weight of the body to the right leg. At the command **march** raise the forearms, fingers closed; to a horizontal position along the waist line; take up an easy run with the step and cadence of double time, allowing a natural swinging motion to the arms inward and upward in the direction of the opposite shoulder.

In marching in quick time, at the command **march**, given as either foot strikes the ground, take one step in quick time, and then step off in double time.

When marching in double time and in running the men breathe as much as possible through the nose, keeping the mouth closed.

A few minutes at the beginning of the setting-up exercises should be devoted to double timing. From lasting only a few minutes at the start it may be gradually increased, so that daily drills should enable the men at the end of five or six months to double time for 15 or 20 minutes without becoming fatigued or distressed.

After the double time the men should be marched for several minutes at quick time; after this the instructor should command:

1. Route step, 2. MARCH.

In marching at route step, the men are not required to preserve silence nor keep the step; if marching at proved intervals, the latter is preserved.

To resume the cadence step in quick time, the instructor commands: 1. Squad, 2. ATTENTION.

Great care must be exercised concerning the duration of the double time and the speed and duration of the run. The demands made upon the men should be increased gradually.

When exercise rather than distance is desired, the running should be done on the balls of the feet, heels raised from the ground.

DOUBLE-TIMING EXERCISES

While the men are double timing the instructor may vary the position of the arms by commanding:

1. 1. Arms forward, 2. RAISE.
2. 1. Arms sideward, 2. RAISE.
3. 1. Arms upward, 2. RAISE.
4. 1. Hands on hips, 2. PLACE.
5. 1. Hands on shoulders, 2. PLACE.
6. 1. Arms forward, 2. CROSS.
7. 1. Arms backward, 2. CROSS.

At the command down, the double-time position for the arms and hands is resumed.

The instructor may combine the following with the double time:

1. 1. Cross step, 2. MARCH.
2. 1. Raise knees, 2. MARCH.

3. 1. **Raise heels**, 2. **MARCH**.
4. 1. **Swing legs forward**, 2. **MARCH**.
5. 1. **Swing legs backward**, 2. **MARCH**.

To continue these exercises, but still continue the double timing, command: 1. **Double time**, 2. **MARCH**. To march in quick time, command: 1. **Quick time**, 2. **MARCH**. Marching in quick or double time, to halt, command: 1. **Squad**, 2. **HALT**.

RIFLE EXERCISES

357. The object of these exercises, which may also be performed with wands or bar bells, is to develop the muscles of the arms, shoulders, and back so that the men will become accustomed to the weight of the piece and learn to wield it with that "handiness" so essential to its successful use. When these exercises are combined with movements of the various other parts of the body, they serve as a splendid, though rather strenuous, method for the all-round development of the men. As the weight of the piece is considerable, instructors are cautioned to be reasonable in their demands. Far better results are obtained if these exercises are performed at commands than when they are grouped and performed for spectacular purposes.

FIG. 1.   [225]   FIG. 2.

357 (contd.)

All the exercises start from the starting position, which is the low extended arm horizontal position in front of the body, arms straight; the right hand grasping the small of the stock and the left hand the barrel; the knuckles turned to the front and the distance between the hands slightly greater than the width of the shoulders. Fig. 1.

This position is assumed at the command: 1. **Starting**, 2. **POSITION**; at the command **position** the piece is brought to the port and lowered to the front horizontal snappily.

To recover the position of order, command: 1. **Order**, 2. **ARMS**; the piece is first brought to the port and then ordered.

ARM EXERCISES

From the starting position (Fig. 1); all exercises are in two counts.

FIG. 3.　　　　　FIG. 4.

1. Raise piece to front, extended arm horizontal.
2. Raise piece to high, overhead, extended arm horizontal.
3. Raise piece to side, horizontal, right or left. Fig. 2.
4. Raise piece to front, perpendicular, right or left hand up. Fig. 3.
5. Raise piece to front, bent arm horizontal, waist high.
6. Raise piece to front, bent arm horizontal, shoulder high. Fig. 4.

FIG. 5.

FIG. 6.

7. Raise piece to rear, bent arm horizontal, on shoulders. Fig. 5.

8. Raise piece to front, bent arm horizontal, shoulder high, arms crossed, left over right or vice versa.

9. Raise piece to low side, perpendicular, right or left, right or left hand up. Fig. 6.

10. Raise piece to high side, perpendicular, right or left. Fig. 7.

In the above exercises the movement begins at the command "**EXERCISE**" and is discontinued at "**HALT.**"

From front bent arm horizontal, shoulder high:

11. Thrust piece forward, upward, downward or sideward, right or left.

12. Thrust piece upward from rear, bent arm horizontal.

From high extended arm horizontal:

13. Circle piece from right to left, or from left to right. Describe complete circle parallel with the front of the body.

ARM COMBINATIONS

All of the following exercises consist of four movements, the third carrying the piece back to the first position, and the fourth to

357 (contd.)

Fig. 7.

the starting position; in other words, the piece is carried back in reverse order at three and four.

1. Raise piece to high, extended arm horizontal; flex to the bent arm horizontal in front of shoulders and return in reverse order.

2. Same as above, except that the piece is brought to the shoulders in rear of head.

3. Raise piece as in 1, lower to right horizontal, and return in reverse order.

4. Same, left.

5. Raise piece to front, bent arm horizontal, shoulder high; thrust piece upward, and return in reverse order.

6. Same, thrusting piece forward or sideward right or left.

7. Raise piece to front, extended arm perpendicular, right hand up; reverse bringing left hand up; reverse again and lower.

8. Raise piece to low side perpendicular, left hand up; change to high side perpendicular, right hand up; and return in reverse order.

9. Same on the left.

[228]

10. Raise piece to front, extended arm horizontal; cross and bend arms to front, bent arm horizontal, right over left; and return in reverse order.

11. Raise piece to front, bent arm horizontal, arms crossed, right over left; change by crossing left over right; reverse and down.

FIG. 8. FIG. 9.

ARM, LEG, AND TRUNK COMBINATIONS

From the starting position: All exercises in two counts.

1. Raise piece to front, extended arm horizontal and bend knees quarter, half, or full.

2. Raise piece to high, extended arm horizontal and raise on toes.

3. Raise piece as in 2 and bend trunk forward.

4. Raise piece to rear, bent arm horizontal on shoulders and bend trunk forward.

5. Raise piece to front perpendicular, left hand up, and bend trunk sideward right. Fig. 8.

6. Same to the left, right hand up.

7. Raise piece to high, right side perpendicular and bend trunk sideward left.

357 (contd.)

0. Same, piece on the left, bending trunk to the right. From front bent arm horizontal, shoulder high.

9. Thrust piece forward or upward and bend knees, quarter, half, or full.

10. Raise piece forward and upward and bend trunk forward.

11. Thrust piece sideward right and bend trunk sideward left.

12. Same reversed.

13. Thrust piece forward and twist body to the left or right.

14. Thrust piece upward and bend trunk backward.

15. Thrust piece upward and hop to side straddle.

16. Thrust piece forward or upward and lunge forward right or left.

17. Thrust piece upward or forward and lunge obliquely forward right or left.

18. Thrust piece sideward left and lunge sideward right or left.

19. Thrust piece upward and lunge backward.

20. Thrust piece downward; lunge forward and bend body forward, Fig. 9.

Fig. 10.   Fig. 11.

21. Thrust piece upward; lunge backward and bend trunk backward. Fig. 10.

22. Thrust piece side right and lunge and bend trunk sideward left.
Fig. 11.

RIFLE DRILL COMBINATION

The following exercises consist of four movements, the third
position always corresponding to the first position and the fourth to the
starting position. They have been grouped and arranged precisely like
the setting-up combination, Fifth Lesson prescribed for trained soldiers.
When performed as a musical drill, the instructions laid down in that
lesson are applicable here.

All exercises begin and end with the first or starting position.
Fig. 1.

FIRST GROUP
First Exercise

FIG. 12.   FIG. 13.

COUNTS.

1—2. Raise piece to bent arm front horizontal, shoulder high, and
stride forward right, Fig. 12;

3—4. Face to the left on both heels and extend piece upward, Fig. 13;

5—6. Resume first position;

7—8. Resume starting position.

Repeat left, right, left.

Second Exercise

Fig. 14. **Fig. 15.**

1—2.. Raise piece to extended high horizontal, and stride sideward right, Fig. 14;

3—4. Bend right knee and lower piece to left horizontal, Fig. 15;

5—6. Resume first position;

7—8. Resume starting position.

Repeat left, right, left.

Third Exercise

FIG. 16.

FIG. 17.

1—2. Raise piece to high side perpendicular on the left, left hand up, and stride backward right, Fig. 16;

3—4. Face about on heels and swing piece down and up to high side perpendicular on the right, Fig 17;

5—6. Resume first position;

7—8. Resume starting position.

Repeat left, right, left.

357 (contd.)

Fourth Exercise

FIG. 18.

FIG. 19.

1—2. Raise piece to extended high horizontal, and stride obliquely forward right, Fig 18;

3—4. Face about on heels and lower piece to horizontal on shoulders; Fig. 19;

5—6. Resume first position;

7—8. Resume starting position.
Repeat left, right, left.

Second Group

First Exercise

FIG. 20.

FIG. 21.

1—2. Lower piece to front extended horizontal and bend trunk forward, Fig. 20;

3—4. Lunge obliquely forward right and raise piece to right oblique, left hand at shoulder, Fig 21;

5—6. Resume first position;

7—8. Resume starting position.

Repeat left, right, left.

Second Exercise

FIG. 22.

FIG. 23.

1—2. Raise piece to high perpendicular on the left, left hand up, and bend trunk sideward right, Fig. 22;

3—4. Lunge sideward right and swing piece down and up to right high perpendicular, right hand up, Fig. 23;

5—6. Resume first position;

7—8. Resume starting position.

Repeat left, right, left.

Third Exercise

FIG. 24.

FIG. 25.

1—2. Raise piece to high extended arm horizontal and bend trunk backward, Fig. 24;

3—4. Lunge forward right, and swing piece to side horizontal, left hand to the rear, Fig. 25;

5—6. Resume first position;

7—8. Resume starting position.
Repeat left, right, left.

Fourth Exercise

FIG. 26.

FIG. 27.

1—2. Raise piece to right high perpendicular and side step position left, Fig. 26;

3—4. Lunge sideward left and swing piece to left high perpendicular, Fig. 27;

5—6. Resume first position;

7—8. Resume starting position.
    Repeat left, right, left.

Third Group

First Exercise

FIG. 28.

FIG. 29.

1—2. Raise piece to front bent horizontal, arms crossed, left over right; lunge sideward right and bend trunk sideward right, Fig. 28;

3—4. Extend right knee and bend trunk to the left, bending left knee and recrossing arms, left over right, Fig 29;

5—6. Resume first position;

7—8. Resume starting position.

Repeat left, right, left.

Second Exercise

FIG. 30.

FIG. 31.

1—2. Raise piece to bent arm horizontal; face right and lunge forward right and bend trunk forward, Fig. 30;

3—4. Raise trunk and turn to the left on both heels and extend piece overhead, Fig. 31;

5—6. Resume first position;

7—8. Resume starting position.
Repeat left, right, left.

Third Exercise

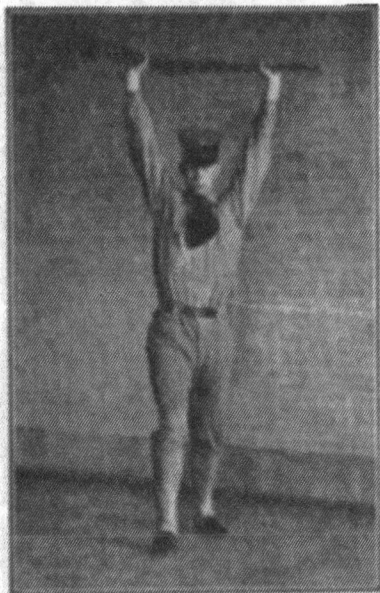

FIG. 32.

FIG. 33.

1—2. Raise piece to left high horizontal; lunge forward right, Fig. 32;
3—4. Bend trunk forward and lower piece to low front horizontal, Fig. 33;
5—6. Resume first position;
7—8. Resume starting position.
    Repeat left, right, left.

Fourth Exercise

FIG. 34.

FIG. 35.

1—2. Raise piece to high extended horizontal and hop to side straddle position, Fig. 34;

3—4. Bend trunk forward and swing piece to extended low horizontal, left hand between legs, right hand forward, Fig. 35;

5—6. Resume first position;

7—8. Resume starting position.

Repeat left, right, left.

GYMNASTIC CONTESTS

358. These exercises are those in which the benefits are lost sight of in the pleasure their attainment provides, which in the case of these contests is the vanquishing of an opponent. The men are pitted against each other in pairs; age, height, weight, and general physical aptitude being the determining factors in the selection.

In the contests in which superiority is dependent upon skill and agility no restrictions need be placed upon the efforts of the contestants;

but in those that are a test of strength and endurance it is well to call a contest a "draw", when the men are equally matched and the contest is likely to be drawn out to the point of exhaustion of one or both contestants.

It is recommended that these contests be indulged in once or twice a month and then at the conclusion of the regular drill.

Contests that require skill and agility should alternate with those that depend upon force and endurance. In order to facilitate the instruction a number of pairs should be engaged at the same time.

1. Cane wrestling: The cane to be about an inch in diameter and a yard long, ends rounded. It is grasped with the right hand at the end, knuckles down, and with the left hand, knuckles up, inside of and close to the opponent's right hand. Endeavor is then made to wrest the cane from the opponent. Loss of grip with either hand loses the bout.

2. Cane twisting: Same cane as in 1. Contestants grasp it as in 1, only the knuckles of both hands are up, and the arms are extended overhead. Object: The contestants endeavor to make the cane revolve in their opponent's hand without allowing it to do so in their own. The cane must be forced down.

3. Cane pulling: Contestants sit on the ground, facing each other, legs straight and the soles of the feet in contact. The cane is grasped as in 2 but close to the feet. Object: To pull the opponent to his feet. The legs throughout the contest must be kept rigid.

4. "Bucked" contest: Contestants sit on the ground "bucked"; i. e., the cane is passed under the knees, which are drawn up, and the arms passed under the cane with the fingers laced in front of the ankles. Object: To get the toes under those of the opponent and roll him over.

5. Single pole pushing: Contestants grasp end of pole, 6 feet long and 2 inches thick, and brace themselves. Object: To push the opponent out of position.

6. Double pole pushing: The poles are placed under the arms close to the arm pits, ends projecting. Object: Same as in 5.

7. Double pole pulling: Position as in 6 but standing back to back. Object: To pull the opponent out of position.

8. "Cock fight:" Contestants hop on one leg with the arms folded closely over the chest. Object: by butting with the fleshy part of the shoulder without raising the arms, or by dodging to make the opponent change his feet or touch the floor with his hand or other part of his body.

9. One-legged tug of war: Contestants hop on one leg and grasp hands firmly. Object: To pull the opponent forward or make him place the raised foot on the floor.

10. The "siege:" One contestant stands with one foot in a circle 14 inches in diameter, the other foot outside, and the arms folded as in 8. Two other contestants, each hopping on one leg, endeavor to dis-

lodge the one in the circle by butting him with the shoulder. The be sieged one is defeated in case he raises the foot in the circle, or removes it entirely from the circle. The besiegers are defeated in case they change feet or touch the floor as in 8. As soon as either of the latter is defeated his place is immediately filled, so that there are always two of them. The besieged should resort to volting, ducking, etc., rather than to depend upon his strength.

11. One-armed tug: Contestants stand facing each other; right hands grasped, feet apart. Object: Without moving feet, to pull the opponent forward. Shifting the feet loses the bout.

12. "Tug royal:" Three contestants stand facing inward and grasp each other's wrists securely with their feet outside a circle about three feet in diameter. Object: by pulling or pushing to make one of the contestants step inside of the circle.

13. Indian wrestling: Contestants lie upon the ground face up, right shoulders in close contact, right elbows locked; at one the right leg is raised overhead and lowered, this is repeated at two, and at three the leg is raised quickly and locked with the opponent's right leg. Object: To roll him over by forcing his leg down.

14. Medicine ball race. Teams of five or six men are organized and a track for each team is marked out. This track consists of marks on the floor or ground at distances of 4 yards. On each of these marks stands a man with legs apart, the team forming a column of files. At "ready," "get set," the contestants prepare for the race, and at "go," the first man in the column rolls a medicine ball, which he has on the floor in front of him, through his legs to No. 2, he in turn rolls it to 3, etc., when it reaches the last man he picks it up and runs to the starting place with it and, the others all having shifted back one mark, the rolling is repeated. This continues until the first man brings the ball back to the starting place and every man is in his original position. The ball should be kept rolling; each man, as it comes to him, pushing it on quickly. Any ball about 9 inches in diameter will answer; it may be made of strong cloth and stuffed with cotton waste.

CHAPTER IV

MANUAL OF INTERIOR GUARD DUTY

(The numbers following the paragraphs are those of the *Manual of Interior Guard Duty*, and references in the text to certain paragraph numbers, refer to these numbers and not to the numbers preceding the paragraphs)

359. [Guard Duty is one of the soldier's most important duties, and in all armies of the world the manner in which it is performed is an index to the discipline of the command and the manner in which other duties are performed.

Upon the guard's vigilance and readiness for action depend not only the enforcement of military law and orders, but also the safety and protection of the post and the quelling of sudden disorder, perhaps even mutiny.

The importance of guard duty is increased during times of war, when the very safety of the army depends upon the vigilance of the sentinels, who are required to watch that others may sleep and thus refresh themselves from the labors of the day. The sentinels are the guardians of the repose, quiet and safety of the camp.

Respect for Sentinels.

360. Respect for the person and office of a sentinel is as strictly enjoined by military law as that required to be paid to an officer. As it is expressed in the *Manual of Guard Duty*, ''All persons of whatever rank in the service are required to observe respect toward sentinels''. Invested as the private soldier frequently is, while on his post, with a grave responsibility, it is proper that he should be fully protected in the discharge of his duty. To permit anyone, of whatever rank, to molest or interfere with him while thus employed, without becoming liable to severe penalty, would clearly establish a precedent highly prejudicial to the interests of the service. *(Davis' Military Law).*

Duty of sentinels.

A sentinel, in respect to the duties with which he is charged, represents the superior military authority of the command to which he belongs, and whose orders he is required to enforce on or in the vicinity of his post. As such he is entitled to the respect and obedience of all persons who come within the scope of operation of the orders, which he is required to carry into effect. Over military persons the authority of the sentinel is absolute, and disobedience of his orders on the part of such persons constitutes a most serious military offence and is prejudicial in the highest degree to the interests of discipline. *(Davis' Military Law).*—Author.]

INTRODUCTION

361. Guards may be divided into four classes: Exterior guards, interior guards, military police, and provost guards. (1)

Exterior guards are used only in time of war. They belong to the domain of tactics and are treated of in the *Field Service Regulations* and in the drill regulations of the different arms of the service.

The purpose of exterior guards is to prevent surprise, to delay attack, and otherwise to provide for the security of the main body.

On the march they take the form of advance guards, rear guards, and flank guards. At a halt they consist of outposts. (2)

Interior guards are used in camp or garrison to preserve order, protect property, and to enforce police regulations. In time of war such sentinels of an interior guard as may be necessary are placed close in or about a camp, and normally there is an exterior guard further out consisting of outposts. In time of peace the interior guard is the only guard in a camp or garrison. (3)

362. Military police differ somewhat from either of these classes. (See *Field Service Regulations*). They are used in time of war to guard prisoners, to arrest stragglers and deserters, and to maintain order and enforce police regulations in the rear of armies, along lines of communication, and in the vicinity of large camps. (4)

363. Provost guards are used in the absence of military police, generally in conjunction with the civil authorities at or near large posts or encampments, to preserve order among soldiers beyond the interior guard. (5)

INTERIOR GUARD

Classification

364. The various elements of an interior guard classified according to their particular purposes and the manner in which they perform their duties are as follows:

(a) The main guard.

(b) Special guards: Stable guards, park guards, prisoner guards, herd guards, train guards, boat guards, watchmen, etc. (6)

Details and Rosters

365. At every military post, and in every regiment or separate command in the field, an interior guard will be detailed and duly mounted.

It will consist of such number of officers and enlisted men as the commanding officer may deem necessary, and will be commanded by the senior officer or noncommissioned officer therewith, under the supervision of the officer of the day or other officer detailed by the commanding officer. (7)

The system of sentinels on fixed posts is of value in discipline and training because of the direct individual responsibility which is imposed and required to be discharged in a definite and precise manner. In order, however, that guard duty may not be needlessly irksome and interfere with tactical instruction, the number of men detailed for guard will be the smallest possible.

Commanding officers are specifically charged with this matter, and, without entirely dispensing with the system of sentinels on fixed posts will, as far as practicable, in time of peace, replace such sentinels with watchmen. (See Par. 221.) (8)

At posts where there are less than three companies the main guard and special guards may all be furnished by one company or by detail from each company.

Where there are three or more companies, the main guard will, if practicable, be furnished by a single company, and, as far as practicable, the same organization will supply all details for that day for special guard, overseer, and fatigue duty. In this case the officer of the day, and the officers of the guard, if there are any, will, if practicable, be from the company furnishing the guard. (9)

At a post or camp where the headquarters of more than one regiment are stationed, or in the case of a small brigade in the field, if but one guard be necessary for the whole command, the details will be made from the headquarters of the command.

If formal guard mounting is to be held, the adjutant, sergeant-major, and band to attend guard mounting will be designated by the commanding officer. (10)

When a single organization furnishes the guard, a roster of organizations will be kept by the sergeant-major under the supervision of the adjutant. (See Appendix B.) (11)

When the guard is detailed from several organizations, rosters will be kept by the adjutant, of officers of the day and officers of the guard by name; by the sergeant-major, under the supervision of the adjutant, of sergeants, corporals, musicians, and privates of the guard by number per organization; and by first sergeants, of sergeants, corporals, musicians, and privates by name. (See Appendix A.) (12)

When organizations furnish their own stable, or stable and park guards, credit will be given each for the number of enlisted men so furnished, as though they had been detailed for main guard. (13)

Special guards, other than stable or park guards, will be credited the same as for main guard, credited with fatigue duty, carried on special duty, or credited as the commanding officer may direct. (Pars. 6, 221, 247, and 300.) (14)

Captains will supervise the keeping of company rosters and see that all duties performed are duly creditded. (See Pars. 355-364, A. R., for rules governing rosters, and Form 342, A. G. O., for instructions as to how rosters should be kept.) (15)

There will be an officer of the day with each guard, unless in the opinion of the commanding officer the guard is so small that his services are not needed. In this case an officer will be detailed to supervise the command and instruction of the guard for such period as the commanding officer may direct. (16)

When more than one guard is required for a command, a field officer of the day will be detailed, who will receive his orders from the brigade or division commander as the latter may direct. When necessary, captains may be placed on the roster for field officer of the day. (17)

The detail of officers of the guard will be limited to the necessities of the service and efficient instruction; inexperienced officers may be detailed as supernumerary officers of the guard for purposes of instruction. (18)

Officers serving in staff departments are, in the discretion of the commanding officer, exempt from guard duty. (19)

Guard details will, if practicable, be posted or published the day preceding the beginning of the tour, and officers notified personally by a written order at the same time. (20)

The strength of guards and the number of consecutive days for which an organization furnishes the guard will be so regulated as to insure privates of the main guard an interval of not less than five days between tours.

When this is not otherwise practicable, extra and special duty men will be detailed for night-guard duty, still performing their daily duties. When so detailed a roster will be kept by the adjutant showing the duty performed by them. (21)

The members of main guards and stable and park guards will habitually be relieved every 24 hours. The length of the tour of enlisted men detailed as special guards, other than stable or park guards, will be so regulated as to permit of these men being held accountable for a strict performance of their duty. (22)

Should the officer of the day be notified that men are required to fill vacancies in the guard, he will cause them to be supplied from the organization to which the guard belongs. If none are available in that organization, the adjutant will be notified and will cause them to be supplied from the organization that is next for guard. (Par. 63.) (23)

The adjutant will have posted on the bulletin board at his office all data needed by company commanders in making details from their companies.

At first sergeant's call, first sergeants will go to headquarters and take from the bulletin board all data necessary for making the details required from their companies; these details will be made from their company rosters. (24)

In order to give ample notice, first sergeants will, when practicable, publish at retreat and post on the company bulletin board all details made from the company for duties to be performed. (25)

Where rosters are required to be kept by this manual, all details will be made by roster. (26)

The Commanding Officer

366. The commanding officer will exact a faithful, vigilant, and correct performance of guard duty in all of its details, giving his orders to the officer of the day, or causing them to be communicated to him with the least practicable delay. He will prescribe the strength of the guard, and the necessary regulations for guard, police, and fatigue duty. (27)

The commanding officer receives the reports of the officers of the day immediately after guard mounting, at his office, or at some other place previously designated; carefully examines the guard report and remarks thereon (questioning the old officer of the day, if necessary, concerning his tour of duty), relieves the old officer of the day and gives the new officer of the day such instructions as may be necessary. (28)

The Officer of the Day

367. The officer of the day is responsible for the proper performance of duty by the guard with which he marches on and for the enforcement of all police regulations. He is charged with the execution of all orders of the commanding officer relating to the safety and good order of the post or camp. His actual tour begins when he receives the instructions of the commanding officer after guard mounting, and ceases when he has been relieved by the commanding officer. In case of emergency during the interval between guard mounting and reporting to the commanding officer, the senior officer of the day will give the necessary instructions for both guards. (29)

In the absence of special instructions from the commanding officer, the officer of the day will inspect the guard and sentinels during the day and at night as such times as he may deem necessary. He will visit them at least once between 12 o'clock midnight and daylight. (30)

He may prescribe patrols(Par. 218) and visits of inspection to be made by officers and noncommissioned officers of the guard whenever he deems it necessary. (31)

He will see that the commander of the guard is furnished with the parole and countersign before retreat in case they are to be used, and will inform him of the presence in post or camp of any person entitled to the compliemnt. (32)

In case of alarm of any kind he will at once take such steps as may be necessary to insure the safety of life and public property and to preserve order in the command, disposing his guard so as best to accomplish this result. (33)

In the performance of his duties as officer of the day he is subject to the orders of the commanding officer only, except that in case of an alarm of any kind, and at a time of great danger, the senior line officer present is competent to give necessary orders to the officer of the day for the employment of the guard. (34)

At the inspections and musters prescribed in *Army Regulations,* the officer of the day will be present at the post of the guard, but all commands to the guard will be given by the commander of the guard. (35)

Both officers of the day together verify the prisoners and inspect the guardhouse and premises. (36)

In the absence of special instructions, the old officer of the day will, at guard mounting, release all garrison prisoners whose sentences expire that day. If there are any prisoners with no record of charges against them, the old officer of the day will report that fact to the commanding officer who will give the necessary instructions. (37)

The old officer of the day signs the report of the commander of the guard. He also enters on it such remarks as may be necessary. (38)

The officers of the day then report to the commanding officer.

On presenting themselves, both salute with the right hand, remaining covered. The old officer of the day, standing on the right of the new, then says: ''Sir, I report as old officer of the day,'' and presents the guard report. As soon as the commanding officer notifies the old officer of the day that he is relieved, the old officer of the day salutes the commanding officer and retires. The new officer of the day again salutes and says: ''Sir, I report as new officer of the day,'' and then receives his instructions. (39)

The officer of the day will always keep the guard informed as to where he may be found at all hours of the day and night. (40)

Commander of the Guard

368. The commander of the guard is responsible for the instruction and discipline of the guard. He will see that all of its members are correctly instructed in their orders and duties, and that they understand and properly perform them. He will visit each relief at least once while it is on post, and at least one of these visits will be made between 12 o'clock midnight and daylight. (41)

He receives and obeys the orders of the commanding officer and the officer of the day, and reports to the latter without delay all orders to the guard not received from the officer of the day; he transmits to his successor all material instructions and information relating to his duties. (42)

He is responsible under the officer of the day for the general safety of the post or camp as soon as the old guard marches away from the guardhouse. In case of emergency while both guards are at the guardhouse, the senior commander of the two guards will be responsible that the proper action is taken. (43)

Officers of the guard will remain constantly with their guards, except while visiting patrols or necessarily engaged elsewhere in the performance of their duties. The commanding officer will allow a reasonable time for meals. (44)

A commander of a guard leaving his post for any purpose will inform the next in command of his destination and probable time of return. (45)

Except in emergencies, the commander of the guard may divide the night with the next in command, but retains his responsibility; the one on watch must be constantly on the alert. (46)

When any alarm is raised in camp or garrison, the guard will be formed immediately. (Par. 234.) If the case be serious, the proper call will be sounded, and the commander of the guard will cause the commanding officer and the officer of the day to be at once notified. (47)

If a sentinel calls: "The Guard," the commander of the guard will at once send a patrol to the sentinel's post. If the danger be great, in which case the sentinel will discharge his piece, the patrol will be as strong as possible. (48)

When practicable, there should always be an officer or noncommissioned officer and two privates of the guard at the guardhouse, in addition to the sentinels there on post. (49)

Between reveille and retreat, when the guard had been turned out for any person entitled to the compliment (See Pars. 222 and 224), the commander of the guard, if an officer, will receive the report of the sergeant, returning the salute of the later with the right hand. He will then draw his saber, and place himself two paces in front of the center of the guard. When the person for whom the guard has been turned out approaches, he faces his guard and commands: 1. **Present**, 2. **ARMS**; faces to the front and salutes. When his salute is acknowledged he resumes the carry, faces about, and commands: 1. **Order**, 2. **ARMS**; and faces to the front.

If it be an officer entitled to inspect the guard, after saluting and before bringing his guard to an order, the officer of the guard reports: "**Sir, all present or accounted for**"; or, "**Sir, (so and so) is absent**"; or, if the roll call has been omitted: "**Sir, the guard is formed**," except that at guard mounting the commanders of the guards present their guards and salute without making any report.

Between retreat and reveille, the commander of the guard salutes and reports, but does not bring the guard to a present. (50)

To those entitled to have the guard turned out but not entitled to inspect it, no report will be made; nor will a report be made to any officer, unless he halts in front of the guard. (51)

When a guard commanded by a noncommissioned officer is turned out as a compliment or for inspection, the noncommissioned officer, standing at a right shoulder on the right of the right guide, commands: 1. Present, 2. ARMS. He then executes the rifle salute. If a report be also required, he will, after saluting, and before bringing his guard to an order, report as prescribed for the officer of the guard. (Par. 50.) (52)

368a. When a guard is in line, not under inspection, and commanded by an officer, the commander of the guard salutes his regimental, battalion, and company commander, by bringing the guard to attention and saluting in person.

For all other officers, excepting those entitled to the compliment from a guard (Par. 224), the commander of the guard salutes in person, but does not bring the guard to attention.

When commanded by a noncommissioned officer the guard is brought to attention in either case, and the noncommissioned officer salutes.

The commander of a guard exchanges salutes with the commanders of all other bodies of troops; the guard is brought to attention during the exchange.

"Present arms" is executed by a guard only when it has turned out for inspection or as a compliment, and at the ceremonies of guard mounting and relieving the old guard. (53)

In marching a guard or a detachment of a guard the principles of paragraph 53 apply. "Eyes right" is executed only in the ceremonies of guard mounting and relieving the old guard. (54)

If a person entitled to the compliment, or the regimental, battalion, or company commander, passes in rear of a guard, neither the compliment nor the salute is given, but the guard is brought to attention while such person is opposite the post of the commander.

After any person has received or declined the compliment, or received the salute from the commander of the guard, official recognition of his presence thereafter while he remains in the vicinity will be taken by bringing the guard to attention. (55)

The commander of the guard will inspect the guard at reveille and retreat, and at such other times as may be necessary, to assure himself that the men are in proper condition to perform their duties and that their arms and equipments are in proper condition. For inspection by other officers, he prepares the guard in each case as directed by the inspecting officer. (56)

The guard will not be paraded during ceremonies unless directed by the commanding officer. (57)

At all formations members of the guard or reliefs will execute inspection arms as prescribed in the drill regulations of their arm. (58)

The commander of the guard will see that all sentinels are habitually relieved every two hours, unless the weather or other cause makes it necessary that it be done at shorter or longer intervals, as directed by the commanding officer. (59)

He will question his noncommissioned officers and sentinels relative to the instructions they may have received from the old guard; he will see that patrols and visits of inspection are made as directed by the officer of the day. (60)

He will see that the special orders for each post and member of the guard, either written or printed, are posted in the guardhouse, and, if practicable, in the sentry box or other sheltered place to which the member of the guard has constant access. (61)

He will see that the proper calls are sounded at the hours appointed by the commanding officer. (62)

Should a member of the guard be taken sick, or be arrested, or desert, or leave his guard, he will at once notify the officer of the day. (Par. 23.) (63)

He will, when the countersign is used (Pars. 210 to 216), communicate it to the noncommissioned officers of the guard and see that it is duly communicated to the sentinels before the hour for challenging; the countersign will not be given to sentinels posted at the guardhouse. (64)

He will have the details for hoisting the flag at reveille, and lowering it at retreat, and for firing the reveille and retreat gun, made in time for the proper performance of these duties. (See Pars. 338, 344, 345, and 346.) He will see that the flags are kept in the best condition possible, and that they are never handled except in the proper performance of duty. (65)

He may permit members of the guard while at the guardhouse to remove their headdress, overcoats, and gloves; if they leave the guardhouse for any purpose whatever he will require that they be properly equipped and armed according to the character of the service in which engaged, or as directed by the commanding officer. (66)

He will enter in the guard report a report of his tour of duty, and, on the completion of his tour, will present it to the officer of the day. He will transmit with his report all passes turned in at the post of the guard. (67)

Whenever a prisoner is sent to the guardhouse or guard tent for confinement, he will cause him to be searched, and will, without unnecessary delay, report the case to the officer of the day. (68)

Under war conditions, if any one is to be passed out of camp at night, he will be sent to the commander of the guard, who will have him passed beyond the sentinels. (69)

The commander of the guard will detain at the guardhouse all suspicious characters or parties attempting to pass a sentinel's post without authority, reporting his action to the officer of the day, to whom persons so arrested will be sent, if necessary. (70)

He will inspect the guard rooms and cells, and the irons of such prisoners as may be ironed, at least once during his tour, and at such other times as he may deem necessary. (71)

He will cause the corporals of the old and new reliefs to verify together, immediately before each relief goes on post, the number of prisoners who should then properly be at the guardhouse. (72)

He will see that the sentences of prisoners under his charge are executed strictly in accordance with the action of the reviewing authority. (73)

When no special prisoner guard has been detailed (Par. 300), he will, as far as practicable, assign as guards over working parties of prisoners sentinels from posts guarded at night only. (74)

The commander of the guard will inspect all meals sent to the guardhouse and see that the quantity and quality of food are in accordance with regulations. (75)

At guard mounting he will report to the old officer of the day all cases of prisoners whose terms of sentence expire on that day, and also all cases of prisoners concerning whom no statement of charges has been received. (See Par. 241.) (76)

The commander of the guard is responsible for the security of the prisoners under the charge of his guard; he becomes responsible for them after their number has been verified and they have been turned over to the custody of his guard by the old guard or by the prisoner guard or overseers. (77)

The prisoners will be verified and turned over to the new guard without parading them, unless the commanding officer or the officer of the day shall direct otherwise. (78)

To receive the prisoners at the guardhouse when they have been paraded and after they have been verified by the officers of the day, the commander of the new guard directs his sergeant to form his guard with an interval, and commands: 1. **Prisoners**, 2. **Right**, 3. **FACE**, 4. **Forward**, 5. **MARCH**. The prisoners having arrived opposite the interval in the new guard, he commands: 1. **Prisoners**, 2. **HALT**, 3. **Left**, 4. **FACE**, 5. **Right** (or left), 6. **DRESS**, 7. **FRONT**.

The prisoners dress on the line of the new guard. (79)

Sergeant of the Guard

369. The senior noncommissioned officer of the guard always acts as sergeant of the guard, and if there be no officer of the guard, will perform the duties prescribed for the commander of the guard. (80)

The sergeant of the guard has general supervision over the other noncommissioned officers and the musicians and privates of the guard, and must be thoroughly familiar with all of their orders and duties. (81)

He is directly responsible for the property under charge of the guard, and will see that it is properly cared for. He will make lists of articles taken out by working parties, and see that all such articles are duly returned. If they are not, he will immediately report the fact to the commander of the guard. (82)

Immediately after guard mounting he will prepare duplicate lists of the names of all noncommissioned officers, musicians, and privates of the guard, showing the relief and post or duties of each. One list will be handed as soon as possible to the commander of the guard; the other will be retained by the sergeant. (83)

He will see that all reliefs are turned out at the proper time, and that the corporals thoroughly understand, and are prompt and efficient in, the discharge of their duties. (84)

During the temporary absence from the guardhouse of the sergeant of the guard, the next in rank of the noncommissioned officers will perform his duties. (85)

Should the corporal whose relief is on post be called away from the guardhouse, the sergeant of the guard will designate a noncommissioned officer to take the corporal's place until his return. (86)

The sergeant of the guard is responsible at all times for the proper police of the guardhouse or guard tent, including the ground about them and the prison cells. (87)

At "first sergeant's call" he will proceed to the adjutant's office and obtain the guard report book. (88)

When the national or regimental colors are taken from the stacks of the color line, the color bearer and guard, or the sergeant of the guard, unarmed, and two armed privates as a guard, will escort the colors to the colonel's quarters, as prescribed for the color guard in the drill regulations of the arm of the service to which the guard belongs. (89)

He will report to the commander of the guard any suspicious or unusual occurrence that comes under his notice, will warn him of the approach of any armed body, and will send to him all persons arrested by the guard. (90)

When the guard is turned out, its formation will be as follows: The senior noncommissioned officer, if commander of the guard, is on the right of the right guide; if not commander of the guard, he is in the line of file closers, in rear of the right four of the guard; the next in rank is right guide; the next left guide; the others in the line of file closers, usually, each in rear of his relief; the field music, with its left three paces to the right of the right guide. The reliefs form in the same order as when the guard was first divided, except that if the guard consists of dismounted cavalry and infantry, the cavalry forms on the left. (91)

The sergeant forms the guard, calls the roll, and, if not in command of the guard, reports to the commander of the guard as prescribed in drill regulations for a first sergeant forming a troop or company;

369 (contd.)

the guard is not divided into platoons or sections, and, except when the whole guard is formed prior to marching off, fours are not counted. (92)

The sergeant reports as follows: ''Sir, all present or accounted for,'' or ''Sir, (so-and-so) is absent''; or if the roll call has been omitted, ''Sir, the guard is formed.'' Only men absent without proper authority are reported absent. He then takes his place, without command. (93)

At night, the roll may be called by reliefs and numbers instead of names; thus, the first relief being on post: **Second relief; No. 1; No. 2,** etc.; **Third relief, Corporal; No. 1,** etc. (94)

Calling the roll will be dispensed with in forming the guard when it is turned out as a compliment, on the approach of an armed body, or in any sudden emergency; but in such cases the roll may be called before dismissing the guard. If the guard be turned out for an officer entitled to inspect it, the roll will, unless he directs otherwise, always be called before a report is made. (95)

The sergeant of the guard has direct charge of the prisoners, except during such time as they may be under the charge of the prisoner guard or overseers, and is responsible to the commander of the guard for their security. (96)

He will carry the keys of the guardroom and cells, and will not suffer them to leave his personal possession while he is at the guardhouse, except as hereinafter provided. (Par. 99.) Should he leave the guardhouse for any purpose, he will turn the keys over to the noncommissioned officer who takes his place. (Par. 85.) (97)

He will count the knives, forks, etc., given to the prisoners with their food, and see that none of these articles remain in their possession. He will see that no forbidden articles of any kind are conveyed to the prisoners. (98)

Prisoners, when paraded with the guard, are placed in line in its center. The sergeant, immediately before forming the guard, will turn over his keys to the noncommissioned officer at the guardhouse. Having formed the guard, he will divide it into two nearly equal parts. Indicating the point of division with his hand, he commands:

1. **Right** (or left), 2. **FACE,** 3. **Forward,** 4. **MARCH,** 5. **Guard,** 6. **HALT,** 7. **Left** (or right), 8. **FACE.**

If the first command be **right face,** the right half of the guard only will execute the movements; if **left face,** the left half only will execute them. The command **halt** is given when sufficient interval is obtained to admit the prisoners. The doors of the guardroom and cells are then opened by the noncommissioned officer having the keys. The prisoners will file out under the supervision of the sergeant, the noncommissioned officer, and sentinel on duty at the guardhouse, and such other sentinels as may be necessary; they will form in line in the interval between the two parts of the guard. (99)

To return the prisoners to the guard room and cells, the sergeant commands:

1. **Prisoners,** 2. **Right** (or left), 3. **FACE,** 4. **Column right,** (or left) 5. **MARCH.**

The prisoners, under the same supervision as before, return to their proper rooms or cells. (100)

To close the guard, the sergeant commands:

1. **Left** (or right), 2. **FACE,** 3. **Forward,** 4. **MARCH,** 5. **Guard,** 6. **HALT,** 7. **Right** (or left), 8. **FACE.**

The left or right half only of the guard, as indicated, executes the movement. (101)

If there be but few prisoners, the sergeant may indicate the point of division as above, and form the necessary interval by the commands:

1. **Right** (or left) **step,** 2. **MARCH,** 3. **Guard,** 4. **HALT,** and close the intervals by the commands:

1. **Left** (or right) **step,** 2. **MARCH,** 3. **Guard,** 4. **HALT.** (102)

If sentinels are numerous, reliefs may, at the discretion of the commanding officer, be posted in detachments, and sergeants, as well as corporals, required to relieve and post them. (103)

Corporal of the Guard

370. A corporal of the guard receives and obeys orders from none but noncommissioned officers of the guard senior to himself, the officers of the guard, the officer of the day, and the commanding officer. (104)

It is the duty of the corporal of the guard to post and relieve sentinels, and to instruct the members of his relief in their orders and duties. (105)

Immediately after the division of the guard into reliefs the corporals will assign the members of their respective reliefs to posts by number, and a soldier so assigned to his post will not be changed to another during the same tour of guard duty, unless by direction of the commander of the guard or higher authority. Usually, experienced soldiers are placed over the arms of the guard, and at remote and responsible posts. (106)

Each corporal will then make a list of the members of his relief, including himself. This list will contain the number of the relief, the name, the company, and the regiment of every member thereof, and the post to which each is assigned. The list will be made in duplicate, one copy to be given to the sergeant of the guard as soon as completed, the other to be retained by the corporal. (107)

371. When directed by the commander of the guard, the corporal of the first relief forms his relief, and then commands: **CALL OFF.**

371 (contd.)

Commencing on the right, the men call off alternately rear and front rank, "one," "two," "three," "four," and so on; if in single rank, they call off from right to left. The corporal then commands:
1. **Right,** 2. **FACE,** 3. **Forward,** 4. **MARCH.**

The corporal marches on the left, and near the rear file, in order to observe the march. The corporal of the old guard marches on the right of the leading file, and takes command when the last one of the old sentinels is relieved, changing places with the corporal of the new guard. (108)

When the relief arrives at six paces from a sentinel (See Par. 168), the corporal halts it and commands, according to the number of the post: **No. (—).**

Both sentinels execute port arms or saber; the new sentinel approaches the old, halting about one pace from him. (See Par. 172.) (109)

The corporals advance and place themselves, facing each other, a little in advance of the new sentinel, the old corporal on his right, the new corporal on his left, both at a right shoulder, and observe that the old sentinel transmits correctly his instructions.

The following diagram will illustrate the positions taken:

R is the relief; A, the new corporal; B, the old; C, the new sentinel; D, the old. (110)

The instructions relative to the post having been communicated, the new corporal commands, **Post;** both sentinels then resume the right shoulder, face toward the new corporal and step back so as to allow the relief to pass in front of them. The new corporal then commands, 1. **Forward,** 2. **MARCH;** the old sentinel takes his place in rear of the relief as it passes him, his piece in the same position as those of the relief. The new sentinel stands fast at a right shoulder until the relief has passed six paces beyond him, when he walks his post. The corporals take their places as the relief passes them. (111)

Mounted sentinels are posted and relieved in accordance with the same principles. (112)

On the return of the old relief, the corporal of the new guard falls out when the relief halts; the corporal of the old guard forms his relief on the left of the old guard, salutes, and reports to the commander of his guard: "**Sir, the relief is present**"; or "**Sir, (so and so) is absent,**" and takes his place in the guard. (113)

To post a relief other than that which is posted when the old guard is relieved, its corporal commands:

1. **(Such) relief,** 2. **FALL IN**; and if arms are stacked, they are taken at the proper commands.

The relief is formed facing to the front, with arms at an **order**; the men place themselves according to the numbers of their respective posts, viz, **two, four, six,** and so on, in the **front rank,** and **one, three, five,** and so on, in the **rear rank.** The corporal, standing about two paces in front of the center of his relief, then commands: **Call off.**

The men call off as prescribed. The corporal then commands: 1. **Inspection,** 2. **ARMS**, 3. Order, 4. **ARMS**; faces the commander of the guard, executes the rifle salute, reports: **"Sir, the relief is present,"** or **"Sir, (so and so) is absent"**; he then takes his place on the right at order arms. (114)

When the commander of the guard directs the corporal: **"Post your relief,"** the corporal salutes and posts his relief as prescribed (Pars. 108 to 111); the corporal of the relief on post does not go with the new relief, except when necessary to show the way. (115)

To dismiss the old relief, it is halted and faced to the front at the guardhouse by the corporal of the new relief, who then falls out; the corporal of the old relief then steps in front of the relief and dismisses it by the proper commands. (116)

Should the pieces have been loaded before the relief was posted, the corporal will, before dismissing the relief, see that no cartridges are left in the chambers or magazines. The same rule applies to sentinels over prisoners. (117)

Each corporal will thoroughly acquaint himself with all the special orders of every sentinel on his relief, and see that each understands and correctly transmits such orders in detail to his successor. (118)

There should be at least one noncommissioned officer constantly on the alert at the guardhouse, usually the corporal whose relief is on post. This noncommissioned officer takes post near the entrance of the guardhouse, and does not fall in with the guard when it is formed. He will have his rifle constantly with him. (119)

Whenever it becomes necessary for the corporal to leave his post near the entrance of the guardhouse, he will notify the sergeant of the guard, who will at once take his place, or designate another noncommissioned officer to do so. (120)

He will see that no person enters the guardhouse, or guard tent, or crosses the posts of the sentinels there posted without proper authority. (121)

Should any sentinel call for the corporal of the guard, the corporal will, in every case, at once and quickly proceed to such sentinel. He will notify the sergeant of the guard before leaving the guardhouse. (122)

371 (contd.)

He will at once report to the commander of the guard any viola
tion of regulations or any unusual occurrence which is reported to him
by a sentinel, or which comes to his notice in any other way. (123)

Should a sentinel call: "The Guard," the corporal will promptly
notify the commander of the guard. (124)

Should a sentinel call: "Relief," the corporal will at once proceed to the post of such sentinel, taking with him the man next for duty
on that post. If the sentinel is relieved for a short time only, the corporal will again post him as soon as the necessity for his relief ceases.
(125)

When the countersign is used, the corporal at the posting of the
relief during whose tour challenging is to begin gives the countersign
to the members of the relief, excepting those posted at the guardhouse.
(126)

He will wake the corporal whose relief is next on post in time
for the latter to verify the prisoners, form his relief, and post it at the
proper hour. (127)

Should the guard be turned out, each corporal will call his own
relief, and cause its members to fall in promptly. (128)

Tents or bunks in the same vicinity will be designated for the
reliefs so that all the members of each relief may, if necessary, be
found and turned out by the corporal in the least time and with the
least confusion. (129)

When challenged by a sentinel while posting his relief, the corporal commands: 1. **Relief,** 2. **HALT**; to the sentinel's challenge he answers "**Relief,**" and at the order of the sentinel he advances alone to
give the countersign, or to be recognized. When the sentinel says, "Advance relief," the corporal commands: 1. **Forward,** 2. **MARCH.**

If to be relieved, the sentinel is then relieved as prescribed. (130)

Between retreat and reveille, the corporal of the guard will challenge all suspicious looking persons or parties he may observe, first halting his patrol or relief, if either be with him. He will advance them in
the same manner that sentinels on post advance like parties (Pars. 191 to
197), but if the route of a patrol is on a continuous chain of sentinels, he
should not challenge persons coming near him unless he has reason to
believe that they have eluded the vigilance of sentinels. (131)

Between retreat and reveille, whenever so ordered by an officer
entitled to inspect the guard, the corporal will call: "**Turn out the
guard,**" announcing the title of the officer, and then, if not otherwise
ordered he will salute and return to his post. (132)

As a general rule he will advance parties approaching the guard
at night in the same manner that sentinels on post advance like parties.
Thus, the sentinel at the guardhouse challenges and repeats the answer
to the corporal, as prescribed hereafter (Par. 200); the corporal, ad-

vancing at "**port arms**," says: "**Advance (so and so) with the counter-sign**," or "**to be recognized**," if there be no countersign used; the countersign being correctly given, or the party being duly recognized, the corporal says: "**Advance (so and so)**"; repeating the answer to the challenge of the sentinel. (133)

When officers of different rank approach the guardhouse from different directions at the same time, the senior will be advanced first, and will not be made to wait for his junior. (134)

Out of ranks and under arms, the corporal salutes with the rifle salute. He will salute all officers whether by day or night. (135)

The corporal will examine parties halted and detained by sentinels, and if he has reason to believe the parties have no authority to cross sentinel's posts, will conduct them to the commander of the guard. (136)

The corporal of the guard will arrest all suspicious looking characters prowling about the post or camp, all persons of a disorderly character disturbing the peace, and all persons taken in the act of committing crime against the Government on a military reservation or post. All persons arrested by corporals of the guard, or by sentinels, will at once be conducted to the commander of the guard by the corporal. (137)

Musicians of the Guard

372. The musicians of the guard will sound call as prescribed by the commanding officer. (138)

Should the guard be turned out for national or regimental colors or standards, uncased, the field music of the guard will, when the guard present arms, sound, "**To the color**" or "**To the standard**"; or, if for any person entitled thereto, the march, flourishes, or ruffles, prescribed in paragraphs 375, 376, and 377, A. R. (139)

Orderlies and Color Sentinels

373. When so directed by the commanding officer, the officer who inspects the guard at guard mounting will select from the members of the new guard an orderly for the commanding officer and such number of other orderlies and color sentinels as may be required. (140)

For these positions the soldiers will be chosen who are most correct in the performance of duty and in military bearing, neatest in person and clothing, and whose arms and accoutrements are in the best condition. Clothing, arms, and equipments must conform to regulations. If there is any doubt as to the relative qualifications of two or more soldiers, the inspecting officer will cause them to fall out at the guardhouse and to form in line in single rank. He will then, by testing them in drill regulations, select the most proficient. The commander of the guard will be notified of the selection. (141)

When directed by the commander of the guard to fall out and report, an orderly will give his name, company, and regiment to the sergenat of the guard, and, leaving his rifle in the arm rack in his company quarters, will proceed at once to the officer to whom he is assigned, reporting: "Sir, Private ————, Company —, reports as orderly." (142)

If the orderly selected be a cavalryman, he will leave his rifle in the arm rack of his troop quarters, and report with his belt on, but without side arms unless specially otherwise ordered. (143)

Orderlies, while on duty as such, are subject only to the orders of the commanding officer and of the officers to whom they are ordered to report. (144)

When on orderly is ordered to carry a message, he will be careful to deliver it exactly as it was given to him. (145)

His tour of duty ends when he is relieved by the orderly selected from the guard relieving his own. (146)

Orderlies are members of the guard, and their name, company, and regiment are entered on the guard report and lists of the guard. (147)

If a color line is established, sufficient sentinels are placed on the color line to guard the colors and stacks. (148)

Color sentinels are posted only so long as the stacks are formed. The commander of the guard will divide the time equally among them. (149)

When stacks are broken, the color sentinels may be permitted to return to their respective companies. They are required to report in person to the commander of the guard at reveille and retreat. They will fall in with the guard, under arms, at guard mounting. (150)

Color sentinels are not placed on the regular reliefs, nor are their posts numbered. In calling for the corporal of the guard, they call: "Corporal of the guard. Color line." (151)

Officers or enlisted men passing the uncased colors will render the prescribed salute. If the colors are on the stacks, the salute will be made on crossing the color line or on passing the colors. (152)

A sentinel placed over the colors will not permit them to be moved, except in the presence of an armed escort. Unless otherwise ordered by the commanding officer, he will allow no one to touch them but the color bearer.

He will not permit any soldier to take arms from the stacks, or to touch them, except by order of an officer or noncommissioned officer of the guard.

If any person passing the colors or crossing the color line fails to salute the colors, the sentinel will caution him to do so, and if the caution be not heeded he will call the corporal of the guard and report the facts. (153)

Privates of the Guard

374. Privates are assigned to reliefs by the commander of the guard, and to posts, usually, by the corporal of their relief. They will not change from one relief or post to another during the same tour of guard duty unless by proper authority. (154)

Orders for Sentinels

Orders for sentinels are of two classes: General orders and special orders. General orders apply to all sentinels. Special orders relate to particular posts and duties. (155)

Sentinels will be required to memorize the following:

My general orders are:

1. To take charge of this post and all Government property in view.

2. To walk my post in a military manner, keeping always on the alert and observing everything that takes place within sight or hearing.

3. To report all violations of orders I am instructed to enforce.

4. To repeat all calls from posts more distant from the guardhouse than my own.

5. To quit my post only when properly relieved.

6. To receive, obey, and pass on to the sentinel who relieves me all orders from the commanding officer, officer of the day, and officers and noncommissioned officers of the guard only.

7. To talk to no one except in line of duty.

8. In case of fire or disorder to give the alarm.

9. To allow no one to commit a nuisance on or near my post.

10. In any case not covered by instructions to call the corporal of the guard.

11. To salute all officers, and all colors and standards not cased.

12. To be especially watchful at night, and, during the time for challenging, to challenge all persons on or near my post, and to allow no one to pass without proper authority. (156)

Regulations Relating to the General Orders for Sentinels

No. 1: To take charge of this post and all Government property in view.

All persons, of whatever rank in the service, are required to observe respect toward sentinels and members of the guard when such are in the performance of their duties. 157)

A sentinel will at once report to the corporal of the guard every unusual or suspicious occurrence noted. (158)

He will arrest suspicious persons prowling about the post or camp at any time, all parties to a disorder occurring on or near his post, and all, except authorized persons, who attempt to enter the camp at night, and will turn over to the corporal of the guard all persons arrested. (159)

374 (contd.)

The number, limits, and extent of his post will invariably constitute part of the special orders of a sentinel on post. The limits of his post should be so defined as to include every place to which he is required to go in the performance if his duties.

No. 2: **To walk my post in a military manner, keeping always on the alert and observing everything that takes place within sight or hearing.** (160)

A sentinel is not required to halt and change the position of his rifle on arriving at the end of his post, nor to execute **to the rear, march,** precisely as prescribed in the drill regulations, but faces about while walking, in the manner most convenient to him, and at any part of his post as may be best suited to the proper performance of his duties. He carries his rifle on either shoulder, and in wet or severe weather, when not in a sentry box, may carry it at a secure. (161)

Sentinels when in sentry boxes stand at ease. Sentry boxes will be used in wet weather only, or at other times when specially authorized by the commanding officer. (162)

In very hot weather, sentinels may be authorized to stand at ease on their posts, provided they can effectively discharge their duties in this position, but they will take advantage of this privilege only on the express authority of the officer of the day or the commander of the guard. (163)

A mounted sentinel may dismount occasionally and lead his horse but will not relax his vigilance.

No. 3: **To report all violations of orders I am instructed to enforce.** (164)

A sentinel will ordinarily report a violation of orders when he is inspected or relieved, but if the case be urgent he will call the corporal of the guard, and also, if necessary, will arrest the offender.

No. 4: **To repeat all calls from posts more distant from the guardhouse than my own.** (165)

To call the corporal, or the guard, for any purpose other than relief, fire, or disorder (Pars. 167 and 173), a sentinel will call, **"Corporal of the guard, No. (—),"** adding the number of his post. In no case will any sentinel call, **"Never mind the corporal;"** nor will the corporal heed such call if given.

No. 5. **To quit my post only when properly relieved.** (166)

If relief becomes necessary, by reason of sickness or other cause, a sentinel will call, **"Corporal of the guard, No. (—), Relief,"** giving the number of his post. (167)

Whenever a sentinel is to be relieved, he will halt, and with arms at a right shoulder, will face toward the relief when it is thirty paces from him. He will come to a port arms with the new sentinel, and in a low tone will transmit to him all the special orders relating to the post, and any other information which will assist him to better perform his duties.

No. 6: **To receive, obey, and pass on to the sentinel who relieves me, all orders from the commanding officer, officer of the day, and officers and noncommissioned officers of the guard only.** (168)

During this tour of duty a soldier is subject to the orders of the commanding officer, officer of the day, and officers and noncommissioned officers of the guard only; but any officer is competent to investigate apparent violations of regulations by members of the guard. (169)

A sentinel will quit his piece on an explicit order from any person from whom he lawfully receives orders while on post; under no cirmumstances will he yield it to any other person. Unless necessity therefor exists, no person will require a sentinel to quit his piece, even to allow it to be inspected. (170)

A sentinel will not divulge the countersign (Pars. 209 to 217) to any one except the sentinel who relieves him, or to a person from whom he properly receives orders, on such person's verbal order given personally. Privates of the guard will not use the countersign except in the performance of their duties while posted as sentinels.

No. 7: **To talk to no one except in line of duty.** (171)

When calling for any purpose, challenging, or holding communication with any person, a dismounted sentinel, armed with a rifle or saber, will take the position of ''port'' arms or saber. At night a dismounted sentinel, armed with a pistol, takes the position of raise pistol in challenging or holding communication. A mounted sentinel does not ordinarily draw his weapon in the daytime when challenging or holding conversation; but if drawn, he holds it at advance rifle, raise pistol, or port saber, according as he is armed with a rifle, pistol, or saber. At night, in challenging and holding conversation, his weapon is drawn and held as just prescribed, depending on whether he is armed with a rifle, pistol, or saber.

No. 8: **In case of fire or disorder to give the alarm.** (172)

In case of fire, a sentinel will call, ''Fire No. (—),'' adding the number of his post; if possible, he will extinguish the fire himself. In case of disorder, he will call: ''The Guard, No (—),'' adding the number of his post. If the danger be great, he will, in either case, discharge his piece before calling.

No. 11: **To salute all officers and all colors and standards not cased.** (173)

When not engaged in the performance of a specific duty, the proper execution of which would prevent it, a member of the guard will salute all officers who pass him. This rule applies at all hours of the day or night, except in the case of mounted sentinels armed with a

374 (contd.)

rifle or pistol, or dismounted sentinels armed with a pistol, after challenging. (See Par. 181.) (174)

Sentinels will salute as follows: A dismounted sentinel armed with a rifle or saber, salutes by presenting arms; if otherwise armed, he salutes with the right hand.

A mounted sentinel, if armed with a saber and the saber be drawn, salutes by presenting saber; otherwise he salutes in all cases with the right hand. (175)

To salute, a dismounted sentinel, with piece at a right shoulder or saber at a carry, halts and faces toward the person to be saluted when the latter arrives within thirty paces.

The limit within which individuals and insignia of rank can be readily recognized is assumed to be about 30 paces, and, therefore, at this distance cognizance is taken of the person or party to be saluted. (176)

The salute is rendered at 6 paces; if the person to be saluted does not arrive within that distance, then when he is nearest. (177)

A sentinel in a sentry box, armed with a rifle, stands at attention in the doorway on the approach of a person or party entitled to salute, and salutes by presenting arms according to the forgoing rules.

If armed with a saber, he stands at a carry and salutes as before. (178)

A mounted sentinel on a regular post halts, faces, and salutes in accordance with the foregoing rules. If doing patrol duty, he salutes, but does not halt unless spoken to. (179)

Sentinels salute, in accordance with the foregoing rules, all persons and parties entitled to compliments from the guard (Par. 244, 227, and 228): officers of the Army, Navy, and Marine Corps; military and naval officers of foreign powers; officers of volunteers, and militia officers when in uniform. (180)

A sentinel salutes as just prescribed when an officer comes on his post; if the officer holds communication with the sentinel, the sentinel again salutes when the officer leaves him.

During the hours when challenging is prescribed, the first salute is given as soon as the officer has been duly recognized and advanced. A mounted sentinel armed with a rifle or pistol, or a dismounted sentinel armed with a pistol, does not salute after challenging.

He stands at advance rifle or raise pistol until the officer passes. (181)

In case of the approach of an armed party of the guard, the sentinel will halt when it is about 30 paces from him, facing toward the party with his piece at the right shoulder. If not himself relieved, he will, as the party passes, place himself so that the party will pass in front of him; he resumes walking his post when the party has reached 6 paces beyond him. (182)

An officer is entitled to the compliments prescribed, whether in uniform or not. (183)

A sentinel in communication with an officer will not interrupt the conversation to salute. In the case of seniors the officer will salute, whereupon the sentinel will salute. (184)

When the flag is being lowered at retreat, a sentinel on post and in view of the flag will face the flag, and, at the first note of the "Star Spangled Banner" or to the color will come to a present arms. At the sounding of the last note he will resume walking his post.

No. 12: **To be especially watchful at night and during the time for challenging, to challenge all persons on or near my post, and to allow no one to pass without proper authority.** (185)

During challenging hours, if a sentinel sees any person or party on or near his post, he will advance rapidly along his post toward such person or party and when within about 30 yards will challenge sharply, "**HALT. Who is there?**" He will place himself in the best possible position to receive or, if necessary, to arrest the person or party. (186)

In case a mounted party be challenged, the sentinel will call, "**HALT. DISMOUNT. Who is there?**" (187)

The sentinel will permit only one of any party to approach him for the purpose of giving the countersign (Pars. 209 to 217), or if no countersign be used, of being duly recognized. When this is done the whole party is advanced, i. e., allowed to pass. (188)

In all cases the sentinel must satisfy himself beyond a reasonable doubt that the parties are what they represent themselves to be and have a right to pass. If he is not satisfied, he must cause them to stand and call the corporal of the guard. So, likewise, if he have no authority to pass persons with the countersign, or when the party has not the countersign, or gives an incorrect one. (189)

A sentinel will not permit any person to approach so close as to prevent the proper use of his own weapon before recognizing the person or receiving the countersign. (190)

When two or more persons approach in one party, the sentinel on receiving an answer that indicates that some one in the party has the countersign, will say, "**Advance one with the countersign,**" and, if the countersign is given correctly, will then say, "**Advance (So and so),**" repeating the answer to his challenge. Thus, if the answer be, "**Relief (friends with the countersign, patrol, etc.),**" the sentinel will say, "**Advance one with the countersign**"; then, "**Advance relief (friends, patrol, etc.).**" (191)

If a person having the countersign approach alone, he is advanced to give the countersign. Thus, if the answer be, "**Friend with the countersign (or officer of the day, or etc.),**" the sentinel will say,

"Advance, friend (or officer of the day, or etc.), with the counter-sign"; then, "Advance, friend (or officer of the day, or etc.)" (192)

If two or more persons approach a sentinel's post from different directions at the same time, all such persons are challenged in turn and required to halt and to remain halted until advanced.

The senior is first advanced, in accordance with the foregoing rules. (193)

If a party is already advanced and in communication with a sentinel, the latter will challenge any other party that may approach; if the party challenged be senior to the one already on his post, the sentinel will advance the new party at once. The senior may allow him to advance any or all of the other parties; otherwise, the sentinel will not advance any of them until the senior leaves him. He will then advance the senior only of the remaining parties, and so on. (194)

The following order of rank will govern a sentinel in advancing different persons or parties approaching his post: Commanding officer, officer of the day, officer of the guard, officers, patrols, reliefs, noncommissioned officers of the guard in order of rank, friends. (195)

A sentinel will never allow himself to be surprised, nor permit two parties to advance upon him at the same time. (196)

If no countersign be used, the rules for challenging are the same. The rules for advancing parties are modified only as follows: Instead of saying "Advance (so and so) with the countersign," the sentinel will say, "Advance (so and so) to be recognized." Upon recognition he will say, "Advance (so and so)." (197)

Answers to a sentinel's challenge intended to confuse or mislead him are prohibited, but the use of such an answer as "Friends with the countersign," is not to be understood as misleading, but as the usual answer made by officers, patrols, etc., when the purpose of their visit makes it desirable that their official capacity should not be announced. (198)

Special Orders For Sentinels at the Post of the Guard.

375. Sentinels posted at the guard will be required to memorize the following:

Between reveille and retreat to turn out the guard for all persons designated by the commanding officer, for all colors or standards not cased, and in time of war for all armed parties approaching my post, except troops at drill and reliefs and detachments of the guard.

At night, after challenging any person or party, to advance no one but call the corporal of the guard, repeating the answer to the challenge. (199)

After receiving an answer to his challenge, the sentinel calls, "Corporal of the guard (So and so)," repeating the answer to the challenge.

He does not in such cases repeat the number of his post. (200)

He remains in the position assumed in challenging until the corporal has recognized or advanced the person or party challenged, when he resumes walking his post, or, if the person or party be entitled thereto, he salutes and, as soon as the salute has been acknowledged, resumes walking his post. (201)

The sentinel at the post of the guard will be notified by direction of the commanding officer of the presence in camp or garrison of persons entitled to the compliment. (Par. 224.) (202)

The following examples illustrate the manner in which the sentinel at the post of the guard will turn out the guard upon the approach of persons or parties entitled to the compliment (Pars. 224, 227, and 228): "Turn out the guard, Commanding Officer"; "Turn out the guard, Governor of a Territory"; "Turn out the guard, national colors"; "Turn out the guard, armed party"; etc.

At the approach of the new guard at guard mounting the sentinel will call "Turn out the guard, armed party." (203)

Should the person named by the sentinel not desire the guard formed, he will salute, whereupon the sentinel will call "Never mind the guard." (204)

After having called "Turn out the guard," the sentinel will never call "Never mind the guard," on the approach of an armed party. (205)

Though the guard be already formed he will not fail to call "Turn out the guard," as required in his special orders, except that the guard will not be turned out for any person while his senior is at or coming to the post of the guard. (206)

The sentinels at the post of the guard will warn the commander of the approach of any armed body and of the presence in the vicinity of all suspicious or disorderly persons. (207)

In case of fire or disorder in sight or hearing, the sentinel at the guardhouse will call the corporal of the guard and report the facts to him. (208)

Countersigns and Paroles

376. **Forty-fourth Article of War.** Any person belonging to the armies of the United States who makes known the watchword to any person not entitled to receive it, according to the rules and discipline of war, or presumes to give a parole or watchword different from that which he received, shall suffer death or such other punishment as a court-martial may direct. (See Par. 171.) (209)

The **countersign** is a word given daily from the principal headquarters of a command to aid guards and sentinels in identifying persons who may be authorized to pass at night.

376 (contd.)

It is given to such persons as may be authorized to pass and re-pass sentinels' posts during the night, and to officers, noncommissioned officers, and sentinels of the guard. (210)

The **parole** is a word used as a check on the countersign in order to obtain more accurate identification of persons. It is imparted only to those who are entitled to inspect guards and to commanders of guards.

The parole or countersign, or both, are sent sealed in the form of an order to those entitled to them. (211)

When the commander of the guard demands the parole, he will advance and receive it as the corporal receives the countersign. (See Par. 133.) (212)

As the communications containing the parole and countersign must at times be distributed by many orderlies, the parole intrusted to many officers, and the countersign and parole to many officers and sentinels, and as both the countersign and parole must, for large commands, be prepared several days in advance, there is always danger of their being lost or becoming known to persons who would make improper use of them; moreover, a sentinel is too apt to take it for granted that any person who gives the right countersign is what he represents himself to be; hence for outpost duty there is greater security in omitting the use of the countersign and parole, or in using them with great caution. The chief reliance should be upon personal recognition or identification of all persons claiming authority to pass.

Persons whose sole means of identification is the countersign, or concerning whose authority to pass there is a reasonable doubt, should not be allowed to pass without the authority of the corporal of the guard after proper investigation; the corporal will take to his next superior any person about whom he is not competent to decide. (213)

The **countersign** is usually the name of a battle; the **parole,** that of a general or other distinguished person. (214)

When they can not be communicated daily, a series of words for some days in advance may be sent to posts or detachments that are to use the same parole or countersign as the main body. (215)

If the countersign be lost, or if a member of the guard desert with it, the commander on the spot will substitute another for it and report the case at once to headquarters. (216)

In addition to the countersign, use may be made of preconcerted signals, such as striking the rifle with the hand or striking the hands together a certain number of times, as agreed upon. Such signals may be used only by guards that occupy exposed points.

They are used before the countersign is given, and must not be communicated to anyone not entitled to know the countersign. Their use is intended to prevent the surprise of a sentinel.

In the daytime signals such as raising a cap or a handkerchief in a prearranged manner may be used by sentinels to communicate with the guard or with each other. (217)

Guard Patrols

377. A guard patrol consists of one or more men detailed for the performance of some special service connected with guard duty. (218)

If the patrol be required to go beyond the chain of sentinels, the officer or noncommissioned officer in charge will be furnished with the countersign, and the outposts and sentinels warned. (219)

If challenged by a sentinel, the patrol is halted by its commander, and the noncommissioned officer accompanying it advances alone and gives the countersign. (220)

Watchmen

Enlisted men may be detailed as watchmen or as overseers over prisoners, and as such will receive their orders and perform their duties as the commanding officer may direct. (221)

Compliments From Guards

378. The compliment from a guard consists in the guard turning out and presenting arms. (See Par. 50.) No compliments will be paid between retreat and reveille except as provided in paragraphs 361 and 362, nor will any person other than those named in paragraph 224 receive the compliment. (222)

Though a guard does not turn out between retreat and reveille as a matter of compliment, it may be turned out for inspection at any time by a person entitled to inspect it. (223)

Between reveille and retreat the following persons are entitled to the compliment: The President, sovereign or chief magistrate of a foreign country, and members of a royal family; Vice-President; President and President pro tempore of the Senate; American and foreign ambassadors; members of the Cabinet; Chief Justice; Speaker of the House of Representatives; committees of Congress officially visiting a military post; governors within their respective States and Territories; governors generala; Assistant Secretary of War officially visiting a military post; all general officers of the Army; general officers of foreign services visiting a post; naval, marine, volunteer, and militia officers in the service of the United States and holding the rank of general officer; American or foreign envoys or ministers; ministers accredited to the United States; chargés d'affaires accredited to the United States; consuls general accredited to the United States; commanding officer of the post or camp; officer of the day. (224)

---

a The term "governors general" shall be taken to mean administrative officers under whom officers with the title of governor are acting.

The relative rank between officers of the Army and Navy is as follows: General with admiral, lieutenant general with the vice admiral, major general with rear admiral, brigadier general with commodore,b colonel with captain, lieutenant colonel with commander, major with lieutenant commander, captain with lieutenant, first lieutenant with lieutenant (junior grade), second lieutenant with ensign. (A. R. 12.) (225)

Sentinels will not be required to memorize paragraph 224, and except in the cases of general officers of the Army, the commanding officer, and the officer of the day, they will be advised in each case of the presence in camp or garrison of persons entitled to the compliment. (226)

Guards will turn out and present arms when the national or regimental colors or standards, not cased, are carried past by a guard or an armed party. This rule also applies when the party carrying the colors is at drill. If the drill is conducted in the vicinity of the guard-house, the guard will be turned out when the colors first pass, and not thereafter. (227)

In case the remains of a deceased officer or soldier are carried past, the guard will turn out and present arms. (228)

In time of war all guards will turn out under arms when armed parties, except troops at drill and reliefs or detachments of the guard, approach their post. (See Par. 53.) (229)

The commander of the guard will be notified of the presence in camp or garrison of all persons entitled to the compliment, except general officers of the Army, the commanding officer, and the officer of the day. Members of the guard will salute all persons entitled to the compliment and all officers in the military or naval service of foreign powers, officers of the Army, Navy and Marine Corps, officers of volunteers, and officers of militia when in uniform. (230)

General Rules Concerning Guard Duty

379. Thirty-sixth Article of War. No soldier shall hire another to do his duty for him. (231)

Thirty-eighth Article of War. Any soldier who is found drunk on his guard, party, or other duty shall suffer such punishment as a court-martial may direct. (232)

All material instructions given to a member of the guard by an officer having authority will be promptly communicated to the commander of the guard by the officer giving them. (233)

Should the guard be formed, soldiers will fall in ranks under arms. At roll call, each man, as his name or number and relief are called, will answer "Here," and come to an order arms. (234)

b The grade of commodore ceased to exist as a grade on the active list of the Navy of the United States on Mar. 3, 1899. By section 7 of the act of Mar. 3, 1899, the nine junior rear admirals are authorized to receive the pay and allowances of a brigadier general of the Army.

380

Whenever the guard or a relief is dismissed, each member not at once required for duty will place his rifle in the arms racks, if they be provided, and will not remove it therefrom unless he requires it in the performance of some duty. (235)

Without permission from the commander of the guard, members of the main guard, except orderlies, will not leave the immediate vicinity of the guard house. Permission to leave will not be granted except in cases of necessity. (236)

Members of the main guard, except orderlies, will not remove their accoutrements or clothing without permission from the commander of the guard. (Par. 66.) (237)

Prisoners

380. Articles of war 66, 67, 68, 69 and 70 have special reference to the confinement of prisoners and should be carefully borne in mind (238)

The commander of the guard will place a civilian in confinement on an order from higher authority only, unless such civilian is arrested while in the act of committing some crime withtin the limits of the military jurisdiction; in which case the commanding officer will be immediately notified. (239)

Except as provided in the twenty-fourth article of war, or when restraint is necessary, no soldier will be confined without the order of on officer, who shall previously inquire into his offense. (A. R. 930.) (240)

An officer ordering a soldier into confinement will send, as soon as practicable, a written statement, signed by himself, to the commander of the guard, setting forth the name, company and regiment of such soldier, and a brief statement of the alleged offense. It is a sufficient statement of the offense to give the number and article of war under which the soldier is charged. (241)

A prisoner, after his first day of confinement, and until his sentence has been duly promulgated, is considered as held in confinement by the commanding officer. After due promulgation of his sentence, the prisoner is held in confinement by authority of the officer who reviews the proceedings of the court awarding sentence. The commander of the guard will state in his report, in the proper place, the name of the officer by whom the prisoner was originally confined. (242)

Enlisted men against whom charges have been preferred will be designated as ''awaiting trial''; enlisted men who have been tried will, prior to the promulgation of the result, be designated as ''awaiting result of trial''; enlisted men serving sentence of confinement, not involving dishonorable discharge, will be designated as ''garrison prisoners.'' Persons sentenced to dismissal or dishonorable discharge and to terms of confinement at military posts or elsewhere will be designated as ''general prisoners.'' (A. R. 928.) (243)

[273]

380 (contd.)

The sentences of prisoners will be read to them when the order promulgating the same is received. The officer of the guard, or the officer of the day, if there be no officer of the guard, will read them unless the commanding officer shall direct otherwise. (244)

When the date for the commencement of a term of confinement imposed by sentence of a court-martial is not expressly fixed by sentence, the term of confinement begins on the date of the order promulgating it. The sentence is continuous until the term expires, except when the person sentenced is absent without authority. (A. R. 969.) (245)

When soldiers awaiting trial or the result of trail, or undergoing sentence, commit offenses for which they are tried, the second sentence will be executed upon the expiration of the first. (246)

Prisoners awaiting trial by, or undergoing sentence of a general court-martial, and those confined for serious offenses, will be kept apart, when practicable, from those confined by sentence of an inferior court, or for minor offenses. Enlisted men in confinement for minor offenses, or awaiting trial or the result of trial for the same, will ordinarily be sent to work under charge of unarmed overseers instead of armed sentinels, and will be required to attend drills unless the commanding officer shall direct otherwise. (247)

Prisoners, other than general prisoners, will be furnished with food from their respective companies or from the organizations to which they may be temporarily attached.

The food of prisoners will, when practicable, be sent to their places of confinement, but post commanders may arrange to send the prisoners, under proper guard, to their messes for meals.

When there is no special mess for general prisoners, they will be attached for rations to companies.

Enlisted men bringing meals for the prisoners will not be allowed to enter the prison room. (See Par. 289.) (248)

With the exception of those specially designated by the commanding officer, no prisoners will be allowed to leave the guard house unless under charge of a sentinel and passed by an officer or noncommissioned officer of the guard. The commanding officer may authorize certain garrison prisoners and paroled general prisoners to leave the guard house, not under the charge of a sentinel, for the purpose of working outside under such surveillance and restrictions as he may impose. (249)

Prisoners reporting themselves sick at sick call, or at the time designated by the commanding officer, will be sent to the hospital under charge of proper guard, with a sick report kept for the purpose. The recommendation of the surgeon will be entered in the guard report. (250)

The security of sick prisoners in the hospital devolves upon the post surgeon, who will, if necessary, apply to the post commander for a guard. (251)

Prisoners will be paraded with the guard only when directed by the commanding officer or the officer of the day. (252)

A prisoner under charge of a sentinel will not salute an officer. (253)

All serviceable clothing which belongs to a prisoner, and his blankets, will accompany him to the post designated for his confinement, and will be fully itemized on the clothing list sent to that post. The guard in charge of the prisoner during transfer will be furnished with a duplicate of this list and will be held responsible for the delivery of all articles itemized therein, with the prisoner. At least one serviceable woolen blanket will be sent with every such prisoner so transferred. (A. R. 939.) (254)

When mattresses are not supplied, each prisoner in the guardhouse will be allowed a bed sack and 30 pounds of straw per month for bedding. So far as practicable, iron bunks will be furnished to all prisoners in post guardhouses and prison rooms. (A. R. 1084.) (255)

If the number of prisoners, including general prisoners confined at a post justifies it, the commanding officer will detail a commissioned officer as "officer in charge of prisoners." At posts where the average number of prisoners continually in confinement is less than 12, the detail of an officer in charge of prisoners will not be made. (256)

Rules and Regulations for the Government of General Prisoners at Posts.

The officer in charge of prisoners, when one is detailed, will make a daily inspection of the cells and prison rooms and will inspect the food and submit to the commanding officer any complaints about the same. (257)

He will have charge of the property, money, and valuables belonging to general prisoners, which they are not permitted to keep in ther possession, and will disburse said money, when desired by the owner, for purposes approved by the commanding officer. If there be no officer in charge of prisoners, this duty will be intrusted to the adjutant. (258)

No general prisoner will be released from confinement except on an order communicated by the commanding officer, who, before giving such order, will verify the date of expiration of the prisoner's sentence by examining all orders fixing or modifying the term of confinement. (259)

The following records and reports will be kept: Record of general prisoner, on blank supplied by the Adjutant General's Department; morning report, and clothing book (ordinary blank book without special ruling furnished by the Quartermaster's Department). (260)

Paragraphs 262 to 295 of this manual will be read to or by every general prisoner as soon as practicable after his confinement, and a copy of these rules and regulations, which will be furnished by the

380 (contd.)

Adjutant General's Department, will be kept posted in each cell and room. (261)

After a general prisoner, who is serving sentence at a post, has served one-half of his sentence, he may submit to the commanding officer of the post an application to be placed upon parole during working hours for the remainder of the term of confinement. Such application will contain a pledge on the part of the applicant to comply with all general conditions under which general prisoners may be paroled, and also with any special requirements that may from time to time be made of him. Upon receipt of such an application, the post commander may, in the exercise of his discretion, parole the prisoner during working hours for work in the Quartermaster Corps upon condition that if the prisoner's conduct is not good the parole status will be forfeited. The granting of the qualified parole here authorized does not constitute a release of the prisoner from military custody or control, but merely authorizes a relaxation of the strict rule which would otherwise require the presence of a guard whenever the prisoner is outside of the guard-house. In determining what constitutes one-half of a sentence the calculation will be based upon the prisoner's term without deduction for good conduct. The authorized abatement for good conduct will continue to accrue during the good conduct of a general prisoner on parole. (A. R. 943.) No paroled general prisoner will be employed about the post exchange or the quarters of any officer except as a mechanic or laborer under the direction of the quartermaster. (262)

Every general prisoner on admission will be minutely searched and will be permitted to retain in his possession only proper clothing and necessary toilet articles. He will then be required to bathe, his hair will be cut close, and his beard, whiskers, and mustache trimmed. (263)

General prisoners will bathe at least once a week and will wear their hair short. The hair and beard of a general prisoner may be allowed to grow during the last month of his confinement. (264)

All articles of personal property taken from a general prisoner will be marked with his name and stored until he is released, when they will be returned to him. (265)

The prison rooms will be properly policed, good order and quiet demeanor maintained, and necessary measures taken for security. The names of occupants of cells will be posted on the doors. Each cell and prison room will be inspected at least once a day for the purpose of detecting contraband articles and of seeing whether any alterations have been made or attempted which might facilitate escape. (266)

The diet of general prisoners shall be determined by the commanding officer. A general prisoner confined on bread-and-water diet will receive an allowance of 18 ounces of bread each day and as much water as he may desire. (267)

Meals will be served in prison rooms or cells when no separate mess is provided. Ample time and a sufficient quanity of food will be allowed for each meal. (268)

Each general prisoner will be furnished with and will have at all times one complete suit of outer clothing, two complete suits of underclothing, one pair of shoes, one hat, and one or two blankets, depending on the temperature. The outer clothing of general prisoners will be conspicuously marked "P" and divested of all ornament. When released such prisoner will have in his possession a serviceable suit of clothing, the outer garments bearing no prison mark. (269)

At the weekly inspection each general prisoner will stand by his bed or bunk, and the inspecting officer will see that the rules for cleanliness have been observed. The bedding and clothing will be folded, clothing on top of the bedding. General prisoners will be held to a strict accountability for clothing in their possession, and they are forbidden to alter it without authority. (270)

General prisoners will be kept at hard labor daily except Sunday, January 1, February 22, May 30, July 4, Labor Day, Thanksgiving Day, and Christmas Day, but in case of pressing necessity they may be employed on these days. So far as practicable, they will perform all scavenger duties at the post. They will not be employed in cultivat-company or private gardens, nor upon ordinary police about stables or barracks. (271)

General prisoners who desire an interview with the commanding officer will make application to the officer in charge, stating the purpose. The officer in charge will receive oral complaints which may be made by them, and will notify them of his action. Complaints in writing will also be addressed to him, and will be laid before the commanding officer with such information as he may possess bearing on the case. If there be no officer detailed in charge of prisoners, the officer of the day will receive application for interviews, complaints, etc., under this paragraph. (272)

Wrongs will be righted, if possible, but those who make frivolous or untruthful complaints will be punished. General prisoners will be permitted to submit explanations for offenses for which reported. No general prisoner will sign any protest or petition in conjunction with other prisoners; each will make his own complaints or requests. (273)

A record will be kept of all reports against general prisoners, with the disciplinary punishment awarded in each case. (274)

Except as otherwise ordered by the commanding officer, general prisoners will be constantly under charge of the guard, and in the event of mutiny, attempted outbreak or escape, or any disorder immediate action will be taken by the guard and enough force used to restore order. The force used in any case will be limited to that necessary to the enforcement of these rules, the preservation of order, and the proper control of prisoners. (275)

380 (contd.)

No disciplinary punishment will be inflicted upon general prisoners unless by direction of the commanding officer, and then only after a full investigation of each case. (276)

A general prisoner who violates any of these rules, who is insolent, insubordinate, disrespectful, or disorderly, or who uses indecent or profane language may be disciplined by—

(a) Being deprived of a meal.
(b) Being locked in his cell when not at work.
(c) Performing extra hard or disagreeable labor.
(d) Solitary confinement on bread-an-water diet.
(e) Forfeiture of good-conduct time.

In addition to being disciplined as indicated he may also be tried by court-martial if the gravity of the offense so demands.

Solitary confinement on bread and water will not exceed 14 consecutive days at any one period, and will not be repeated until an interval of 14 days shall have elapsed and shall not exceed 84 days in one year. (277)

No good-conduct time can be forfeited in advance. When it is necessary to discipline a general prisoner who has none to his credit, the punishment must take some other form. (278)

Any general prisoner who attempts to escape will forfeit all good-conduct time previously earned. A recaptured prisoner will suffer the same forfeiture. In either case, the prisoner may, in addition, be tried by court-martial. (279)

A general prisoner who refuses to work may, for the first offense, be closely confined and deprived of his next meal, but food will be allowed him as soon as he consents to resume work; and he may be further punished for his offense by loss of not more than 20 days' good-conduct time, or by being locked in his cell for not more than 30 days, except when at work. (280)

Letters will be sent out by general prisoners through the officer in charge or officer of the day. Each prisoner will be permitted to write to his family or friends once in each month, all letters to be submitted unsealed (without stamp or envelope) for inspection. Paper will be furnished to prisoners for official as well as private communications. (281)

Prison authorities without the consent of a general prisoner will not open and inspect letters addressed to him. Such letters may, however, be retained unopened until the prisoner is released, or his letters otherwise disposed of under judicial process. (282)

General prisoners will be permitted to make application for clemency as soon after their arrival at a post for confinement as they may desire, but thereafter not until six months shall have elapsed since the date of final action upon the last application. Applications should be addressed to the officer in charge (or the officer of the day), but applicants may state to what authority they wish to appeal. (283)

Applications for clemency should be based on reasonable grounds. Good conduct is rewarded by an allowance of good-conduct time, but does not of itself furnish any claim to clemency or further mitigation of sentence. It will aid, however, in obtaining favorable consideration for applications based upon other grounds. (284)

General prisoners, other than those confined in penitentiaries, will be allowed in abatement of their terms of confinement, when serving sentences of over 3 months and not over 12 months, 5 days for each complete period of 25 days during the whole of which their conduct has been good; but the abatement of 5 days so authorized shall not have the effect in any case of reducing the confinement below 3 months. On sentences exceeding 1 year there will be allowed the foregoing abatement for the first year of the sentence, including abatement, and thereafter 10 days for each complete period of 20 days during the whole of which the conduct of the prisoners has been good. Abatements thus authorized may be forfeited wholly or in part by subsequent misconduct, such forfeiture to be determined by the commanding officer of the post where the prisoner is confined. A general prisoner serving sentence in a penitentiary will receive the abatement authorized for convicts in that penitentiary. (A. R. 942.) (285)

In order to secure uniformity in computing abatement of terms of confinement, the following method of computation will be used:

A general prisoner will be credited at the beginning of his confinement with all the good-conduct time that can be earned in his case during the entire period of his sentence. All months will be assumed to consist of 30 days. When forfeitures of good-conduct time are imposed they will be deducted from the amount of the prisoner's credit, but care will be taken not to impose or deduct a forfeiture in excess of the amount of good-conduct time that has actually been earned at date of forfeiture. (A. R. 942.) Except when the loss of good-conduct time is prescribed for specific offenses, the other minor penalties enumerated in paragraph 275 will ordinarily be inflicted before resort is had to loss of good-conduct time. (286)

Talking, gazing about, or laughing in ranks is prohibited. General prisoners who are not at work will stand at attention when addressed by an officer or concommissioned officer. Those at work will, under no circumstances, leave their places of employment without the permission of the noncommissioned officer or sentinel in charge of the party. (287)

A general prisoner desiring to speak to a sentinel will hold up his hand as a signal for the desired permission. (288)

380 (contd.)

No persons will be permitted to enter the prison rooms without authority from the commanding officer, the officer of the day, or the officer in charge. (289)

The beds will be neatly made up as soon as the cells are unlocked. The night buckets will be emptied, cleaned, and put in the place provided for them during the day. A small quantity of disinfecting fluid will be placed in each bucket, and the buckets will be taken into the cells immediately after supper. (290)

Spitting on the walls or floors of cells and prison rooms, or defacing them, is forbidden. Any general prisoner who makes unnecessary litter or dirt in the prison will be reported to the officer in charge or officer of the day. (291)

Trafficking with general prisoners is forbidden. (292)

General prisoners will be in bed at taps. Loud talking or loud noises of any kind will not be permitted at any time. Strict silence is enjoined after tattoo. (293)

General prisoners will be respectful in their treatment of one another. They are forbidden to hold any conversation with soldiers or citizens, except on a matter of duty, without authority from the commanding officer, officer of the day, or officer in charge. (294)

A record of all violations of these rules will be kept by the provost sergeant or commander of the guard, and report of the same will be made to the officer in charge of prisoners or the officer of the day, in time to accompany the morning report of general prisoners. (295)

The foregoing rules will be enforced with reference to garrison prisoners so far as applicable. (296)

Garrison prisoners will be allowed in abatement of their terms of confinement when serving sentences of 1 month, 5 days for good conduct. On sentences exceeding 1 month they will be allowed the foregoing abatement for the first month of the sentence, and thereafter 10 days for each complete period of 20 days during the whole of which their conduct has been good. Abatements thus authorized may be forfeited, wholly or in part, by subsequent misconduct, such forfeiture to be determined by the commanding officer of the post where the prisoner is confined. (A. R. 942.) (297)

After a garrison prisoner has served one-half of his sentence he may, if his enlistment has not expired, submit to the commander of the post where the sentence is being executed a request to be put on probation for the remainder of the term of confinement adjudged, and upon the request being granted the soldier will be restored to duty upon condition that if his conduct is not good while on probation he will be required to serve the remainder of his sentence. In determining what constituted one-half of a sentence the calculation will be based upon the prisoner's term without deduction for good conduct. The authorized abatement for good conduct will continue to accrue during the good conduct of a garrison prisoner on probation. (A. R. 943.) (298)

Guarding Prisoners.

381. The sentinel at the post of the guard has charge of the prisoners except when they have been turned over to the prisoner guard or overseers. (Par. 247 and 300 to 304.)

(a) He will allow none to escape.

(b) He will allow none to cross his post leaving the guardhouse except when passed by an officer or noncommissioned officer of the guard.

(c) He will allow no one to communicate with prisoners without permission from proper authority.

(d) He will promptly report to the corporal of the guard any suspicious noise made by the prisoners.

(e) He will be prepared to tell whenever asked how many prisoners are in the guardhouse and how many are out at work or elsewhere.

Whenever prisoners are brought to his post returning from work or elsewhere, he will halt them and call the corporal of the guard, notifying him of the number of prisoners returning. Thus: "Corporal of the guard, (so many) prisoners."

He will not allow prisoners to pass into the guardhouse until the corporal of the guard has responded to the call and ordered him to do so. (299)

Whenever practicable special guards will be detailed for the particular duty of guarding working parties composed of such prisoners as cannot be placed under overseers. (Par. 247.) (300)

The prisoner guard and overseers will be commanded by the police officer; if there be no police officer, then by the officer of the day. (301)

The provost sergeant is sergeant of the prisoner guard and overseers, and as such receives orders from the commanding officer and the commander of the prisoner guard only. (302)

Details for prisoner guard are marched to the guardhouse and mounted by being inspected by the commander of the main guard, who determines whether all of the men are in proper condition to perform their duties and whether their arms and equipments are in proper condition, and rejects any men found unfit. (303)

When prisoners have been turned over to the prisoner guard or overseers, such guards or overseers are responsible for them under their commander, and all responsibility and control of the main guard ceases until they are returned to the main guard. (Par. 306.) (304)

If a prisoner attempts to escape, the sentinel will call "Halt". If he fails to halt when the sentinel has once repeated his call, and if

381 (contd.)

there be no other possible means of preventing his escape, the sentinel will fire upon him.

The following will more fully explain the important duties of a sentinel in this connection:

(CIRCULAR.)

WAR DEPARTMENT,
ADJUTANT GENERAL'S OFFICE,
Washington, November 1, 1887.

By direction of the Secretary of War the following is published for the information of the Army:

United States Circuit Court, Eastern District of Michigan, August 1, 1887.
The United States *v.* James Clark.

The circuit court has jurisdiction of a homicide committed by one soldier upon another within a military reservation of the United States

If a homicide be committed by a military guard without malice and in the performance of his supposed duty as a soldier, such homicide is excusable, unless it was manifestly beyond the scope of his authority or was such that a man of ordinary sense and understanding would know that it was illegal.

It seems that the sergeant of the guard has a right to shoot a military convict if there be no other possible means of preventing his escape

The common-law distinction between felonies and misdemeanors has no application to military offenses.

While the finding of a court of inquiry acquitting the prisoner of all blame is not a legal bar to a prosecution, it is entitled to weight as an expression of the views of the military court of the necessity of using a musket to prevent the escape of the deceased.

*   *   *   *   *

By order of the Secretary of War:

R. C. DRUMM,
Adjutant General

The following is taken from Circular No. 3 of 1883, from Headquarters Department of the Columbia:

VANCOUVER BARRACKS, W. T.,
April 20, 1883.

TO THE ASSISTANT ADJUTANT GENERAL,
DEPARTMENT OF THE COLUMBIA.
SIR:

*   *   *   *   *   *

A sentinel is placed as guard over prisoners to prevent their escape, and for this purpose he is furnished a musket, with ammunition. To prevent escape is his first and most important duty.

*   *   *   *   *

I suppose the law to be this: That a sentinel shall not use more force or violence to prevent the escape of a prisoner than is necessary to effect that object, but if the prisoner, after being ordered to halt, continues his flight, the sentinel may maim or even kill him, and it is his duty to do so

A sentinel who allows a prisoner to escape without firing upon him and firing to hit him, is, in my judgment, guilty of a most serious military offense, for which he should and would be severely punished by a general court-martial.

*   *   *   *

[Signed]   HENRY A. MORROW,
Colonel 21st Infantry, Commanding Post.

[Third indorsement.]

OFFICE JUDGE ADVOCATE,
MILITARY DIVISION OF THE PACIFIC,
May 11, 1883.

Respectfully returned to the Assistant Adjutant General, Military Division of the Pacific, concurring fully in the views expressed by Colonel Morrow. I was not aware that such a view had ever been questioned That the period is a time of *peace* does not affect the authority and duty of the sentinel or guard to fire upon the escaping prisoner, if this escape can not otherwise be prevented. He should, of course, attempt to stop the prisoner before firing, by ordering him to halt, and will properly warn him by the words, "Halt, or I fire," or words to such effect.

W. WINTHROP,
Judge Advocate.

[Fourth indorsement.]
HEADQUARTERS MILITARY DIVISION OF THE PACIFIC,
May 11, 1883.
Respectfully returned to the Commanding General, Department of the Columbia, approving the opinion of the commanding officer, Twenty-first Infantry, and of the Judge Advocate of the Division, in respect to the duty of and method to be adopted by sentinels in preventing prisoners from escaping. * * *

By command of Major General Schofield:

J. C. KELTON,
Assistant Adjutant General.
See also Circular No. 53, A. G. O., December 22, 1900. (305)

On approaching the post of the sentinel at the guardhouse, a sentinel of the prisoner guard or an overseer in charge of prisoners will halt them and call, "No. 1, (so many prisoners.)" He will not allow them to cross the post of the sentinel until so directed by the corporal of the guard. (306)

Members of the prisoner guard and overseers placed over prisoners for work will receive specific and explicit instructions covering the required work; they will be held strictly responsible that the prisoners under their charge properly and satisfactorily perform the designated work. (307)

Stable Guards.

382. Under the head of stable guards will be included guards for cavalry stables, artillery stables and parks, mounted infantry stables, machine-gun organization stables and parks, and quartermaster stables and parks. Where the words "troop" and "cavalry" are used they will be held to include all of these organizations. (308)

When troop stable guards are mounted they will guard the stables of the cavalry (See Par. 13). When no stable guards are mounted, the stables will be guarded by sentinels posted from the main guard, under the control of the officer of the day.

The instructions given for troop stable guard will be observed as far as applicable by the noncommissioned officers and sentinels of the main guard when in charge of the stables. (309)

Troop Stable Guards.

383. Troop stable guards will not be used except in the field, or when it is impracticable to guard the stables by sentinels from the main guard. (310)

Troop stable guards will be under the immediate control of their respective troop commanders; they will be posted in each cavalry stable, or near the picket line, and will consist of not less than one noncommissioned officer and three privates.

Stable guards are for the protection of the horses, stables, forage, equipments, and public property generally. They will in addition enforce the special regulations in regard to stables, horses, and parks. (311)

Sentinels of stable guards will be posted at the stables or at the picket lines when the horses are kept outside. The troop stable guard may be used as a herd guard during the day time or when grazing is practicable. (312)

The troop stable guard, when authorized by the post commander, will be mounted under the supervision of the troop commander. It will be armed, at the discretion of the troop commander, with either rifle or pistol. (313)

The tour continues for 24 hours, or until the guard is relieved by a new guard. (314)

The employment of stable guards for police and fatigue duties at the stables is forbidden; but this will not prohibit them from being required to assist in feeding grain before reveille. (315)

The troop stable guard will attend stables with the rest of the troop and groom their own horses, the sentinels being taken off post for the purpose. (316)

Neither the noncommissioned officer nor the members of the stable guard will absent themselves from the immediate vicinity of the stables except in case of urgent necessity, and then for no longer time than is absolutely necessary. No member of the guard will leave for any purpose without the authority of the noncommissioned officer of the guard. (317)

The noncommissioned officer and one member of the stable guard will go for meals at the proper hour; upon their return the other members of the guard will be directed to go by the noncommissioned officer. (318)

When the horses are herded each troop will furnish its own herd guard. (Par. 14.) (319)

Smoking in the stables or their immediate vicinity is prohibited. No fire or light, other than electric light or stable lanterns, will be permitted in the stables. A special place will be designated for trimming, filling, and lighting lanterns. (320)

Noncommissioned Officer of the Troop Stable Guard

384. The noncommissioned officer receives his orders from his troop commander, to whom he will report immediately after posting his first relief, and when relieved will turn over all his orders to his successor. He instructs his sentinels in their general and special duties; exercises general supervision over his entire guard; exacts order and cleanliness about the guardroom; prevents the introduction of intoxicants into the guardhouse and stables; receives, by count, from his predecessor, the animals, horse equipments, and all property (both private and public) pertaining thereto; examines, before relieving his predecessor, all locks, windows, and doors, and should any be found insecure he will report the fact to his troop commander when he reports for orders. He will personally

post and relieve each sentinel, taking care to verify the property responsibility of the sentinel who comes off post, and see that the sentinel who goes on post is aware of the property responsibility that he assumes. (321)

That the noncommissioned officer may be more thoroughly informed of his responsibility, all horses returning, except those from a regular formation, will be reported to him. He will then notify the sentinel on post, and, in the absence of the stable sergeant, will see that the horses are promptly cared for.

In case of abuse, he will promptly report to the troop commander. Should the horse be the private property of an officer, he will report such abuse to the owner. (322)

The noncommissioned officer will report any unusual occurrence during his tour direct to his troop commander. (323)

Horses and other property for which the noncommissioned officer is responsible will not be taken from the stables without the authority of the post or troop commander. (324)

The noncommissioned officer must answer the sentinel's calls promptly. (325)

In case of fire, the noncommissioned officer will see that the requirements of paragraph 334 are promptly carried out. (326)

Whenever it becomes necessary for the noncommissioned officer to leave his guard, he will designate a member of it to take charge and assume his responsibility during his absence. (327)

Sentinels of the Troop Stable Guard

385. The sentinel in the discharge of his duties will be governed by the regulations for sentinels of the main guard whenever they are applicable—such as courtesies to officers, walking post in a soldierly manner, challenging, etc.; he will not turn out the guard except when ordered by proper authority. (328)

The sentinel will receive orders from the commanding officer, the troop commander, and the noncommissioned officers of the stable guard only, except when the commanding officer directs the officer of the day to inspect the stable guard. (329)

In the field and elsewhere when directed by the commanding officer the sentinel when posted will verify the number of horses for which he is responsible, and when relieved will give the number to his successor. (330)

The sentinel will not permit any horse or equipments to be taken from the stables, except in the presence of the noncommissioned officer. (331)

Should a horse get loose, the sentinel will catch him and tie him up. If he be unable to catch the horse, the noncommissioned officer will at once be notified. In case a horse be cast, or in any way entangled, he will relieve him, if possible; if unable to relieve him, he will call the noncommissioned officer. Sentinels are forbidden to punish or maltreat a horse. (332)

When a horse is taken sick, the sentinel will notify the noncommissioned officer, who in turn will call the farrier, and see that the horse is properly attended to. (333)

In case of fire the sentinel will give the alarm by stepping outside the stable and firing his pistol or piece repeatedly, and calling out at the same time, "Fire, stables, Troop (———)."

As soon as the guard is alarmed, he will take the necessary precautions in opening or closing the doors so as to prevent the spreading of the fire and make it possible to remove the horses; he will drop the chains and bars, and, with the other members of the guard, proceed to lead out the horses and secure them at the picket line or such other place as may have been previously designated. (334)

Sentinels over horses, or in charge of prisoners, receive orders from the stable sergeant, so far as the care of the horses and the labor of prisoners are concerned. (335)

In field artillery and machine-gun organizations, the guard for the stables has charge of the guns, caissons, etc., with their ammunition and stores, as well as the horses, harness, and forage. (336)

Flags

386. The garrison, post and storm flags are national flags and shall be of bunting. The union of each is as described in paragraph 216, *Army Regulations*, and shall be of the following proportions: Width, seven-thirteenths of the hoist of the flag; length, seventy-six one-hundredths of the hoist of the flag.

The garrison flag will have 38 feet fly and 20 feet hoist. It will be furnished only to posts designated in orders from time to time from the War Department, and will be hoisted only on holidays and important occasions.

The post flag will have 19 feet fly and 10 feet hoist. It will be furnished for all garrison posts and will be hoisted in pleasant weather.

The storm flag will have 9 feet 6 inches fly and 5 feet hoist. It will be furnished for all occupied posts for use in stormy and windy weather. It will also be furnished to national cemeteries. (A. R. 223.) (337)

At every military post or station the flag will be hoisted at the sounding of the first note of the reveille, or of the first note of the march, if a march be played before the reveille. The flag will be lowered at the sounding of the last note of the retreat, and while the flag is being lowered the band will play "The Star Spangled Banner," or, if there be no band present, the field music will sound "to the color." When "to the color" is sounded by the field music while the flag is being

387

lowered the same respect will be observed as when ''The Star Spangled Banner'' is played by the band, and in either case officers and enlisted men out of ranks will face toward the flag, stand at attention, and render the prescribed salute at the last note of the music. (A. R. 437)

The lowering of the flag will be regulated as to be completed at the last note of ''The Star Spangled Banner'' or ''to the color.'' (338)

The national flag will be displayed at a seacoast or lake fort at the beginning of and during an action in which a fort may be engaged, whether by day or by night. (A. R. 437.) (339)

The national flag will always be displayed at the time of firing a salute. (A. R. 397.) (340)

The flag of a military post will not be dipped by way of salute or compliment. (A. R. 405.) (341)

On the death of an officer at a military post the flag is displayed at half-staff and so remains, between reveille and retreat until the last salvo or volley is fired over the grave; or if the remains are not interred at the post, until they are removed therefrom. (A. R. 422.) (342)

During the funeral of an enlisted man at a military post the flag is displayed at half-staff. It is hoisted to the top after the final volley or gun is fired or after the remains are taken from the post. The same honors are paid on the occasion of the funeral of a retired enlisted man. (A. R. 423.) (343)

When practicable, a detail consisting of a noncommissioned officer and two privates of the guard will raise or lower the flag. This detail wears side arms or, if the special equipments do not include side arms, then belts only.

The noncommissioned officer, carrying the flag, forms the detail in line, takes his post in the center, and marches it to the staff. The flag is then securely attached to the halyards and rapidly hoisted. The halyards are then securely fastened to the cleat on the staff and the detail marched to the guardhouse. (344)

When the flag is to be lowered, the halyards are loosened from the staff and made perfectly free. At retreat the flag is lowered at the last note of retreat. It is then neatly folded and the halyards made fast. The detail is then reformed and marched to the guardhouse, where the flag is turned over to the commander of the guard.

The flag should never be allowed to touch the ground and should always be hoisted or lowered from the leeward side of the staff, the halyards being held by two persons. (345)

Reveille and Retreat Gun

387. The morning and evening gun will be fired by a detachment of the guard, consisting, when practicable, of a corporal and two privates. The morning gun is fired at the first note of reveille, or, if marches be played before the reveille, it is fired at the beginning of the first march.

The retreat gun is fired at the last note of retreat.

The corporal marches the detachment to and from the piece, which is fired, sponged out, and secured under his direction. (846)

Guard Mounting

388. Guard mounting will be formal or informal as the commanding officer may direct. It will be held as prescribed in the drill regulations of the arm of the service to which the guard belongs; if none is prescribed, then as for infantry. In case the guard is composed wholly of mounted organizations, guard mounting may be held mounted. (347)

When infantry and mounted troops dismounted are united for guard mounting, all details form as prescribed for infantry. (348)

Formal Guard Mounting for Infantry

Formal guard mounting will ordinarily be held only in posts or camps where a band is present. (349)

At the assembly, the men designated for the guard fall in on their company parade grounds as prescribed in paragraph 106, I. D. R. The first sergeant then verifies the detail, inspects it, replaces any man unfit to go on guard, turns the detail over to the senior noncommissioned officer, and retires. The band takes its place on the parade ground so that the left of its front rank shall be 12 paces to the right of the front rank of the guard when the latter is formed. (350)

At adjutant's call, the adjutant, dismounted, and the sergeant-major on his left, marches to the parade ground. The adjutant halts and takes post so as to be 12 paces in front of and facing the center of the guard when formed; the sergeant-major continues on, moves by the left flank, and takes post, facing to the left, 12 paces to the left of the front rank of the band; the band plays in quick or double quick time; the details are marched to the parade ground by the senior noncommissioned officers; the detail that arrives first is marched to the line so that, upon halting, the breast of the front-rank man shall be near to and opposite the left arm of the sergeant-major; the commander of the detail halts his detail, places himself in front of and facing the sergeant-major, at a distance equal to or a little greater than the front of his detail, and commands: 1. Right, 2. DRESS. The detail dresses up to the line of the sergeant-major and its commander, the right front-rank man placing his breast against the left arm of the sergeant-major; the noncommissioned officers take post two paces in rear of the rear rank of the detail. The detail aligned, the commander of the detail commands: FRONT, salutes, and then reports: "The detail is correct;" or "So many sergeants, corporals, or privates are absent;" the sergeant-major returns the salute with the right hand after the report is made; the commander then passes by the right of the guard and takes post in the line of noncommissioned officers in rear of the right file or his detail.

Should there be more than one detail, it is formed in like manner on the left of the one preceding; the privates, noncommissioned officers, and commander of each detail dress on those of the preceding details in the same rank or line; each detail commander closes the rear rank to the right and fills blank files, as far as practicable, with the men from his front rank.

Should the guard from a company not include a noncommissioned officer, one will be detailed to perform the duties of commander of the detail. In this case the commander of the detail, after reporting to the sergeant-major, passes around the right flank between the guard and the band and retires. (351)

When the last detail has formed, the sergeant-major takes a side step to the right, draws sword, verifies the detail, takes post two paces to the right and two paces to the front of the guard, facing to the left, causes the guard to count off, completes the left squad, if necessary, as in the school of the company, and if there be more than three squads, divides the guard into two platoons, again takes post as described above and commands: 1. **Open ranks**, 2. **MARCH**.

At the command **march**, the rear rank and file closers march backward four steps, halt, and dress to the right. The sergeant major aligns the ranks and file closers and again, taking post as described above, commands: **FRONT**, moves parallel to the front rank until opposite the center, turns to the right, halts midway to the adjutant, salutes, and reports: "**Sir, the details are correct;**" or, "**Sir, (so many) sergeants, corporals, or privates are absent;** ' the adjutant returns the salute, directs the sergeant-major: **Take your post**, and then draws saber; the sergeant-major faces about, approaches to within two paces of the center of the front rank, turns to the right, moves three paces beyond the left of the front rank, turns to the left, halts on the line of the front rank, faces about, and brings his sword to the order. When the sergeant-major has reported, the officer of the guard takes post, facing to the front, three paces in front of the center of the guard, and draws saber.

The adjutant then commands: 1. **Officer (or officers) and noncommissioned officers**, 2. **Front and Center**, 3. **MARCH**.

At the command **center**, the officers carry saber. At the command **march**, the officer advances and halts three paces from the adjutant, remaining at the carry; the noncommissioned officers pass by the flanks, along the front, and form in order of rank from right to left, three paces in rear of the officer, remaining at the right shoulder; if there is no officer of the guard the noncommissioned officers halt on a line three paces from the adjutant; the adjutant then assigns the officers and noncommissioned officers according to rank, as follows: **Commander of the guard, leader of first platoon, leader of second platoon, right guide of first platoon, left guide of second platoon, left guide of first platoon,**

388 (contd.)

right guide of second platoon, and file closers, or, if the guard is not divided into platoons: Commander of the guard, right guide, left guide, and file closers.

The adjutant then commands: 1. **Officer (or officers) and noncommissioned officers**, 2. **POSTS**, 3. **MARCH**.

At the command **posts**, all, except the officer commanding the guard, face about. At the command **march**, they take the posts prescribed in the school of the company with open ranks. The adjutant directs: **Inspect your guard, sir**; at which the officer commanding the guard faces about, commands: **Prepare for inspection**, returns saber, and inspects the guard.

During the inspection, the band plays; the adjutant returns saber, observes the general condition of the guard, and falls out any man who is unfit for guard duty or does not present a creditable appearance. Substitutes will report to the commander of the guard at the guardhouse. (352)

The adjutant, when so directed, selects orderlies and color sentinels, as prescribed in paragraphs 140 and 141, and notifies the commander of the guard of his selection. (353)

If there be a junior officer of the guard he takes post at the same time as the senior, facing to the front, 3 paces in front of the center of the first platoon; in going to the front and center he follows and takes position on the left of the senior and is assigned as leader of the first platoon; he may be directed by the commander of the guard to assist in inspecting the guard.

If there be no officer of the guard, the adjutant inspects the guard. A noncommissioned officer commanding the guard takes post on the right of the right guide, when the guard is in line; and takes the post of the officer of the guard, when in column or passing in review. (354)

The inspection ended, the adjutant places himself about 30 paces in front of and facing the center of the guard, and draws saber; the new officer of the day takes post in front of and facing the guard, about 30 paces from the adjutant; the old officer of the day takes post 3 paces to the right of and 1 pace to the rear of the new officer of the day; the officer of the guard takes post 3 paces in front of its center, draws saber with the adjutant and comes to the order; thereafter he takes the same relative positions as a captain of a company.

The adjutant then commands: 1. **Parade**, 2. **REST**, 3. **SOUND OFF**, and comes to the order and parade rest.

The band, playing, passes in front of the officer of the guard to the left of the line, and back to its post on the right, when it ceases playing.

[290]

The adjutant then comes to attention, carries saber, and commands: 1. **Guard**, 2. **ATTENTION**, 3. **Close ranks**, 4. **MARCH**.

The ranks are opened and closed as in paragraph 745, I. D. R.

The adjutant then commands: 1. **Present**, 2. **ARMS**, faces toward the new officer of the day, salutes, and then reports: **Sir, the guard is formed.** The new officer of the day, after the adjutant has reported, returns the salute with the hand and directs the adjutant: **March the guard in review, sir.**

The adjutant carries saber, faces about, brings the guard to an order, and commands: 1. **At trail, platoons (or guard) right**, 2. **MARCH**, 3. **Guard**, 4. **HALT**.

The platoons execute the movements; the band turns to the right and places itself 12 paces in front of the first platoon.

The adjutant places himself 6 paces from the flank and abreast of the commander of the guard; the sergeant major, 6 paces from the left flank of the second platoon.

The adjutant then commands: 1. **Pass is review**, 2. **FORWARD**, 3. **MARCH**.

The guard marches in quick time past the officer of the day, according to the principles of review, and is brought to **eyes right** at the proper time by the commander of the guard; the adjutant, commander of the guard, leaders of platoons, sergeant-major, and drum major salute.

The band, having passed the officer of the day, turns to the left out of the column, places itself opposite and facing him, and continues to play until the guard leaves the parade ground. The field music detaches itself from the band when the latter turns out of the column, and, remaining in front of the guard, commences to play when the band ceases.

Having passed 12 paces beyond the officer of the day, the adjutant halts; the sergeant-major halts abreast of the adjutant and 1 pace to his left; they then return saber, salute, and retire; the commander of the guard then commands: 1. **Platoons, right by squads**, 2. **MARCH**, and marches the guard to its post.

The officers of the day face toward each other and salute; the old officer of the day turns over the orders to the new officer of the day.

While the band is sounding off, and while the guard is marching in review, the officers of the day stand at parade rest with arms folded. They take this position when the adjutant comes to parade rest, resume the attention with him, again take the parade rest at the first note of the march in review, and resume attention as the head of the column approaches.

The new officer of the day returns the salute of the commander of the guard and the adjutant, making one salute with the hand. (355)

If the guard be not divided into platoons, the adjutant commands: 1. **At trail, guard right**, 2. **MARCH**, 3. **Guard**, 4. **HALT**, and it passes in review as above; the commander of the guard is 3 paces in front of its center; the adjutant places himself 6 paces from the left flank and abreast of the commander of the guard; the sergeant covers the adjutant on a line with the front rank. (356)

Informal Guard Mounting for Infantry

889. Informal guard mounting will be held on the parade ground of the organization from which the guard is detailed. If it is detailed from more than one organization, then at such place as the commanding officer may direct. (357)

At assembly, the detail for guard falls in on the company parade ground. The first sergeant verifies the detail, inspects their dress and general appearance, and replaces any man unfit to march on guard. He then turns the detail over to the commander of the guard and retires. (358)

At adjutant's call, the officer of the day takes his place 15 paces in front of the center of the guard and commands: 1. **Officer (or officers) and noncommissioned officers,** 2. **Front and center,** 3. **MARCH;** whereupon the officers and noncommissioned officers take their positions, are assigned and sent to their posts as prescribed in formal guard mounting. (Par. 352.)

The officer of the day will then inspect the guard with especial reference for its fitness for the duty for which it is detailed, and will select as prescribed in paragraphs 140 and 141, the necessary orderlies and color sentinels. The men found unfit for guard will be returned to quarters and will be replaced by others found to be suitable, if available in the company. If none are available in the company, the fact will be reported to the adjutant immediately after guard mounting.

When the inspection shall have been completed, the officer of the day resumes his position and directs the commander of the guard to march the guard to its post. (359)

Relieving the Old Guard

390. As the new guard approaches the guardhouse, the old guard is formed in line, with its field music 3 paces to its right; and when the field music at the head of the new guard arrives opposite its left, the commander of the new guard commands: 1. **Eyes,** 2. **RIGHT;** the commander of the old guard commands: 1. **Present,** 2. **ARMS;** commanders of both guards salute. The new guard marches in quick time past the old guard.

When the commander of the new guard is opposite the field music of the old guard, he commands: **FRONT;** the commander of the old guard commands: 1. **Order,** 2. **ARMS,** as soon as the new guard shall have cleared the old guard.

The field music having marched 3 paces beyond the field music of the old guard, changes direction to the right, and, followed by the guard, changes direction to the left when on a line with the old guard; the changes of direction are without command. The commander of the guard halts on the line of the front rank of the old guard, allows his guard to march past him, and when its rear approaches forms it in line to the left, establishes the left guide 3 paces to the right of the field music of the old guard, and on a line with the front rank, and then dresses his guard to the left; the field music of the new guard is 3 paces to the right of its front rank. (360)

The new guard being dressed, the commander of each guard, in front of and facing its center, commands: 1. **Present**, 2. **ARMS**, resumes his front, salutes, carries saber, faces his guard and commands: 1. **Order**, 2. **ARMS**.

Should a guard be commanded by a noncommissioned officer, he stands on the right or left of the front rank, according as he commands the old or new guard, and executes the rifle salute. (361)

After the new guard arrives at its post, and has saluted the old guard, each guard is presented by its commander to its officer of the day; if there be but one officer of the day present, or if one officer acts in the capacity of old and new officer of the day, each guard is presented to him by its commander. (362)

If other persons entitled to a salute approach, each commander of the guard will bring his own guard to attention if not already at attention. The senior commander of the two guards will then command "1. **Old and new guards**, 2. **Present**, 3. **ARMS**."

The junior will salute at the command "**Present Arms**" given by the senior. After the salute has been acknowledged, the senior brings both guards to the order. (363)

After the salutes have been acknowledged by the officers of the day, each guard is brought to an order by its commander; the commander of the new guard then directs the orderly or orderlies to fall out and report, and causes bayonets to be fixed if so ordered by the commanding officer; bayonets will not then be unfixed during the tour except in route marches while the guard is actually marching, or when specially directed by the commanding officer.

The commander of the new guard then falls out members of the guard for detached posts, placing them under charge of the proper noncommissioned officers, divides the guard into three reliefs, **first**, **second**, and **third**, from right to left, and directs a list of the guard to be made by reliefs. When the guard consists of troops of different arms combined, the men are assigned to reliefs so as to insure a fair division of duty, under rules prescribed by the commanding officer. (364)

The sentinels and detachments of the old guard are at once relieved by members of the new guard; the two guards standing at ease or at rest while these changes are being made. The commander of the old transmits to the commander of the new guard all his orders, instructions, and information concerning the guard and its duties. The commander of the new guard then takes possession of the guardhouse and verifies the articles in charge of the guard. (365)

If considerable time is required to bring in that portion of the old guard still on post, the commanding officer may direct that as soon as the orders and property are turned over to the new guard, the portion of the old guard at the guardhouse may be marched off and dismissed. In such a case, the remaining detachment or detachments of the old guard will be inspected by the commander of the new guard when they reach the guardhouse. He will direct the senior noncommissioned officer present to march these detachments off and dismiss them in the prescribed manner. (366)

In bad weather, at night, after long marches, or when the guard is very small, the field music may be dispensed with. (367)

Appendix A

391. When the guard for the day is supplied by more than one organization, the details due from the several companies will be determined as follows: Take the number of privates for duty in each company from its morning report for the day next preceding that on which the tour of duty is to commence, deducting details for detached service of over 24 hours, made after the morning report has been received; the total of these gives the total number of privates available. Then: The total strength is to the strength of a company as the total detail is to the detail from the company. Multiply the total detail by the strength of the company, and divide the result by total strength; carry out to two places of decimals, disregarding all smaller fractions. This rule is applied for each company.

The whole numbers in the results thus obtained are added together, and if the total is less than the total detail required add one to the whole number in the result that has the largest fraction, and so on for each company till the required total is obtained.

There will thus be a difference between the exact proportion and the number detailed from each company; this difference is entered in the credit column and the next day is carried forward and added or subtracted from the first proportion.

FIRST DAY

Company.	Strength.	Privates for guard required, including 3 for stable guard.		Total strength.		Proportion	Detail	Credits.
A	25	×	14 ÷	160	=	−2.18	+ 2	− 18
B	24	×	14 ÷	160	=	−2.10	+ 2	−.10
C	30	×	14 ÷	160	=	−2.62	+ 3	+.38
D	22	×	14 ÷	160	=	+1 92	+ 2	+.08
E	22	×	14 ÷	160	=	+1.92	+ 2	+.08
F (Cav.)	37	×	14 ÷	160	=	−3.23	+*3	−.23
	160					11	14	

* Troop F. Furnishes 3 stable and no other guard

Note.—The proportion due from a company is always given a minus sign and the detail furnished given a plus sign

SECOND DAY

Company.	Strength.	Privates for guard including 3 for stable guard.		Total strength.		First proportion.	Credits brought forward.	Final proportion.	Detail.	Credits.
A	27	×	14 ÷	160	=	−2.36	−.18	−2.54	+ 2	−.54
B	23	×	14 ÷	160	=	−2.01	−.10	−2.11	+ 2	−.11
C	28	×	14 ÷	160	=	−2.45	+.38	−2 07	+ 2	−.07
D	23	×	14 ÷	160	=	−2.01	+.08	−1.93	+ 2	+.07
E	21	×	14 ÷	160	=	−1.83	+.08	−1.75	+ 2	+.25
F (Cav.)	38	×	14 ÷	160	=	−3 32	−.23	−3.55	+*4	+.45
	160								11	14

* Troop F furnishes 3 stable and 1 main guard.

The number of sergeants, corporals, and musicians will be determined in like manner.

A convenient form for the roster is as follows.

ROSTER I. PRIVATES

Enlisted strength of guard, 14 privates.

	Guard required		
	Jan 1, 14.	Jan 2, 14	Jan 3, 14
A Company:			
Strength	25	27	27
First proportion	—2.18	—2 36	—2 36
Final proportion	....	—2.54	—2 90
Detail	+2	+2	+3
Credits	— .18	— 54	+ 10
B Company:			
Strength	24	23	23
First proportion	—2.10	—2 01	—2 01
Final proportion	....	—2.11	—2 12
Detail	+2	+2	+2
Credits	— .10	— 11	— 12
C Company:			
Strength	30	28	28
First proportion	—2.62	—2.45	—2 45
Final proportion	....	—2.07	—2 52
Detail	+3	+2	+3
Credits	+ .38	— 07	+ .48
D Company:			
Strength	22	23	23
First proportion	—1.92	—2 01	—2 01
Final proportion	....	—1 93	—1 94
Detail	+2	+2	+2
Credits	+ .08	+ 07	+ 08
E Company:			
Strength	22	21	21
First proportion	—1 92	—1 84	—1.84
Final proportion	....	—1 76	—1 59
Detail	+2	+2	+1
Credits	+ .08	+ .24	— .59
F Troop:			
Strength	37	38	38
First proportion	—3 23	—3 32	—3 32
Final proportion	....	—3 55	—2 87
Detail	+3	+4	+3
Credits	— .23	+ 45	+ .13

Appendix B

392. When details for guard and fatigue are made as prescribed in paragraph 11, no account will be taken of very small disproportions in the strength of companies.

When the disproportion is considerable a roster will be kept by the sergeant-major under the supervision of the adjutant as follows: In accordance with the method explained in Appendix A, determine the proportion of privates each company would be required to furnish.

In the credit column, charge each company, except the one furnishing the guard, with this proportion, i. e., with the number of men it was due to furnish but did not furnish. Enter this number or proportion with a minus sign.

Then credit the company furnishing the guard with the number of men furnished, less the proportion it was due to furnish. The difference is the number of men it furnished in excess, and is entered in the credit column with a plus sign.

Whether the same or different companies furnish the guard on consecutive days, the debits and credits will be determined for each day and added algebraically to the credit or debit brought forward.from the preceding day. The result will then be entered in the credit column for the day.

When a new company is to relieve the one furnishing the guard, that one will ordinarily be detailed which has the largest minus number in the credit column.

The following table indicates the form of the roster.

The order in which companies are shown in this table as furnishing the guard has no especial significance, as many reasons may enter into determination of this matter.

Roster II.—Privates

Date	Guard furnished Company	Number of privates	A Company Strength	A Company Proportion	A Company Credits	B Company Strength	B Company Proportion	B Company Credits	C Company Strength	C Company Proportion	C Company Credits	D Company Strength	D Company Proportion	D Company Credits	E Company Strength	E Company Proportion	E Company Credits
Jan. 1	A	13	47	3.39	+9.61	29	2.09	−2.09	34	2.46	−2.46	34	2.46	−2.46	36	2.60	−2.60
Jan. 2	A	13	47	3.39	+19.22	29	2.09	−4.18	34	2.46	−4.92	34	2.46	−4.92	36	2.60	−5.20
Jan. 3	E	13	47	3.39	+15.82	29	2.09	−6.27	34	2.46	−7.38	34	2.46	−7.38	36	2.60	+11.27
Jan. 4	D	10	47	2.51	+13.30	31	1.65	−7.92	36	1.93	−11.24	37	1.98	−9.36	36	1.93	+1.34
Jan. 5	C	10	47	2.51	+10.79	31	1.65	−9.57	36	1.93	−3.17	37	1.98	−1.34	36	1.93	+.41
Jan. 6	B	10	47	2.51	+8.28	31	1.65	−11.22	36	1.93	−5.10	37	1.98	−3.32	36	1.93	+1.48
Jan. 7	D	10	47	2.51	+5.77	31	1.65	−2.87	36	1.93	−7.03	37	1.98	−5.30	36	1.93	+1.55
Jan. 8	C	10	47	2.51	+3.26	31	1.65	−4.52	36	1.93	−1.04	37	1.98	−2.72	36	1.93	+1.62
Jan. 9	B	10	47	2.51	+.75	31	1.65	−6.17	36	1.93	−.89	37	1.98	+.84	36	1.93	+1.69
Jan. 10	A	10	47	2.51	−1.76	31	1.65	+2.18	36	1.93	+2.82	37	1.98	−1.14	36	1.93	−.24
Jan. 11	D	10	47	2.51	+9.25	31	1.65	−.53	36	1.93	−4.75	37	1.98	−3.12	36	1.93	−2.17
Jan. 12	C	10	47	2.51	+16.74	31	1.65	+1.12	36	1.93	−6.68	37	1.98	−5.10	36	1.93	−4.10
Jan. 13	D	10	47	2.51	+14.23	31	1.65	−2.77	36	1.93	+4.25	37	1.98	+6.88	36	1.93	−6.03
Jan. 14	C	10	47	2.51	+11.72	31	1.65	−4.42	36	1.93	−2.32	37	1.98	+4.90	36	1.93	+2.04
Jan. 15	E	10	47	2.51	+9.21	31	1.65	−6.07	36	1.93		37	1.98	+2.92	36	1.93	

CHAPTER V

SIGNALING

Signals and Codes

General Service Code. (International Morse Code.)

393. Used for all visual and sound signaling, radiotelegraphy, and on cables using siphon recorders, used in communicating with Navy.

A	.-		N	-.
B	-...		O	---
C	-.-.		P	.--.
D	-..		Q	--.-
E	.		R	.-.
F	..-.		S	...
G	--.		T	-
H	....		U	..-
I	..		V	...-
J	.---		W	.--
K	-.-		X	-..-
L	.-..		Y	-.--
M	--		Z	--..

NUMERALS

1	.----		6	-....
2	..---		7	--...
3	...--		8	---..
4	....-		9	----.
5	.....		0	-----

PUNCTUATION

Period ......
Comma .-.-.-
Interrogation ..--..

THE MORE IMPORTANT CONVENTIONAL FLAG SIGNALS

For communication between the firing line and the reserve or commander in rear. In transmission, their concealment from the enemy's view should be insured. In the absence of signal flags the headdress or other substitute may be used.

(See Par. 72 for the signals.—Author.)

Wigwag

Signaling by flag, torch, hand lantern, or beam of searchlight (without shutter)[1]

394. 1. There is one position and there are three motions. The position is with flag or other appliance held vertically, the signalman facing directly toward the station with which it is desired to communicate. The first motion (the dot) is to the right of the sender, and will embrace an arc of 90°, starting with the vertical and returning to it, and will be made in a plane at right angles to the line connecting the two stations. The second motion (the dash) is a similar motion to the left of the sender. The third motion (front) is downward directly in front of the sender and

---

[1] Extracts from Signal Book, United States Army.

ERROR
A
1
F
6
NEGATIVE
K
B
2
G
7
L
REPEAT
C
3
H
8
M
D
4
I
9
N
E
5
J
0
INTERROGATORY
O

THE TWO-ARM SEMAPHORE CODE.

AFFIRMATIVE	U	Z
Q	V	ATTENTION
ACKNOWLEDGE R	W	INTERVAL
S	X	NUMERALS
T	Y	

394 (contd.)

instantly returned upward to the first position. This is used to indicate a pause or conclusion.

2. The beam of the searchlight, though ordinarily used with the shutter like the heliograph, may be used for long-distance signaling, when no shutter is suitable or available, in a similar manner to the flag or torch, the first position being a vertical one. A movement of the beam 90° to the right of the sender indicates a dot, a similar movement to the left indicates a dash; the beam is lowered vertically for front.

3. To use the torch or hand lantern, a footlight must be employed as a point of reference to the motion. The lantern is more conveniently swung out upward to the right of the footlight for a dot, to the left for a dash, and raised vertically for front.

4. To call a station, make the call letter until acknowledged, at intervals giving the call or signal of the calling station. If the call letter of a station is unknown, wave flag until acknowledged. In using the search light without shutter throw the beam in a vertical position and move it through an arc of 180° in a plane at right angles to the line connecting the two stations until acknowledged. To acknowledge a call, signal "Acknowledgment (or) I understand (- - - - front)" followed by the call letter of the acknowledging station.

Signaling with heliograph, flash lantern, and searchlight (with shutter)[1]

1. The first position is to turn a steady flash on the receiving station. The signals are made by short and long flashes. Use a short flash for dot and a long steady flash for dash. The elements of a letter should be slightly longer than in sound signals.

2. To call a station, make the call letter until acknowledged, at intervals the call or signal of the calling station.

3. If the call letter of a station be unknown, signal a series of dots rapidly made until acknowledged. Each station will then turn on a steady flash and adjust. When the adjustment is satisfactory to the called station, it will cut off its flash, and the calling station will proceed with its message.

4. If the receiver sees that the sender's mirror needs adjustment, he will turn on a steady flash until answered by a steady flash. When the adjustment is satisfactory, the receiver will cut off his flash and the sender will resume his message.

5. To break the sending station for other purposes, turn on a steady flash.

---

[1] Extracts from Signal Book, United States Army

APPENDIX 8

Sound Signals.[1]

1. Sound signals made by the whistle, foghorn, bugle, trumpet, and drum may be used in a fog, mist, falling snow, or at night. They may be used with the dot and dash code.

2. In applying the code to whistle, foghorn, bugle, or trumpet, one short blast indicates a dot and one long blast a dash. With the drum, one tap indicates a dot and two taps in rapid succession a dash. Although these signals can be used with a dot and dash code, they should be so used in connection with a preconcerted or conventional code.

Morse Code. (American Morse Code)[1]

395. Used only by the army on telegraph lines, on short cables, and on field lines, and on all commercial lines in the United States.

A .—		O . .
B — . . .		P . . . . .
C . . .		Q . . — .
D — . .		R . . .
E .		S . . .
F . — .		T —
G — — .		U . . —
H		V . . . —
I . .		W . — —
J — . — .		X . — . .
K — . —		Y . . . .
L ———		Z . . . .
M — —		& . . . .
N — .		

NUMERALS

1 . — — .		6 . . . . . .
2 . . — . .		7 — — . .
3 . . . — .		8 — . . . .
4 . . . . —		9 — . . —
5 — — —		0 ———

PUNCTUATION

Period .... — — . .
Comma .... — . — .
Interrogation .... — . . — .

[1] Extracts from Signal Book, United States Army.

CHAPTER VI

SMALL-ARMS FIRING MANUAL

(To include Changes No. 2, June 26, 1914.)

EXTRACTS

(The numbers following the paragraphs are those of the Small-Arms Firing Manual, and references in the text to certain paragraph numbers refer to these numbers, and not to the numbers preceding the paragraphs.)

INDIVIDUAL INSTRUCTION

Preliminary Instruction and Sighting Drills for the Rifle

396. **Nomenclature and Care of the Rifle.**—Although each recruit is required to be instructed in the nomenclature, care, use, and preservation of the rifle, this instruction will be repeated as the initial step for each season's known distance practice.

The precautions necessary to avoid accidents will also be thoroughly impressed upon the soldier at this time. (12)

Sighting Drills

397. **Value.**—The value of the sighting drills and the position and aiming drills can not be too strongly emphasized. By means of them the fundamental principles of shooting may be inculcated before the soldier fires a shot. (13)

To Whom Given.—The sighting drills will be given to all soldiers who have not qualified as "marksman" or better in the preceding target year. (14)

Purpose:

(a) To show how to align the sights properly on the mark.

(b) To discover and demonstrate errors in sighting.

(c) To teach uniformity in sighting. (15)

Apparatus and Its Use—Sighting Bar.—(See Pl. I.) To consist of:

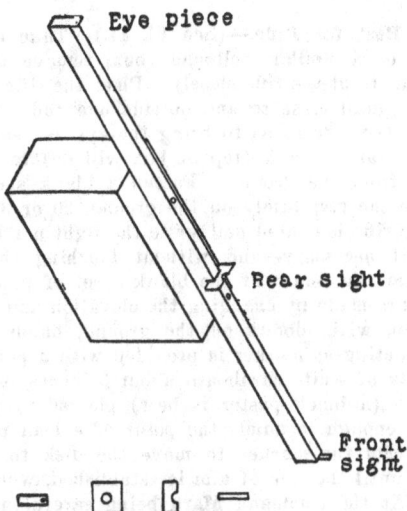

PLATE I.

(a) A bar of wood about 1 by 2 inches by 4 feet, with a thin slot 1 inch deep cut across the edge about 20 inches from one end.

(b) A front sight of tin or cardboard ½ by 3 inches tacked to the end nearer the slot and projecting 1 inch above bar.

(c) An eyepiece of tin or cardboard 1 by 3 inches tacked to the other end of, and projecting 1 inch above, the bar, with a very small hole (0.03 inch) ½ inch from top of part projecting above the bar.

(d) An open rear sight of tin or cardboard 1½ by 3 inches, with a U-shaped notch ¾ inch wide cut in the middle of one of the long edges. This is placed in the slot on the bar. A slight bend of the part of the tin fitting in the slot will give enough friction to hold the sight in any part of slot in which it is placed.

(e) A peep rear sight of tin or cardboard 3 by 3 inches, with a peep hole ¾ inch in diameter cut in the center. This replaces the open sight when the peep sight is shown.

Carefully blacken all pieces of tin or cardboard and the top of the bar. Nail the bar to a box about 1 foot high and place on the ground, table, or other suitable place. Then adjust the open or peep rear sight in the slot and direct the bar upon a bull's-eye (preferably a Y target) placed about 5 yards from the bar. No other than the sight desired can be seen. Errors, etc., are shown by manipulating the open and peep rear sights. (16)

Sighting Rest for Rifle.—(See Pl. II.) Take an empty pistol ammunition box or a similar well-made box, remove the top and cut notches in the ends to fit the rifle closely. Place the rifle in these notches with the trigger guard close to and outside one end. (The stock may be removed from the rifle so as to bring the eye as near the rear sight 'as in shooting.) Nail a plank (top of box will do) to a stake or wall about 12 inches from the ground. Fasten a blank sheet of paper to the plank. Place the rest firmly on the ground, 20 or 30 feet from the plank, so that the rifle is canted neither to the right nor left—weight the box with sand if necessary—and without touching the rifle or rest, sight the rifle near the center of the blank sheet of paper. Changes in the line of sight are made by changing the elevation and windage. Take the prone position with elbows on the ground, hands supporting the head. A soldier acting as marker is provided with a pencil and a small rod bearing a disk of white cardboard about 3 inches in diameter, with a black bull's-eye (a black paster is best) pierced in the center with a hole just large enough to admit the point of a lead pencil. The soldier sighting directs the marker to move the disk to the right, left, higher, or lower, until the line of aim is established, when he commands **Mark** or **Hold**. At the command **Mark** being careful not to move the disk, the marker records through the hole in its center the position of the disk and then withdraws it. At the command **Hold** the marker holds the disk carefully in place without marking until the position is verified by the instructor, and the disk is not withdrawn until so directed. (17)

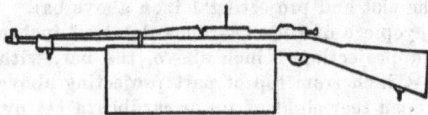

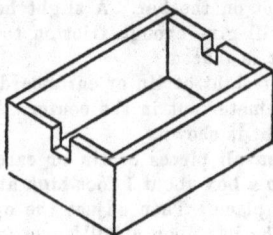

PLATE II.

Line of Sight.—With the open sight the line of sight is determined by a point on the middle line of the notch of the rear sight and the top of the front sight. With the peep sight, the line of sight is determined by the **center** of the peep and the top of the front sight. (18)

Point of Aim.—The soldier will be informed that to give the greatest uniformity a point just below the mark, and not the mark, is taken as **the point of aim**, as it is impossible to always know, if touching the mark with the top of the front sight, how much of the front sight is seen; that the term ''on the mark or bull's-eye'' will be understood to mean an aim, taken just below the mark, showing a fine line of light between the mark and the top of the front sight. (19)

The Normal Sight.—Look through the rear-sight notch at the bull's-eye or mark and bring the top of the front sight on a line with the top of and in the center of the rear-sight notch and aligned upon the point of aim. (See Fig. 1, Pl. III.) (20)

FIG. 1

FIG 2

PLATE III.

[307]

FIG. 1.

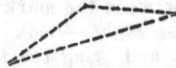

FIG. 3.

FIG. 2.

TRIGGER.

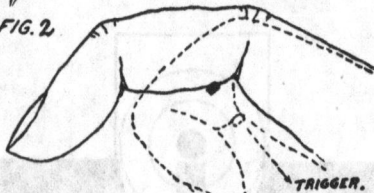

FIG. 4.

PLATE IV.

The Peep Sight.—Look through the peep hole at the bull's-eye or mark and bring the top of the front sight to the center of the aperture and aligned upon the point of aim. (See Fig. 2, Pl. III.)

The soldier should be informed that regular results in firing can be obtained only when the same amount of front sight is taken each time, and that this can be done only by using the normal sight with the open notch or the peep sight in the manner described above. He should understand that the effect of taking less than the normal amount of sight is to cause a point lower than that aimed at to be struck, and that taking too much of the front sight causes a higher point to be struck.

Although men will be found occasionally who can get excellent results by using the fine sight (Fig. 1, Pl. IV), the average man can not, and this form of sighting is not recommended. The so-called full sight should not be taught under any circumstances. If shown to the men at all, it should be for the purpose of pointing out a fault to be carefully avoided.

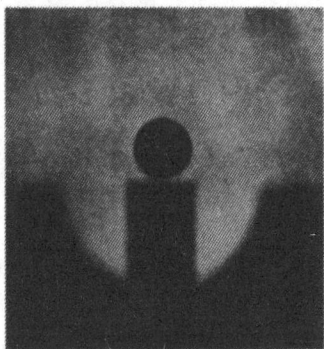

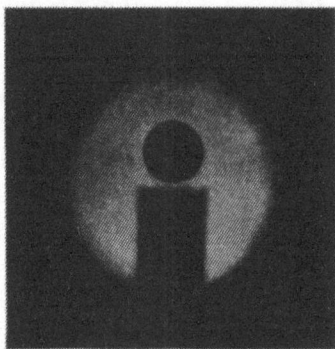

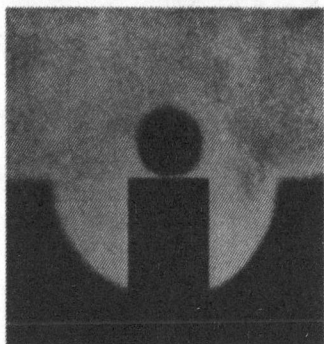

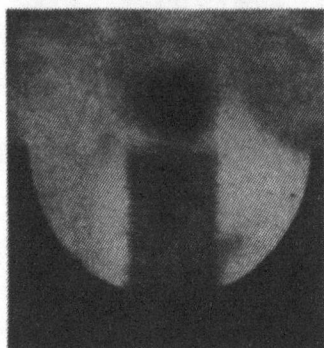

PLATE V.

Remarks.—The eye can be focused accurately upon objects at but one distance at a time; all other objects in the field of view will appear more or less blurred, depending on their distance from the eye. This can readily be seen if a pencil is placed in the field of view near the eye while looking at some distant object. The pencil will appear blurred. This is the condition met with by the normal eye in sighting a rifle. If the eye is focused on one of the three points—the bull's-eye, the front sight, or the rear sight—the other two will appear blurred. This blurring effect is best overcome by using the ''peep sight,'' as though looking through a window, and focusing the eye on the bull's-eye. The blurring of the peep hole will be concentric, giving a clear and easily defined center. The blurring of the front sight will be less, but symmetrical on both sides with very little blur on the top. It can be readily and naturally brought to the center of the peep hole. Variations in light have less effect on the peep than on the open sight.

But the limited field of view and lack of readiness in getting a quick aim with the peep sight limit its use to those stages of the combat when comparative deliberation will be possible. In the later stages of battle—especially when a rapid fire is to be delivered—the open sight will, in most cases, be used. In this case the normal sight should be used, as the horizontal line at the top of the notch of the rear sight affords a good guide for regularity.

Whatever sight is used, the eye must be focused on the bull's-eye, or mark, not on the front or rear sight.   (21)

First Sighting Exercise

Using illustrations, describe the normal sight and the peep sight.   (22)

Using the sighting bar, represent the normal sight and the peep sight and require each man in the squad to look at them.   (23)

Using the sighting bar, describe and represent the usual errors of sighting and require each man in the squad to look at them.   (24)

Second Sighting Exercise

Using the sighting rest for the rifle, require each man to direct the marker to move the disk until the rifle is directed on the bull's-eye with the normal sight and command **Hold.** The instructor will verify this line of sight. Errors, if any, will be explained to the soldier and another trial made. If he is still unable to sight correctly, the first exercise will be repeated.

Soldiers will sometimes be found who do not know how to place the eye in the line of sight; they often look over or along one side of the notch of the rear sight and believe that they are aiming through the notch because they see it at the same time that they do the front sight. This error will probably be made evident by the preceding exercise.

397 (contd.)

Some men in sighting will look at the front sight and not at the object. As this often occasions a blur, which prevents the object from being distinctly seen and increases both the difficulties and inaccuracies of sighting, it should be corrected. (25)

Repeat the above, using the peep sight. (26)

Third Sighting Exercise

Using the sighting rest for the rifle, require each man to direct the marker to move the disk until the rifle is directed on the bull's-eye with the normal sight and command Mark; then, being careful not to move the rifle or sights, repeat the operation until three marks have been made.

(a) **The Triangle of Sighting.**—Join the three points determined as above by straight lines, mark with the soldier's name, and call his attention to the triangle thus formed. The shape and size of this triangle will indicate the nature of the variations made in aiming.

(b) **Abnormal Shape, Causes.**—If the triangle is obtuse angled, with its sides approaching the vertical (See Fig. 2, Pl. IV), the soldier has not taken a uniform amount of front sight. If the sides of the triangle are more nearly horizontal (See Fig. 3, Pl. IV) the errors were probably caused by not looking through the middle of the notch or not over the top of the front sight. If any one of the sides of the triangle is longer than one-half inch, the instructor directs the exercise to be repeated, verifying each sight and calling the soldier's attention to his errors. The instructor will explain that the sighting gains in regularity as the triangle becomes smaller.

(c) **Verifying the Triangle.**—If the sides of the triangle are so small as to indicate regularity in sighting, the instructor will mark the center of the triangle and then place the center of the bull's-eye on this mark. The instructor will then examine the position of the bull's-eye with reference to the line of sight. If the bull's-eye is properly placed with reference to the line of sight, th soldier aims correctly and with uniformity. If not so placed, he aims in a regular manner but with a constant error.

(d) **Causes of Errors.**—If the bull's-eye is directly above its proper position, the soldier has taken in aiming too little front sight, or if directly below too much front sight. If directly to the right or left, the soldier has not sighted through the center of the rear-sight notch and over the top of the front sight. If to the right, he has probably either sighted along the left of the rear sight notch or the right side of the front sight, or has committed both of these errors. If the bull's-eye is too far to the left, he has probably sighted along the right of the rear sight notch or to the left of the front sight, or has combined both of these errors.

If the bull's-eye is placed with reference to its proper position diagonally above and to the right, the soldier has probably combined the errors which placed it too high and too far to the right. Any other diagonal position would be produced by a similar combination of vertical and horizontal errors.

As the errors thus shown are committed when the rifle is fixed in position, while that of the bull's-eye or target is altered, the effect will be directly opposite to the changes in the location of a hit in actual fire, occasioned by the same errors, when the target will be fixed and the rifle moved in aiming.

After the above instruction has been given to one man, the line of sight will be slightly changed by moving the sighting rest or by changing the elevation and windage, and the exercises similarly repeated with the other men in the squad. (27)

Repeat the third sighting exercise, using the peep sight. (28)

Fourth Sighting Exercise

This exercise is a demonstration of the effect of canting the piece. The soldier must be impressed with the necessity of keeping the sights vertical when aiming, and not canting the piece to the right or left. Explain to the soldier that if the piece is canted to the right, the bullet will strike to the right and below the point aimed at, even though the rifle be otherwise correctly aimed and the sights correctly set. Similarly, if the piece is canted to the left, the bullet will strike to the left and low. This can be explained by showing that the elevation fixes the height of the point where the bullet will hit the target, and that the windage fixes the point to the right or left; i. e., the elevation gives vertical effects and the windage horizontal effects. Let a pencil (or rod) held vertical represent the elevation; now if the pencil is turned to the right 90°, or horizontal, all of the elevation has been taken off, causing the shot to strike low and changed into windage, causing the shot to strike to the right. (29)

This effect may be demonstrated as follows: Use the sighting rest with the rifle firmly held in the notches, the bolt removed. Paste a black paster near the center of the bottom line of the target. Sight the rifle on this mark, using about 2,000 yards elevation, then, being careful not to move the rifle, look through the bore and direct the marker to move the disk until the bull's-eye is in the center of the field of view and command **Mark**. Next turn the rest with the rifle on its side, and with the same elevation sight on the same paster as above, then, being careful not to move the rifle, look through the bore and again direct the marker to move the disk until the bull's-eye is in the center of the field of view and command **Mark**. Not considering the fall of the bullet, the first mark represents the point struck with the sight vertical, the second mark represents the point struck, low and to the right, using the same elevation and the same point of aim, when the piece is canted 90° to the right.

Different degrees of canting the piece can be represented by drawing an arc of a circle through the two marks with the paster as a center. The second mark will be at a point on this arc corresponding to the degree of canting the piece. Emphasis will be laid upon the fact that this effect of canting increases with the distance from the target. (30)

Other Exercises.—If time permits, the instructor may devise other exercises which suggest themselves as useful and beneficial to his men. The following are examples:

(a) In strong sunlight make a triangle of sighting, using a rifle having sights worn bright. Then, being careful not to move the rifle, blacken sights and make another triangle. Use dotted lines for the triangle made with bright sights and full lines for the triangle made with blackened sights. The position and size of the two triangles will plainly show the advantage of the blackened sights.

(b) In strong sunlight make a triangle of sighting; then, being careful not to move the rifle, make another triangle, having first shaded the target and the man sighting. The relative position of the triangles will show the importance of knowing the effects of varying degrees of light. (31)

Position and Aiming Drills

398. Purpose.—These drills are intended to so educate the muscles of the arm and body that the piece, during the act of aiming, shall be held without restraint, and during the operation of firing shall not be deflected from the target by any convulsive or improper movement of the trigger finger or of the body, arms, or hands. They also establish between the hand and eye such prompt and intimate connection as will insure that the finger shall act upon the trigger, giving the final pressure at the exact moment when the top of the front sight is seen to be directed upon the mark.

The fact, though simple, can not be too strongly impressed upon the recruit that if, at the moment of discharge, the piece is properly supported and correctly aimed, the mark will surely be hit. Since any intelligent man can be taught to aim correctly and to hold the sights aligned upon the mark with a fair amount of steadiness, it follows that bad shooting must necessarily arise from causes other than bad aiming. The chief of these causes is known to be the deflection given to the rifle when it is discharged, due to the fact that the soldier, at the moment of firing, instead of squeezing the trigger, jerks it. This convulsive action is largely due to lack of familiarity with the methods of firing and to a constrained position of the muscles of the body, arm, and hands, which constrained position it is the purpose of the position and aiming drills to correct.

To become a good shot, constant, careful, and patient practice is required. Systematic aiming and squeezing the trigger will do much to make a rifleman. The men will be taught to take advantage of every opportunity for practicing aiming, and squeezing the trigger. For this purpose the barracks and ground in the vicinity of the barracks should be furnished with aiming targets, which the men will be encouraged to use at odd moments, as when waiting for a formation or during a rest. At drill the soldier will be cautioned never to squeeze the trigger without selecting an object and taking careful aim. When on the range waiting for his turn to fire, the soldier should use part of his time in position and aiming exercises, aiming at the target or at objects outside of the range, and he should be made to understand that this practice previous to firing will tend to prevent nervousness and will have a marked effect upon his score. (32)

To Whom Given.—The position and aiming drills will be given to all soldiers who have not qualified as "marksman" or better in the preceding target year. Some practice in these drills (especially in the trigger-squeeze exercise) is recommended for those who have qualified as "marksman" or better. The amount to be given is left to the discretion of the company commander. (33)

Drills; General Instructions.—These drills are divided into four progressive exercises. The first exercise teaches the position; the second exercise teaches the position and the aim; the third exercise teaches the aim and the manner of squeezing the trigger; and the fourth exercise teaches the methods of rapid fire. The exercises should be taught by the numbers at first; when fully understood, without numbers.

To correct any tendency to cant the piece, the rear sight will be raised. A black paster at which to aim will be placed on the wall opposite each man. The squad being formed in single rank, with an interval of 1 yard between files, the instructor directs the men to take the position of **Ready** except that the position of the feet will be such as to insure the greatest firmness and steadiness of the body. The instructor then cautions **"Position and aiming drill."**

The exercise which is being taught should be repeated frequently and made continuous. The instructor prefaces the preparatory command by **Continue the motion** or **At will** and gives the command **Halt** at the conclusion of the exercise, when the soldier will return to the position of **Ready.** Or the soldier may be made to repeat the first and second motions by the command **One, Two,** the exercise concluding with the command **Halt.**

Care should be taken by the instructor not to make the position and aiming drills tedious. Thirty minutes daily should be spent in this practice during the period of preliminary instruction. After gallery practice is taken up, however, five or ten minutes daily should be sufficient for these exercises.

In order that the instructor may readily detect and correct errors, the squads for these drills should not consist of more than eight men.

The instructor should avoid holding the squad in tiresome positions while making explanations or correcting errors. (34)

Position Exercise

The instructor commands: 1. **Position**, 2. **EXERCISE**. At the last command, without moving the body or eyes, raise the rifle smartly to the front of the right shoulder to the full extent of the left arm, elbow inclined downward, the barrel nearly horizontal, muzzle slightly depressed, heel of the butt on a line with the top of the shoulder.

(Two.) Bring the piece smartly against the hollow of the shoulder, without permitting the shoulder to give way, and press the rifle against it, mainly with the right hand, only slightly with the left, the forefinger of the right hand resting lightly against the trigger, the rifle inclined neither to the right nor left.

(Three.) Resume the position of **Ready**. (35)

Remarks.—The instructor should especially notice the position of each soldier in this exercise, endeavoring to give to each man an easy and natural position. He should see that the men avoid drawing in the stomach, raising the breast, or bending the small of the back.

The butt of the piece must be pressed firmly, but not too tightly, into the hollow of the shoulder and not against the muscles of the upper arm. If held too tightly, the pulsations of the body will be communicated to the piece; if too loosely, the recoil will bruise the shoulder. If only the heel or toe touches the hollow of the shoulder, the recoil may throw the muzzle down or up, affecting the position of the hit. While both arms are used to press the piece to the shoulder, the left arm should be used to direct the piece and the sight forefinger must be left free to squeeze the trigger. (36)

Aiming Exercise

The instructor will first direct the sights to be adjusted for the lowest elevation and subsequently for the different longer ranges.

The instructor commands: 1. **Aiming**, 2. **EXERCISE**.

At the last command execute the first and second motion of the position exercise.

(Two.) Bend the head a little to the right, the cheek resting against the stock, the left eye closed, the right eye looking through the notch of the rear sight at a point slightly below the mark.

(Three.) Draw a moderately long breath, let a portion of it escape, then, with the lungs in a state of rest, slowly raise the rifle with the left hand, being careful not to incline the sight to either side, until the line of sight is directly on the mark; hold the rifle steadily directed on the mark for a moment; then, without command and just before the power to hold the rifle steadily is lost, drop the rifle to the position of **Ready** and resume the breathing. (37)

Remarks.—Some riflemen prefer to extend the left arm. Such a position gives greater control over the rifle when firing in a strong wind or at moving objects. It also possesses advantages when a rapid as well as accurate delivery of fire is desired. Whatever the position, whether standing, kneeling, sitting, or prone, the piece should rest on the palm of the left hand, never on the tips of the fingers, and should be firmly grasped by all the fingers and the thumb.

The eye may be brought to the line of sight either by lowering the head or by raising the shoulder; it is best to combine somewhat these methods; the shoulder to be well raised by raising the right elbow and holding it well to the front and at right angles to the body.

If the shoulder is not raised, it will be necessary for the soldier to lower the head to the front in order to bring the eye in to the line of sight. Lowering the head too far to the front brings it near the right hand, which grasps the stock. When the piece is discharged, this hand is carried by the recoil to the rear and, when the head is in this position, may strike against the nose or mouth. This often happens in practice, and as a result of this blow often repeated many men become gun-shy, or flinch, or close their eyes at the moment of firing. Much bad shooting, ascribed to other causes, is really due to this fault. Raising the right elbow at right angles to the body elevates the right shoulder, and lifts the piece so that it is no longer necessary to incline the head materially to the front in order to look along the sights.

As the length of the soldier's neck determines greatly the exact method of taking the proper position, the instructor will be careful to see that the position is taken without restraint. (38)

As changes in the elevation of the rear sight will necessitate a corresponding change in the position of the soldier's head when aiming, the exercise should not be held with the sight adjusted for the longer ranges until the men have been practiced with the sights as the latter would generally be employed for offhand firing. (39)

The soldier must be cautioned that while raising the line of sight to the mark he must fix his eyes on the mark and not on the front sight; the latter can then be readily brought into the line joining the rear-sight notch and mark. If this plan be not followed when firing is held on the range at long distances the mark will generally appear blurred and indistinct. The front sight will always be plainly seen, even though the eye is not directed particularly upon it. (40)

The rifle must be raised slowly, without jerk, and its motion stopped gradually. In retaining it directed at the mark, care must be taken not to continue the aim after steadiness is lost; this period will probably be found to be short at first, but will quickly lengthen with practice. No effort should be made to prolong it beyond the time that

398 (contd.)

breathing can be easily restrained. Each soldier will determine for himself the proper time for discontinuing the aim. (41)

The men must be cautioned not to hold the breath too long, as a trembling of the body will result in many cases. (42)

Some riflemen prefer, in aiming, to keep both eyes open but, unless the habit is fixed, the soldier should be instructed to close the left eye. (43)

Trigger-squeeze Exercise

The instructor commands: 1. **Trigger squeeze**, 2. **EXERCISE**. At the command **EXERCISE**, the soldier will execute the first motion of the aiming exercise.

(Two.) The second motion of the aiming exercise.

(Three.) Draw a moderately long breath, let a portion of it escape, hold the breath and slowly raise the rifle with the left hand until the line of sight is on the mark, being careful not to incline the sights to either side. Contract the trigger finger gradually, slowly and steadily increasing the pressure on the trigger, while the aim is being perfected; continue the gradual increase of pressure so that when the aim has become exact the additional pressure required to release the point of the sear can be given almost insensibly and without causing any deflection of the rifle. Continue the aim a moment after the release of the firing pin, observe if any change has been made in the direction of the line of sight, and then resume the position of **Ready**, cocking the piece by raising and lowering the bolt handle. (44)

Remarks.—Poor shooting is often the result of lack of proper coördination of holding the breath, the maximum steadiness of aim, and the squeeze of the trigger. By frequent practice in this exercise, each man may come to know the exact instant his firing pin will be released. He must be taught to hold the breath, bring the sights to bear upon the mark, and squeeze the trigger all at the same time. (45)

The Trigger Squeeze.—The trigger should be squeezed, not pulled, the hand being closed upon itself as a sponge is squeezed, the forefinger sharing in this movement. The forefinger should be placed as far around the trigger as to press it with the second joint. (See Fig.4, Pl. IV.) By practice the soldier becomes familiar with the trigger squeeze of his rifle, and knowing this, he is able to judge at any time, within limits, what additional pressure is required for its discharge. By constant repetition of this exercise he should be able finally to squeeze the trigger to a certain point beyond which the slightest movement will release the sear. Having squeezed the trigger to this point, the aim is corrected and, when true, the additional pressure is applied and the discharge follows. (46)

Rapid-fire Exercise

Object.—The object of this exercise is to teach the soldier to aim quickly and at the same time accurately in all the positions he will be called upon to assume in range practice. (47)

The instructor commands: 1. **Rapid-fire exercise**, 2. **COMMENCE FIRING**. At the first command the first and second motions of the trigger-squeeze exercise are performed. At the second command the soldier performs the third motion of the trigger-squeeze exercise, squeezing the trigger without disturbing the aim or the position of the piece, but at the same time without undue deliberation. He then, without removing the rifle from the shoulder, holding the piece in position with the left hand, grasps the handle of the bolt with the right hand, rapidly draws back the bolt, closes the chamber, aims, and again squeezes the trigger. This movement is repeated until the trigger has been squeezed five times, when, without command, the piece is brought back to the position of **Ready**.

When the soldier has acquired some facility in this exercise, he will be required to repeat the movement ten times, and finally by using dummy cartridges, he may, by degrees, gain the necessary quickness and dexterity for the execution of the rapid fire required in range firing. (48)

Methods.—The methods of taking position, of aiming, and of squeezing the trigger, taught in the preceding exercises, should be carried out in the rapid-fire exercise, with due attention to all details taught therein; the details being carried out as prescribed except that greater promptness is necessary. In order that any tendency on the part of the recruit to slight the movements of aiming and of trigger squeeze shall be avoided, the rapid-fire exercise will not be taught until the recruit is thoroughly drilled and familiar with the preceding exercises. The recruit will be instructed that with practice in this class of fire the trigger can be squeezed promptly without deranging the piece. (49)

Repetition.—If the recruit seems to execute the exercise hurriedly or carelessly, the instructor will require him to repeat it at a slower rate. (50)

Manipulation of the Breech Mechanism.—To hold the piece to the shoulder and, at the same time, manipulate the breech mechanism with the proper facility, are learned only after much practice. Some riflemen, especially men who shoot from the left shoulder, find it easier, in rapid firing, to drop the piece to the position of load after each shot. While at first trial this method may seem easier, it is believed that, with practice, the advantage of the former method will be apparent. (51)

Position and Aiming Drill, Kneeling

These exercises will be repeated in the kneeling position by causing the squad to kneel by the commands prescribed in the *Drill Regulations*. The exercises will be executed as prescribed for standing, except that at the command Two in the position exercise, the soldier will rest the left elbow on the left knee, the point of the elbow in front of the kneecap. The pasters for the kneeling exercise should be at 2½ feet from the floor or ground. (52)

Remarks.—Frequent rests will be given during practice in these exercises kneeling, as the position, if long continued, becomes constrained and fatigues the soldier unnecessarily.

In raising the rifle to the mark in the second and third exercises, the position of the left hand should not be changed, but the left forearm should be brought toward the body and at the same time the body bent slightly to the rear.

When aiming kneeling there is, from the nature of the position, a tendency to press the butt of the rifle against the upper arm instead of against the hollow of the shoulder; this will necessitate inclining the head considerably to the right to get the line of sight, and by bringing the rifle so far to the rear will, if the thumb is placed across the stock, cause it to give by the recoil a blow upon the nose or mouth.

These difficulties may be avoided by advancing the right elbow well to the front, at the same time raising it so that the arm is about parallel with the ground. The hollow of the shoulder will then be the natural place for the rifle butt, and the right thumb will be brought too far from the face to strike it in the recoil.

Some riflemen prefer, by bending the ankle, to rest the instep flat on the ground, the weight of the body coming more on the upper part of the heel; this obviates any tendency of the right knee to slip; or, by resting the right side of the foot on the ground, toe pointing to the front, to bring the weight of the body on the left side of the foot. These positions are authorized. (53)

Choice of Position.—In firing kneeling, the steadiness obtained depends greatly upon the position adopted. The peculiarities of conformation of the individual soldier exert when firing kneeling a greater influence than when firing either standing, sitting, or prone; the instructor should, therefore, carefully endeavor, noticing the build of each soldier, to place him in the position for which he is best adapted and which will exert the least tension or strain upon the muscles and nerves. It should be remembered, however, that without the rest of the left elbow on the knee this position possesses no advantage of steadiness over the standing position. (54)

Kneeling Position; When Taken.—The kneeling position can be taken more quickly than either the sitting or the prone position. It is, therefore, the position naturally assumed when a soldier, who is standing or advancing, has to make a quick shot at a moving or disappearing object and desires more steadiness than can be obtained standing. (55)

Position and Aiming Drill, Sitting Down

In many cases the men, while able to kneel and hold the piece moderately steady, can obtain in a sitting position much better results. All should, therefore, be instructed in aiming sitting down as well as kneeling.

To practice the soldier in the preceding exercises in a sitting position, the squad being formed in a single rank, with an interval of one pace between files, the rifle should first be brought to **Order arms**; the instructor then commands: **Sit Down.**

At this command make a half face to the right and, assisted by the left hand on the ground, sit down, facing slightly to the right, the left leg directed toward the front, right leg inclined toward the right, both heels, but not necessarily the bottoms of the feet, on the ground, the right knee slightly higher than the left; body erect and carried naturally from the hips; at the same time drop the muzzle of the piece to the front, and to the position of the first motion of load, right hand upon the thigh, just in front of the body, the left hand slightly above, but not resting upon, the left leg.

The exercise will be executed as heretofore prescribed, except that at the command **Two** (position exercise) the soldier will rest the left elbow on the left knee, the point of the elbow in front of the kneecap, and the right elbow against the left or inside of the right knee, at the same time inclining the body from the hips slightly forward.

For the aiming and trigger-squeeze exercises the pasters, used as aiming points, will be 2½ feet from the floor or the ground.

To afford the men rest, or on the completion of the kneeling or sitting down exercises, the instructor will command **RISE**, when the men rise, face to the front, and resume the **Order arms.** (56)

Remarks.—If the preceding position is carefully practiced, steadiness is quickly attained. The right leg should not be carried so far to the right as not to afford a good support or brace for the right elbow.

This position may be modified, but, in general, not without impairing the steadiness of the man, by crossing the legs at the ankle, the outside of each foot resting upon the ground, body more erect, and the knees slightly more raised than in the previous position. (57)

Position and Aiming Drill, Prone

From the nature of the position it is not practicable to execute these exercises according to the method followed when standing or kneeling. Instruction will, however, always be given with reference to

398 (contd.)

the position, to the manner of assuming it, and to aiming and squeezing the trigger.

For this purpose the squad being formed as specified in paragraph 56 (the black pasters therein mentioned being about 12 inches from the ground), the squad will be brought to **Order arms.**

Then (the squad either standing or kneeling), the instructor commands: **LIE DOWN,** which will be executed as prescribed in the *Drill Regulations*; the legs may be spread apart and the toes turned out if found to give a steadier position.

After the squad has taken the position as prescribed above, the legs should be inclined well to the left, and either crossed or separated as the soldier prefers or as his particular conformation appears to render most desirable, and the body at the same time inclined slightly to the right.

With care and practice the soldier may acquire an easy position which he is able to assume with great facility.

Being at **Ready** the instructor then commands: 1. **Trigger squeeze,** 2. **EXERCISE.**

At the latter command carry the left elbow to the front and slightly to the right, the left hand under the barrel at the balance, weight of the body mainly supported by the left elbow, the right resting lightly on the floor or ground.

(Two.) Slide the rifle with the right hand through the left hand to the front until the left hand is a little in front of the trigger guard; at the same time raise the rifle with both hands and press it against the hollow of the shoulder.

(Three.) Direct the rifle upon the mark and carry out the further details of aiming and squeezing the trigger as prescribed in paragraph 44.

Then resume the position, lying down.

As soon as the men have acquired with accuracy the details of the position they will be practiced, without the numbers, in aiming and squeezing the trigger at will; after which the rapid-fire exercise in the prone position will be practiced, the necessary skill and dexterity being acquired by degrees.

To afford the men rest, or on completion of the exercise, the instructor will command: **RISE,** which is executed as prescribed in the *Drill Regulations.* (58)

Remarks.—The preceding position for firing lying down possesses in a greater degree than any other position the merit of adaptability to the configuration of the ground; it enables the soldier to deliver fire over low parapets or improvised shelters, thus making the best use of cover. The importance of training the soldier in firing from the other positions should not, however, be lost sight of, since from the prone position it will frequently be impossible to see the objective.

Back positions are not authorized.

In the prone position, when aiming, the left elbow should be well under the barrel, the other elbow somewhat to the right, but not so far as to induce any tendency to slip on the floor or ground.

The greater changes in elevation required in first directing the rifle on the object should be given by altering the position of the left hand under the barrel, the slighter changes only by advancing or withdrawing the shoulder.

As the body does not yield to the recoil, as when firing standing or kneeling, the force of recoil, if the rifle is not properly held, may severely bruise the soldier. It is one of the objects of this exercise to so teach him that this will be prevented by assuming a correct position. Care must be exercised that the butt is not brought against the collar bone. By moving the shoulder slightly to the front or rear, and by moving the right elbow from the body or toward it, each soldier may determine the position in which the shoulder gives to the butt of the rifle the easiest rest. This will probably be the one in which the force of the recoil will be least felt.

The soldier should persist in this exercise until he obtains a position in which he feels no constraint, which will not subject him to bruises from the recoil, and from which the mark appears plainly through the sights. Having secured such a position, he must not change it when firing, as a variation in the points of support of the rifle, the distance of the eye from the rear sight, or the tension of the hold has a decided effect, especially at the longer ranges, upon the location of the point struck. (59)

Use of Sling.—After the soldier has been drilled in the proper standing, kneeling, sitting, and prone positions in the foregoing exercises, the use of the sling will be taught. Its use is described in paragraph 91. Adjustments and their advantages will be taught with the idea on noninterference with quickness and freedom of action. The trigger-squeeze exercises will then be continued in the different positions, using the sling. (60)

General Remarks on the Preceding Drills

The importance of sighting and position and aiming drills can not be too persistently impressed upon the soldier. If these exercises are carefully practiced, the soldier, before firing a shot at a target, will have learned to correctly aim his piece, to hold his rifle steadily, to squeeze the trigger properly, to assume that position best adapted to the particular conformation of his body, and will have acquired the quickness and manual dexterity required for handling the piece in rapid fire. This knowledge can not be successfully acquired upon the target ground. At that place the time that can be given to instruction is limited and should be devoted to the higher branches of the subject.

Even, if the desired amount of attention could be given to each soldier, nevertheless, from the circumstances of the firing, his errors can not be readily determined. It is more than likely that the soldier would never discover the reasons for his failures and would, therefore, be unable to properly correct them.

Under such conditions the knowledge that he may have of the many other requisites for good marksmanship can not be utilized to full advantage, and, in fact, can but in a limited degree compensate for the neglect of these first principles and for the failure to lay, by assiduously practicing them, the only firm foundation for future proficiency.

If, in the instruction practice on the range, it is found that the soldier makes errors in his position, he should be required to stop firing and to practice the third exercise for 10 or 15 minutes. He should be encouraged to go through these exercises frequently at other than drill hours, care being taken that, in the aiming and trigger-squeeze exercises, he always has some definite object for a mark. (61)

399. Deflection and Elevation Correction Drills.

Sight Correction.—The soldier may find when firing at a target that the first shot has missed the bull's-eye or figure, and in order to cause the second to hit, two methods may be used: The point of aim may be changed or the sights may be moved and the same point be aimed at. In order to do accurate shooting it is essential to have a well-defined mark at which to aim; consequently, except for very slight corrections, the method of moving the sights, involving changes in elevation and windage, is devised. (62)

Elevation.—The instructor will show the men the graduations on the rear-sight leaf, and will explain to them the value of the different divisions. He will explain how to adjust their sights for different distances. He will make it clear that raising or lowering the slide on the rear-sight leaf has the effect of raising or lowering the point struck. The amount of change which a given amount of elevation will cause in the point struck varies with the range and with the rifle and the ammunition used. (63)

Deflection.—The instructor will explain how to move the movable base by use of the windage screw; that the graduations on the rear end of the movable base are for convenience in setting the sights and applying corrections; that each division is called a point of windage; that turning the movable base of the rear sight to the right or left changes the point struck to the right or left; that, to overcome the drifting effect of a wind from the right, the movable base must be moved to the right, and, if the wind be from the left, the movable sight base must be moved to the left. (64)

Adjusting the Sights.—(a) **Elevation.**—The graduations on the rear sight will be found correct for but few rifles. This is due to slight variations in the parts of the rifle, especially the barrel, which occur under the most exact methods of fabrication. Not all rifles are tested at the arsenal, and when the graduations for the rear sight have been experimentally determined, they are correct only for the particular conditions existing when they were so determined. The correction necessary for each particular rifle at any range is found by shooting it at that range, and is constant with the same ammunition and when firing under the same conditions. If no correction is necessary, the rifle is said to "shoot on the mark."

(b) **The zero of a rifle.**—That reading of the wind gauge necessary to overcome the drift of a rifle at a particular range is called the "zero" of that rifle for that range, and all allowances for wind should be calculated from this reading.

The "zero" of a rifle is found by shooting it on a perfectly calm day. (65)

The following table gives the approximate corrections on the rear-sight leaf and the wind gauge necessary to move the point struck 1 foot at ranges from 100 to 1,000 yards:

Range.	Correction on wind gauge necessary to move the point struck 1 foot.	Correction in elevation necessary to change the point struck 1 foot.
Yards.	*Points.*	*Yards.*
100	3	415
200	1.5	185
300	1	105
400	.75	70
500	.6	48
600	.5	35
700	.43	25
800	.375	20
900	.333	15
1,000	.3	12

(66)

Exercise.—To give the soldier practice in correcting elevation and windage, a target should be placed on the wall facing the squad and a blank paster attached a foot or more from the bull's-eye, at first directly above or below the bull's-eye, then on a horizontal line with it, and finally in an oblique direction.

For this drill the rifle of each soldier who has not determined by actual firing the "zero" and the correct elevations for the different ranges will be assumed to shoot on the mark and to require no windage.

Announce the range and tell the men that the paster represents the position of an assumed hit and require each man to correct his sight so as to bring the next hit into the bull's-eye. This exercise should be repeated daily during gallery practice, using the A, B, C, and D tar-

399 (contd.)

gets, until the men have acquired accuracy in making corrections for all ranges up to 1,000 yards.

When the men have learned how to adjust their sights, this exercise should be carried on in connection with gallery practice. The rear sight on each rifle is given an incorrect setting in elevation and windage by the instructor, and the soldier required to find the correct adjustment by firing. (67)

The Effect of Wind.—It is important that before going on the range the soldier should be taught to estimate the force and direction of the wind and the amount of correction necessary to apply to the movable rear-sight base to overcome the effect of the wind on the bullet in its flight.

The direction of the wind, for convenience, is expressed by a clock-face notation, the clock being supposed to lie on the ground with the hour XII toward the target or mark and the hour III at the firer's right hand. A wind blowing from the front (that is, from the direction of the target) is called a "XII-o'clock wind," one directly from the left across the field of fire is called a "IX-o'clock wind," and so on. The direction of the wind can be obtained by observing its effect upon smoke, on trees, or grass, or dust, or by wetting the finger and holding it up.

The force of the wind is designated in miles per hour. An anemometer should be placed near the barracks, where it will not be exposed to cross currents, and so that the dial can be readily seen. The force of the wind can then be read from the dial and at the same time the effect of the wind on the boughs of trees, flags, and streamers, and the smoke from chimneys should be observed. The soldier should be required to estimate the force of the wind and then verify his estimation by anemometer readings. This exercise should be repeated frequently until the soldier has learned to estimate roughly the force of the wind without the aid of an anemometer.

Heat waves, when present, are an important aid in estimating the force of the wind. (68)

In the following table are shown the points of windage necessary to correct for a 10-mile-an-hour wind:

Range	III or IX.	II, IV, VIII, X	I, V, VII, XI.
Yards	Points	Points	Points
100	0.23	0 2	0 1
200	.34	31	.17
300	.61	53	30
400	86	75	43
500	1.11	96	55
600	1.39	1 2	.69
700	1.68	1 45	.84
800	2 00	1.73	1.00
900	2.34	2 03	1 17
1,000	2 67	2 30	1.33

Winds blowing from XII and VI o'clock directions have no deviating influence on the flight of the bullet, but these winds have the effect of shortening (in the case of XII-o'clock wind) or lengthening the range (in the case of VI-o'clock wind). The correction necessary to apply in the case of such winds is inappreciable. (69)

Remarks.—If the soldier is well drilled in applying the windage and elevation corrections necessary to bring an assumed hit into the bull's-eye or figure, using in turn each of the targets at which he fires on the range, he will need very little further instruction in applying the necessary corrections. The instructor should assure himself that the men understand the reasons for these corrections, and they should never forget that they must move the rear-sight movable base **into the wind and in the same direction** they wish to move the point struck. (70)

Gallery Practice

400. After the soldier has been thoroughly instructed in sighting, and in the position, aiming, deflection, and elevation-correction drills, he will be exercised in firing at short ranges with reduced charges. (71)

Value of Gallery Practice.—Notwithstanding the value of the position and aiming drills, it is impossible to keep up the soldier's interest if these exercises are unduly prolonged. By gallery practice, however, the interest is easily maintained and further progress, especially in teaching the trigger squeeze, is made. Many of the external influences, which on the range affect the firing, being absent, the soldier is not puzzled by results for which, at this stage of his education, he could not account were he advanced to firing with full charges. Furthermore, as there is no recoil to induce nervousness or flinching, the soldier soon finds that he can make good scores, and this success is the surest stimulus to interest.

Not only to the beginner is gallery practice of value; to the good shot it is a means of keeping, to a certain extent, in practice, and practice in shooting, as much as in anything else, is essential. Since it can be carried on throughout the year, gallery practice is of much value in fixing in the men the **habit of aimed fire,** than which nothing in his training is of more importance. (72)

Gallery Practice.—During the month preceding range practice the minimum number of scores given in the following table must be fired by all who are required to fire under the provisions of paragraph 89:

401

Range (feet).	Target.	Position.	Scores.
50		Kneeling.	
50	The iron gallery target issued by the Ordnance Department, or one similar thereto, or the paper "X" target.	Standing.	A minimum of four (4) at each range in each of the positions prescribed
75		Prone.	
75		Sitting.	(73)

Scores.—Gallery practice will be conducted in scores of five shots. The number of scores to be fired by any man at a single practice is determined by the company commander.

No reports of the results of the firing will be required, although a record of it should be kept in the company for the instruction and guidance of the soldier. (74)

Additional Practice.—In addition to the minimum number of scores prescribed in paragraph 73, practice in this class of firing should be carried on throughout the year when practicable, the amount and details of the practice being left to the discretion of the company commander. The practice should be varied as much as possible. It should include exercises in slow and rapid fire, in assuming the various firing positions and opening fire quickly, and in finding the correct adjustment of the rear sight without unnecessary loss of time. Moving or disappearing targets can easily be improvised and the instruction made as interesting as possible. (75)

Matches.—Matches in gallery firing between the men, particularly the recruits, and between teams of the same or different companies, should be promoted and encouraged. While such matches increase the interest of the men in their practice, they at the same time afford experience in the conditions of competitive firing. (76)

Estimating-Distance Test

401. Importance.—Ability to estimate distances correctly is an important element in the education of the soldier.

While it is true that fire on the battle field will usually be by groups and the ranges given by officers or noncommissioned officers, the battle field is reached only after a long series of experiences in scout, patrol, and outpost duty, in which the soldier is frequently placed in positions where it is essential that he shall determine for himself the range to be used in order that the fire may be effective. It is, therefore, here made a prerequisite to qualification that the soldier shall be proficient in estimating distances by eye.

[326]

During the estimating-distance drills advantage should be taken of every opportunity to train the soldier in observing his surroundings from positions and when on a march. He should be practiced in pointing out and naming different features of the ground; in discovering and describing different objects; in counting different objects or beings. Especially should noncommissioned officers be trained in describing the location, with reference to other objects, of objects difficult to see and in imparting information of this kind quickly and accurately. (77)

Distances can be estimated by the eye or by sound; they can be determined by range-finding instruments, by trial shots or volleys, or from maps. (78)

Estimation of Distance by Eye.—To estimate distance by the eye with accuracy, it is necessary to be familiar with the appearance, as to length, of a unit of measure which can be compared mentally with the distance which is to be estimated. The most convenient unit of length is 100 yards. To impress upon the soldier the extent of a stretch of 100 yards two posts 100 yards apart, with short stakes between to mark each 25 yards, should be placed near the barracks, or on the drill ground, and the soldier required to pace off the marked distance several times counting his steps. He will thus learn how many of his steps make 100 yards and will become familiar with the appearance of the whole distance and of its fractional parts.

Next a distance of more than 100 yards will be shown him and he will be required to compare this distance with the 100-yard unit and to estimate it. Having made this estimate, he will be required to verify its accuracy by pacing the distance.

A few minutes each day should be spent in this practice, the soldier often being required to make his estimate by raising his rearsight leaf and showing it to the instructor. After the first drills the soldier should be required to pace the distance only when the estimate is unusually inaccurate.

The soldier should be taught that, in judging the distance from the enemy, his estimate may be corrected by a careful observation of the clearness with which details of dress, the movements of limbs or of the files in a line may be seen. In order to derive the benefit of this method, the soldier will be required to observe closely all the details noted above in single men or squads of men posted at varying distances, which will be measured and announced.

Although the standing and kneeling silhouettes used in field practice afford good objects upon which to estimate distances, the instructor should make frequent use of living figures and natural objects, as this is the class of targets from which the soldier will be compelled to estimate his range in active service. (79)

401 (contd.)

Methods of Estimating Long Distances by the Eye.—The following methods are found useful:

(a) The soldier may decide that the object can not be more than a certain distance away nor less than a certain distance; his estimates must be kept within the closest possible limits and the mean of the two taken as the range.

(b) The soldier selects a point which he considers the middle point of the whole distance, estimates this half distance and doubles it, or he similarly divides the distance into a certain number of lengths which are familiar to him.

(c) The soldier estimates the distance, along a parallel line, as a road on one side, having on it well-defined objects.

(d) The soldier takes the mean of several estimates made by different persons. This method is not applicable to instruction. (80)

Appearance of Objects: How Modified by Varying Conditions of Light; Difference of Level, etc.—During instruction the men should be taught the effect of varying conditions of light and terrain upon the apparent distance of an object.

Objects seem nearer—

(a) When the object is in a bright light.

(b) When the color of the object contrasts sharply with the color of the background.

(c) When looking over water, snow, or a uniform surface like a wheat field.

(d) When looking from a height downward.

(e) In the clear atmosphere of high altitudes.

Objects seem more distant—

(a) When looking over a depression in the ground.

(b) When there is a poor light or a fog.

(c) When only a small part of the object can be seen.

(d) When looking from low ground upward toward higher ground. (81)

Estimating Distance by Sound.—Sound travels at the rate of about 1,100 feet, or 366 yards, per second. If a gun is fired at a distance, a certain time elapses before the sound is heard. If the number of seconds or parts of seconds between the flash and the report be carefully taken and multiplied by 366, the product will be approximately the distance in yards to the gun. This method will be of doubtful use of the battle field, owing to the difficulty of distinguishing the sound of the gun, whose flash is seen, from that of any other. It will probably be useful in determining the range to a hostile battery when it first opens fire. (82)

Determining Distance by Range-Finding Instruments.—Accuracy in determining distance by range-finding instruments depends upon care and facility in use of the instrument and clearness of definition of the objective. Knowledge of the use of the instrument issued is essential to all company officers and should be imparted to sergeants when time is available. (83)

Determination of Distance by Trial Shots or Volleys.—If the ground is so dry and dusty that the fall of the bullets is visible through a glass or with the naked eye, a method of determining the distance is afforded by using a number of trial shots or volleys.

The method of using trial volleys is as follows:

The sights are raised for the estimated range and one volley is fired. If this appears to hit but little short of the mark an increase of elevation of 100 yards will be used for the next volley. When the object is inclosed between two volleys, a mean of the elevations will be adopted as the correct range.

The range may be obtained from a near-by battery or machine gun. This is the best method where available. (84)

Estimating Distance Test.—When instruction shall in the opinion of the company commander, have progressed to such an extent as to enable the soldier to judge distances with the eye with fair accuracy, he will be tested for proficiency.

As the danger space is continuous for a man kneeling within a range of 547 yards (battle-sight range), and as individual fire and the fire of small squads will ordinarily be limited to 1,200 yards, the soldier will be tested for proficiency at distances between these two ranges.

The rules governing this test are as follows:

(a) The test will be supervised by an officer.

(b) Each soldier will be tested separately.

(c) The ground shall be other than that over which he fired or has previously estimated distances.

(d) The use of any device to mark the limits within which distances are tested (550 and 1,200 yards), at the time the test is given, so that this device can be seen from the estimating point is prohibited.

(e) The objectives will be natural objects, men standing, kneeling, or prone, or silhouettes.

(f) For objectives, five or more natural objects will be selected, or single men or groups of men stationed or silhouettes placed within the ranges indicated above. The distances of the objectives will not be measured until all who are to estimate on them have made their estimates.

The men to be tested are conducted to a point near that from which the estimates are to be made, and remain facing away from the objectives or hidden therefrom by some feature of the ground. They are not permitted to know what objects are to be used in the test until they are called up to the estimating point.

[329]

The officer conducting the test calls up one man at a time, points out to him an objective and causes him to estimate the distance thereto. This is continued until the soldier's estimates on five objectives have been obtained. When the test for the day is completed by any man, he will not be allowed to join the squad awaiting test. After all the men have made estimates of distances to a given series of objectives, ranges thereto will be measured.

(g) Proficiency for the expert rifleman and for the sharpshooter shall consist in making in five consecutive estimates an average degree of accuracy of 90 per cent.

Similarly for proficiency, marksmen, first-class men, and second-class men are required to make in five consecutive estimates an average degree of accuracy of 85 per cent.

Not more than three trials will be given, and should the soldier fail three times to make the required percentage, his final qualification will be reduced one grade below that obtained in firing. (85)

Range Finders.—The estimating test having been completed, five or six enlisted men, selected by the company commander from the most accurate estimators, will be designated as "Range finders." These men will be given practice in estimating distances throughout the year. The practice will be on varied ground and at distances up to 2,000 yards. (86)

Known Distance Practice

402. General Description.—When gallery practice has been completed as required, the soldier is advanced to known-distance firing. The general scheme for firing is as follows:

KNOWN DISTANCE PRACTICE.

Regular courses.

Qualification course —
- Instruction practice —
  - Slow fire, targets A and B.
  - Slow fire, target D.
  - Rapid fire, target D.
- Record practice —
  - Slow fire, targets A and B.
  - Rapid fire, target D.

Long-distance practice. — Slow fire —
- 800 yards, target C.
- 1,000 yards, target C.

Practice with telescopic sights. —
- Target B.
- Target C.

Supplementary Course—Special Course A, Instruction Practice.

Special courses.

Special course A. —
- Instruction practice —
  - Slow fire, target A.
  - Slow fire, target D.
  - Rapid fire, target D.
- Record practice —
  - Slow fire, target A.
  - Rapid fire, target D.

Courses for Organized Militia .... —
- Qualification course. —
  - Instruction practice. —
    - Slow fire, targets A and B.
    - Slow fire, target D.
    - Rapid fire, target D.
  - Record practice. —
    - Slow fire, targets A and B.
    - Rapid fire, target D.
- Long-distance practice, slow fire ... —
  - 800 yards, target C.
  - 1,000 yards, target C.
- Practice with telescopic sights ... —
  - Target B.
  - Target C.

(87)

Special Courses

403. Preliminary Drills.—Special courses will be preceded by the required preliminary instruction and sighting drills. (119)

Special Course A

When Used.—When a complete range is not provided and a range of 200 and 300 yards is available, practice may be conducted, if authorized by the department commander, as prescribed in the following special course A. This practice is also prescribed for the Coast Artillery Corps and for bands in the Philippine Islands when required to fire. The instruction practice, special course A, will be followed for the supplementary firing for recruits and others who join too late to fire in the regular season. (120)

Special course A shall consist of instruction and record practice as follows:

Instruction practice.—Slow fire, target A; slow fire, target D; rapid fire, target D.

Record practice.—Slow fire, target A; rapid fire, target D.

The details of this practice are as prescribed in the tables given below, which have the force of written regulations.

Instruction Practice

TABLE 1.—*Slow fire, target A*

Range.	Time	Shots.	Position
200	No limit	15	5 prone 5 kneeling 5 standing
300		10	5 prone 5 sitting

TABLE 2.—*Slow fire, target D*
(Battle sight only will be used)

Range.	Time	Shots	Position
200	No limit.	10	5 kneeling 5 standing
300		10	5 prone. 5 sitting.

TABLE 3.—*Rapid fire, target D*
(Battle sight only will be used)

Range.	Time.	Shots	Position
200	1 minute.	10	Kneeling from standing
300	1 minute, 10 seconds.	10	Prone from standing.

RECORD PRACTICE
TABLE 4.—*Slow fire, target A*

Range.	Time.	Shots.	Position
200	No limit.	10	5 kneeling 5 standing.
300		10	5 prone. 5 sitting.

Rapid fire as given in Table 3.

All practice in special course A will be conducted according to the rules prescribed for the firing at the same ranges in the instruction and record firing in the qualification course.

For qualification see paragraph 242. (121)

The courses for Organized Militia are prescribed in Part V. (122)

Except as indicated in paragraphs 115 and 117 the Philippine Scouts will fire the course prescribed for the Regular Army, and will be subject to the same rules in regard to qualification and classification. (123)

Advice to Riflemen

404. For purposes of instruction, all firing may be divided into three classes, viz:

1. Slow fire at 600 yards and under.
2. Slow fire at ranges over 600 yards.
3. Rapid fire. (124)

Short-Range Practice.—In the first class of fire, slight changes of wind, light, and temperature may be almost disregarded. The principal things to be learned are: Setting the sight properly at the beginning of a score; aiming properly; squeezing the trigger properly; holding the rifle. These constitute probably 90 per cent of the work at ranges under 600 yards, and if the soldier performs these actions correctly with each shot, he will make a good score regardless of small changes in the atmospheric conditions. It is very important that the piece be held firmly and sighted uniformly. (125)

Long Range Practice.—In the second class of firing (at distances greater than 600 yards) a large part of the work is in the holding, but changes of wind, temperature, and light must be studied in order to make good scores. (126)

Wind.—Wind is the most important factor to be considered in long range known distance practice. It is unnecessary to teach recruits and others who never shoot beyond 600 yards more than the adjustment of the wind gauge for a right or left wind and how to change the wind gauge when a hit is made.

The direction of the wind is shown by considering the range as a clock face, the firer being in the center and the target at 12 o'clock. The direction is then indicated as a 10 o'clock wind, 2 o'clock wind, etc.

The force of the wind is indicated in miles per hour and is shown accurately by the anemometer, and is estimated by observation of flags, by throwing up leaves, grass, or bits of paper, and by the "feel" of the wind on the hands or face.

At long-distance ranges, after firing a shot and before firing again, the firer should look carefully for any change in direction and force of the wind. A change of 4 miles in force or of one hour in direction will make a decided difference in the location of a hit.

Any wind deflects the bullet from its course in the direction the wind is blowing. The amount of deflection varies with the direction and force of the wind.

The wind guage is graduated in points, and 1 point will move the bullet approximately 4 inches for each 100 yards of distance the firer is from the target and in the direction the movable base of the wind guage is moved:

At 200 yards 1 point equals 8 inches.

At 600 yards 1 point equals 24 inches.

At 800 yards 1 point equals 32 inches.

At 1,000 yards 1 point equals 40 inches.

The amount of windage to be taken is determined by estimating the force and direction of the wind.

A simple rule for determining the approximate windage at any range is as follows:

$$\frac{\text{Range} \times \text{Velocity}}{10}$$ equals quarter points required for 3 or 9 o'clock

winds. Winds one hour away from 3 and 9 o'clock require only slightly less windage. Winds one hour away from 12 or 6 o'clock require half as much windage as 3 or 9 o'clock winds.

Example: Range 800 yards: 5-mile wind blowing from 9 o'clock.

$$\frac{8 \times 5}{10}$$ equals 4 quarter points, or 1 point of windage. (127)

Temperature.—After the proper adjustment of the sight has been determined, it will rarely happen while firing a single, or even several consecutive scores, that such changes can occur in the temperature as to make further corrections necessary. If the first shot has been fired from a clean, cool gun, the subsequent fouling and heating of the barrel and the different vibrations of the latter, which are caused by the heating, will generally make necessary a slight increase in elevation for the second shot, and often an additional increase for the third shot. This should be followed, in some cases, where a number of shots are fired without cleaning or without any considerable interval, by a slight lowering of the elevation after additional shots.

A decided increase in the temperature will cause the bullet to strike high; a decided drop in temperature will cause the bullet to strike low. (128)

Light.—Changes of light do not effect the flight of the bullet; they do affect the manner in which the aim is taken. As all men are not affected alike by changes of light, each man must determine for himself how changes of light affect him.

Using the peep sight, the bull's-eye of a bright target is more clearly defined than the bull's-eye of a dark one, and the firer will usually hold closer to the bright bull's-eye than to a dark one. If the target changes from bright to dark, the next shot will usually go low.

With the open sight, as the light changes from bright to dark and the rear notch fills with shadow, more front sight is seen, and the shot goes high.

If occasional shadows drift across the face of the target, do not fire until the target is bright.

In a permanent change of light, let the eye accustom itself to the change before firing. Then the aim will be the same as before and there will be no change in the position of the hit.

In rapid firing with open sights, on very dark days shots seem to go high, due entirely to the firer taking more front sight than on bright days. (129)

Mirage.—This is the term applied in target practice to heated air in motion, as seen through telescopes or field glasses on clear days with winds of from 2 to 14 miles per hour. Through the telescope, waves appear to be moving across the face of the target in the direction the wind is blowing.

These waves indicate the general direction and speed of the wind. As to direction, they indicate a right or left wind only, and not one from 11, 1, 5, or 7 o'clock.

In a light 6 o'clock wind or with no wind at all, the waves will go straight up or "boil."

With a light wind the mirage moves slowly across the face with a decided vertical motion, giving a saw-tooth appearance. As the wind increases, the vertical motion of the mirage decreases until, with a 12 to 14 mile wind, the waves seem nearly flat and run across the target with very little vertical motion.

On hot days, with no wind, or a very light wind from 6 o'clock, the mirage will rise straight from the bottom to the top of the target. This condition seldom lasts long, and in a very short time the mirage will run from one side to the other.

Never fire while the mirage is "boiling," for there is usually a slight drift toward one side or the other, invisible to the firer, and if a shot is fired with no windage in a "boil," it will usually be out of the bull's-eye. Wait for the mirage to move from one side. (130)

Rapid Firing.—Success in rapid firing depends upon catching a quick and accurate aim, holding the piece firmly and evenly and in squeezing the trigger without a jerk.

In order to give as much time as possible for aiming accurately, the soldier must practice taking position, loading with the clip, and working the bolt so that no time will be lost in these operations. With constant practice all these movements may be made quickly and without false motions.

When the bolt handle is raised it must be done with enough force to start the shell from the chamber; and when the bolt is pulled back it must be with sufficient force to throw the empty shell well away from the chamber, and far enough to engage the next cartridge.

In loading, use force enough to load each cartridge with one motion.

The aim must be caught quickly, and once caught must be held, and the trigger squeezed steadily. Rapid firing, as far as holding the aim and squeezing the trigger are concerned, should be done with all the precision of slow fire. The gain in time should be in getting ready to fire, loading, and working the bolt.

Constant practice will increase the accuracy of aim, and any exercise that will strengthen arms and hands will enable one to hold better through a long string of shots. (131)

Firing at Moving Targets.—In firing at moving targets, the rifle must move with the target. If the target moves across the front, the aim must be a certain distance in front of it, depending on the distance of the target and its speed. If it moves toward the firer he must hold below it; if away from him, he must hold over it.

The following table shows the approximate distance necessary to aim ahead of the body of a man or horse moving across the range at various distances and various rates of speed:

Distance (yards).	Man walking.		Man double timing.		Horse walking.		Horse trotting.		Horse running.	
	Ft	*In*	*Ft.*	*In*	*Ft*	*In*	*Ft*	*In*	*Ft.*	*In*
100	Front	edge	..	6	Front edge of body		Front of body		Front edge of body.	
200	..	8	1	8	Front edge of body.		Front of body.		1	4
300	1	5	3	..	Front of body		1	6	3	10
400	2	2	4	5	Front of body.		3	4	6	6
500	3	1	5	11	1	..	5	4	9	6
600	4	..	7	7	2	..	7	7	12	10

(132)

Firing with Rests.—In the ordinary positions for firing with piece supported by hands, arms, and shoulder, the explosion of the powder charge sets up in the barrel of the rifle certain vibrations which become disturbed and altered somewhat when the rifle is fired with the additional support of a solid rest applied at some point of the barrel.

Using the same elevations and aiming point the effect of a rest is exhibited in a changed point of strike of the bullet.

The vertical vibrations of the barrel are the more pronounced, and as these are interfered with by a point of rest under the barrel this species of support will usually change the point of strike more than in the case of a side rest against a vertical surface. In the latter case the piece is steadied rather than rested.

With a rest beneath the balance, or near that point, the tendency is to shoot above and to the right of the point of strike that would be attained without a rest, using the same elevation and point of aim.

The tendency is the same and more pronounced when the rest is under a point near the muzzle. The change in the point of strike in any case is slight and insufficient to carry the shot off the target from the center of target D at 600 yards.

In order that the shooting may be uniform, the piece should always be rested at the same point.

A side rest will cause no appreciable change in the point of strike.

In firing with the bayonet fixed, usually a lower point on the target will be struck corresponding to a reduction in the range of about 50 yards. (133)

Cleaning the Rifle.—After firing, the bore of the rifle is covered with fouling. This is of two kinds, a black deposit covering the entire bore, caused by the burning powder and easily removed with rags, and a metallic fouling, caused by particles of the metal jacket of the bullet adhering to the barrel and which can be removed only by the use of ammonia solution.

The powder fouling must be removed first. Then the metallic fouling can be seen in patches on the lands.

To remove the powder fouling use a cleaning rod long enough to clean from the breech; Hoppe's Powder Solvent No. 9; rags about 1¼ inches square, of thin flannel or any other soft material.

A cleaning rack should be provided for every barrack.

Rifles should always be cleaned from the breech, thus avoiding any possible injury to the muzzle. Any injury to the rifling at the muzzle will affect the shooting adversely. If the bore for a length of 6 inches at the muzzle is perfect, a minor injury near the chamber will have little effect on the accuracy of the rifle.

The rifle should be cleaned as soon as the firing for the day is completed. The fouling is easier to remove then, and if left longer it will corrode the barrel.

Take a couple of rags soaked in No. 9 and run them through the barrel until they have removed all the powder fouling; run clean rags through to dry the barrel; clean with ammonia solution as directed and finish by wiping out with a greased rag or a clean rag soaked in No. 9. For grease, use vaseline, cosmic, or "3 in 1" oil. After the barrel is cleaned, wipe out the chamber, the cams, bolt, and all visible working parts. Occasionally clean out the magazine and wipe off magazine spring, then wipe all working parts with a greased rag.

After cleaning the working parts, wipe off stock and outside of barrel with oiled rag.

405

Before firing again, wipe all oil out of barrel, but leave chamber and working parts slightly oily. This will prevent shells binding in chamber and will make parts work easier. Wipe all oil from outside of barrel and stock.

To remove metallic fouling, use ammonia solution. This is made as follows: Take ammonia persulphate, 1 ounce; ammonium carbonate, 200 grains; ammonia (28 per cent), 6 ounces; water, 4 ounces. One rounded tablespoonful equals 1 ounce of persulphate or 200 grains of carbonate.

Powder the persulphate and carbonate separately. Dissolve persulphate in the ammonia and the carbonate in the water and then pour the mixture in a strong bottle, and cork. If mixed in this manner, it may be used in an hour.

To use.—After the barrel has been cleaned with No. 9 and wiped dry, cork up breech with a small cork, put a piece of rubber tubing about an inch long on the muzzle, and fill the barrel with the solution. It will boil up instantly with a white foam, very slightly blue. Let the solution stay in the barrel not more than 10 minutes and then pour out. If there was any metal fouling, the solution will be dark blue.

Fill the barrel with water to remove any remaining ammonia, pour out, and then remove the cork and rubber tube, wipe barrel perfectly dry, and then rub with oiled rag.

Care should be used in mixing and using this solution, for if improperly mixed or used it will injure the rifle. If the solution, after being used, is brown, it is bad and should be thrown away. The proportions of persulphate and carbonate should be the same in bulk. Too much persulphate will injure the barrel.

Keep the barrel filled. If the solution evaporates, it will leave a deposit of persulphate on the surface of the bore and will injure it.

An experienced noncommissioned officer should mix the solution and supervise its use.

Care should be taken not to spill the solution on the barrel or in the mechanism. (134)

Pistol and Revolver Practice; Preliminary Drills; Position and Aiming Drills[1]

405. **Nomenclature and Care of the Weapon; Handling and Precautions against Accidents.**—The soldier will first be taught the nomenclature of those parts of the weapon necessary to an understanding of its action and use and the proper measures for its care and preservation. Ordnance pamphlets Nos. 1866 (description of the Colt's Automatic Pistol), 1919, and 1927 (descriptions of the Colt's revolver, calibers .38

[1] Whenever in these regulations the word "pistol" appears, the regulation applies with equal force to the revolver, if applicable to that weapon.

and .45, respectively), contain full information on this subject and are furnished to organizations armed with these weapons.

Careless handling of the pistol or revolver is the cause of many accidents and results in broken parts of the mechanism. The following rules will, if followed, prevent much trouble of this character:

(a) On taking the **pistol** from the armrack or holster, take out the magazine and see that it is empty before replacing it; then draw back the slide and make sure that the piece is unloaded. Observe the same precaution after practice on the target range, and again before replacing the pistol in the holster or in the armrack. When taking the **revolver** from the armrack or holster and before returning it to the same, open the cylinder and eject empty shells and cartridges. Before beginning a drill and upon arriving on the range observe the same precaution.

(b) Neither load nor cock the weapon until the moment of firing, nor until a run in the mounted course is started.

(c) Always keep the pistol or revolver in the position of **Raise pistol** (Par. 156, *Cavalry Drill Regulations*), except when it is pointed at the target. (The position of **Lower pistol** is authorized for mounted firing only.)

(d) Do not place the weapon on the ground where sand or earth can enter the bore or mechanism.

(e) Before loading the **pistol,** draw back the slide and look through the bore to see that is is free from obstruction. Before loading the **revolver,** open the cylinder and look through the bore to see that it is free from obstruction. When loading[1] **the pistol** for target practice place 5 cartridges in the magazine and insert the magazine in the handle; draw back the slide and insert the first cartridge in the chamber and carefully lower[2] the hammer fully down.

[1] *TO LOAD PISTOL: Being at Raise Pistol* (Right hand grasping stock at the height of and 6 inches in front of the point of the right shoulder, forefinger alongside barrel, barrel to the rear and inclined forward about 30°).

Without deranging position of the hand, rotate the pistol so the sights move to the left, the barrel pointing to the right front and up.

With the thumb and forefinger of the *Left* hand (thumb to the right), grasp the slide and pull it towards the body until it stops and then release it. The pistol is thus loaded, and the hammer at full cock.

. If the pistol is to be kept in the hand and not to be fired at once, engage the safety lock with the thumb of the *Right hand.*

If the pistol is to be carried in the holster, remove safety lock, if on, and lower the hammer *fully* down.

[2] *TO LOWER THE HAMMER: Being at the loading position at full cock.*

I. Firmly seat thumb of *Right* hand on the hammer; insert forefinger inside trigger guard.

II. With *thumb* of *Left* hand exert a momentary pressure on the grip-safety to release hammer from sear.

III. At the same instant exert pressure on the trigger and carefully and slowly lower the hammer fully down.

IV. Remove finger from trigger.

V. Insert pistol in holster.

(Caution) The pistol must *never* be placed in the holster until hammer is *fully down.*

405 (contd.)

In loading the **revolver** place 5 cartridges in the cylinder and let the hammer down on the **empty chamber**.

(f) Whenever the pistol is being loaded or unloaded, the muzzle must be kept up.

(g) Do not point the weapon in any direction where an accidental discharge might do harm.

(h) After loading do not cock the pistol or the revolver until ready to fire.

(i) Keep the working parts properly lubricated. (135)

Position, Dismounted.—Stand firmly on both feet, body perfectly balanced and erect and turned at such an angle as is most comfortable when the arm is extended toward the target; the feet far enough apart to insure firmness and steadiness of position (about 8 to 10 inches); weight of body borne equally upon both feet; right arm fully extended, left arm hanging naturally.

Remarks.—The right arm may be slightly bent, although the difficulty of holding the pistol uniformly and keeping it as well as the forearm in the same vertical plane makes this objectionable. (136)

The Grip.—Grasp the stock as high as possible with the thumb and last three fingers, the forefinger alongside the trigger guard, the thumb extended along the stock. The barrel, hand, and forearm should be as nearly in one line as possible when the weapon is pointed toward the target. The grasp should not be so tight as to cause tremors of the hand or arm to be communicated to the weapon, but should be firm enough to avoid losing the grip when the recoil takes place.

Remarks.—The force of recoil of the pistol or revolver is exerted in a line above the hand which grasps the stock. The lower the stock is grasped the greater will be the movement or "jump" of the muzzle caused by the recoil. This not only results in a severe strain upon the wrist, but in loss of accuracy.

If the hand be placed so that the grasp is on one side of the stock, the recoil will cause a rotary movement of the weapon toward the opposite side.

The releasing of the sear causes a slight movement of the muzzle, generally to the left. The position of the thumb along the stock overcomes much of this movement. The soldier should be encouraged to practice this method of holding until it becomes natural.

To do uniform shooting the weapon must be held with exactly the same grip for each shot. Not only must the hand grasp the stock at the same point for each shot, but the tension of the grip must be uniform. (137)

(a) **The Trigger Squeeze.**—The trigger must be squeezed in the same manner as in rifle firing. (See Pars. 44 and 46.) The pressure of the forefinger on the trigger should be steadily increased and should be straight back, not sideways. The pressure should continue to that point beyond which the slightest movement will release the sear. Then, when the aim is true, the additional pressure is applied and the pistol fired.

Only by much practice can the soldier become familiar with the trigger squeeze. This is essential to accurate shooting. It is the most important detail to master in pistol or revolver shooting.

(b) **Self-Cocking Action.**—The force required to squeeze the trigger of the revolver when the self-cocking device is used is considerably greater than with the single action. To accustom a soldier to the use of the self-cocking mechanism, and also to strengthen and develop the muscles of the hand, a few minutes' practice daily in holding the unloaded revolver on a mark and snapping it, using the self-cocking mechanism, is recommended. The use of the self-cocking device in firing is not recommended except in emergency. By practice in cocking the revolver the soldier can become sufficiently expert to fire very rapidly, using single action, while his accuracy will be greater than when using double action. (138)

Aiming.—Except when delivering rapid or quick fire, the rear and front sights of the pistol are used in the same manner as the rifle sights. The normal sight is habitually used (See Pl. VI), and the line of sight is directed upon a point just under the bull's-eye at "6 o'clock." The front sight must be seen through the middle of the rear-sight notch, the top being on a line with the top of the notch. Care must be taken not to cant the pistol to either side.[1]

---

[1] The instructor should take cognizance of the fact that the proper aiming point is often affected by the personal and fixed peculiarities of the firer, and if unable to correct such abnormalities, permit firer to direct sight at such point as promises effective results.

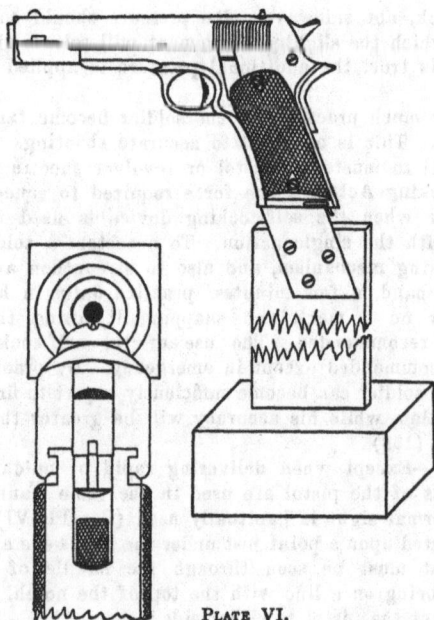

PLATE VI.

If the principles of aiming have not been taught, the soldier's instruction will begin with sighting drills as prescribed for the rifle so far as they may be applicable. The sighting bar with open sight will be used to teach the normal sight and to demonstrate errors likely to be committed.

To construct a sighting rest for the pistol (See Pl. VI) take a piece of wood about 10 inches long, 1¼ inches wide, and 9/16 inch thick. shape one end so that it will fit snugly in the handle of the pistol when the magazine has been removed. Screw or nail this stick to the top of a post or other object at such an angle that the pistol when placed on the stick will be approximately horizontal. A suitable sighting rest for the revolver may be easily improvised. (139)

(a) **How to Cock the Pistol.**—The pistol should be cocked by the thumb of the right hand and with the least possible derangement of the grip. The forefinger should be clear of the trigger when cocking the pistol. Some men have difficulty at first in cocking the pistol with the right thumb. This can be overcome by a little practice. Jerking the pistol forward while holding the thumb on the hammer will not be permitted.

(b) **How to Cock the Revolver.**—The revolver should be cocked by putting the thumb on the hammer at as nearly a right angle to the hammer as possible, and by the action of the thumb muscles alone bringing the hammer back to the position of full cock. Some men with large hands are able to cock the revolver with the thumb while holding it in the position of aim or raise pistol. Where the soldier's hand is small this can not be done, and in this case it assists the operation to give the revolver a slight tilt to the right and upward (to the right). Particular care should be taken that the forefinger is clear of the trigger or the cylinder will not revolve. Jerking the revolver forward while holding the thumb on the hammer will not be permitted. (140)

Position and Aiming Drills, Dismounted.—For this instruction the squad will be formed with an interval of 1 pace between files. Black pasters to simulate bull's-eyes will be pasted opposite each man on the barrack or other wall, from which the squad is 10 paces distant.

The squad being formed as described above, the instructor gives the command, 1. **Raise,** 2. **PISTOL** (Par 156, *Cavalry Drill Regulations*) and cautions, "Position and Aiming Drill, Dismounted." The men take the positions described in paragraph 136, except that the pistol is held at **Raise pistol.**

The instructor cautions, "Trigger squeeze exercise." At the command **Ready,** cock the weapon as described in paragraph 140. At the Command, 1. **Squad,** 2. **FIRE,** slowly extend the arm till it is nearly horizontal, the pistol directed at a point about 6 inches below the bull's-eye. At the same time put the forefinger inside the trigger guard and gradually "feel" the trigger. Inhale enough air to comfortably fill the lungs and gradually raise the piece until the line of sight is directed at the point of aim, i. e., just below the bull's-eye at 6 o'clock. While the sights are directed upon the mark, gradually increase the pressure on the trigger until it reaches that point where the slightest additional pressure will release the sear. Then, when the aim is true, the additional pressure necessary to fire the piece is given so smoothly as not to derange the alignment of the sights. The weapon will be held on the mark for an instant after the hammer falls and the soldier will observe what effect, if any, the squeezing of the trigger has had on his aim.

It is impossible to hold the arm perfectly still, but each time the line of sight is directed on the point of aim a slight additional pressure is applied to the trigger until the piece is finally discharged at one of the moments when the sights are correctly aligned upon the mark.

When the soldier has become proficient in taking the proper position, the trigger squeeze should be executed at will. The instructor prefaces the preparatory command by "At will" and gives the command **Halt** at the conclusion of the exercise, when the soldier will return to the position of "**Raise pistol.**"

At first this exercise should be executed with deliberation, but gradually the soldier will be taught to catch the aim quickly and to lose no time in beginning the trigger squeeze and bringing it to the point where the slightest additional pressure will release the sear.

Remarks.—In service few opportunities will be offered for slow aimed fire with the pistol or revolver, although use will be made of the weapon under circumstances when accurate pointing and rapid manipulation are of vital importance.

In delivering a rapid fire, the soldier must keep his eyes fixed upon the mark and, after each shot, begin a steadily increasing pressure on the trigger, trying at the same time to get the sights as nearly on the mark as possible before the hammer again falls. The great difficulty in quick firing with the pistol lies in the fact that when the front sight is brought upon the mark, the rear sight is often found to be outside the line joining the eye with the mark. This tendency to hold the pistol obliquely can be overcome only by a uniform manner of holding and pointing. This uniformity is to be attained only by acquiring a grip which can be taken with certainty each time the weapon is fired. It is this circumstance which makes the position and aiming drills so important. The soldier should constantly practice pointing the pistol until he acquires the ability to direct it on the mark in the briefest interval of time and practically without the aid of sights.

The soldier then repeats the exercises with the pistol in the left hand, the left side being turned toward the target. (141)

To Draw and Fire Quickly—Snap Shooting.—With the squad formed as described in paragraph 141 except that the pistol is in the holster and the flap, if any, buttoned, the instructor cautions "Quick fire exercise." And gives the command, 1. **Squad,** 2. **FIRE.** At this command, each soldier, keeping his eye on the target, quickly draws his pistol, cocks it as in paragraph 140, thrusts it toward the target, squeezes the trigger and at the instant the weapon is brought in line with the eye and the objective, increases the pressure, releasing the sear. To enable the soldier to note errors in pointing, the weapon will be momentarily held in position after the fall of the hammer. Efforts at deliberate aiming in this exercise must be discouraged.

Remarks under paragraph 141 are specially applicable also to this type of fire. When the soldier has become proficient in the details of this exercise, it should be repeated at will; the instructor cautions "At will; quick fire exercise." The exercise should be practiced until the mind, the eye, and trigger finger act in unison.

To simulate this type of fire mounted, the instructor places the squad so that the simulated bull's-eyes are in turn, to the **Right,** to the **Left,** to the **Right Front,** to the **Left Front,** to the **Right Rear.** With the squad in one of these positions, the instructor cautions "Position and aiming drill, mounted.'' At this caution the right foot is carried 20 inches to the right and the left hand to the position of the bridle hand (Par. 246, *Cavalry Drill Regulations*). The exercise is carried out as described for the exercise dismounted, using the commands and means laid down in paragraphs 161 to 168, inclusive, *Cavalry Drill Regulations*, for firing in the several directions. The exercise is to be executed at will when the squad has been sufficiently well instructed in detail.

When firing to the left the pistol hand will be about opposite the left shoulder and the shoulders turned about 45° to the left; when firing to the right rear the shoulders are turned about 45° to the right.

When the soldier is proficient in these exercises with the pistol in the right hand, they are repeated with the pistol in the left hand. (142)

Position and Aiming Drill—Mounted

Preliminary Training of Horses.—This course must be preceded necessarily by much work during the year, having for its object the training of the horses to the sight of the targets and to the noise of discharge of pistols. In addition to work on the riding track, much can be accomplished in this line by having blank cartridges fired while the horses are being groomed, and by placing targets just outside the corral or in such position that the horses will pass near them when being led to water or to the stables. During the dismounted practice the horses may be picketed near the firing point. (143)

Quick Aim Drill.—The soldier must be instructed and practiced in taking rapid aim while the horse is in motion. To this end, frequent practice should be had with the pistol throughout the year when drilling on the riding track, going through the motions of aiming and firing (at will) at silhouette targets and other objects placed along the track and 5 yards from it. This practice should be conducted at a walk, trot, and gallop. In quick-aimed fire at a gallop the soldier must endeaver to discharge his pistol at the moment when the horse is in the act of rising in the leap. This can best be done by holding the pistol pointed toward the targets and moving the arm up and down in unison with the motion of the horse. With the eyes fixed on the target, point the pistol just as it starts on the upward motion and squeeze the trigger. The soldier will be taught to exchange magazines and to load his weapon with facility at all gaits. (144)

Preliminary Range Practice Drills.—The pointing and snapping exercises outlined above for the riding school will, in the target season, be extended to and amplified on the range where a track will be laid out, as illustrated in Plate VIII, with a barrier in front of each target

406

to preserve a uniform distance from the horse to the targets. Parallel to, 10 yards distant from, and facing this track will be placed 5 standing silhouette figures 10 yards apart. The squad in column of troopers, with a distance between troopers of about 10 yards, will move around the track at a walk, trot, and gallop, each trooper pointing and snapping his pistol at each target as he arrives opposite it. (145)

(a) **Practice as With Ball Cartridges, Mounted.**—As soon as the horses have become sufficiently accustomed to the targets and to the noise of firing, the trooper mounted will be practiced in the details of procedure laid down for the several types of fire prescribed for Range Practice with ball cartridges. In firing to the left and left front, the weapon may be held in the left hand.

(b) **Quick Fire.**—Target: Silhouette of standing figure arranged to revolve as a bobbing target and to be operated by means of ropes laid under the track. This type of target is used for firing at the halt and at the walk. For any given individual, the target is turned before firing commences, edge of target toward the firing point. The targets are operated by revolving the target through an arc of 90° so that face and edge are alternately turned toward the firing point. Exposures are for a specified length of time with an interval 3 to 5 seconds between exposures. The individual who is to fire takes position at the firing point with his weapon in the holster and loaded as prescribed in paragraph 135 (e); flap of holster, if any, buttoned. At the first appearance of the target the trooper draws and fires, or attempts to fire, one shot at his target before it disappears. Prior to firing of this type the soldier should have been thoroughly instructed in quick pointing to eliminate, so far as practicable, personal errors. For purpose of instruction, the instructor may cause each shot to be marked and the procedure of firing the first shot to be repeated for each shot of the score. (Plate XI.) (146)

COURSES FOR ORGANIZED MILITIA

406. The following courses in small-arms firing are prescribed for the Organized Militia:

General Scheme

The general scheme of instruction for the Organized Militia embraces: First, a certain amount of instruction in the preliminary drills and exercises, followed by gallery practice, with a prescribed test before the soldier can be advanced to practice on the target range; second, a definite course of instruction practice, under which, by selected scores of five shots each, a soldier must attain a certain proficiency before he can be advanced to fire the record practice, Organized Militia, or the qualification course, Regular Army; third, a definite test, either the qualification course, Organized Militia, or the qualification course, Regular Army, at the discretion of the State authorities, under which the soldier attains a certain grade in marksmanship; fourth, long-range practice. (345)

Preliminary Drills and Instruction

(a) The period for indoor instruction will be determined by the State authorities and may extend into or include the entire range practice season:

(b) The essentials of indoor instruction will include—

Nomenclature, covering the most important parts and elements.

Manipulation and use of the various working parts.

Care of the arm.

Sighting, aiming, positions, and trigger squeeze.

Gallery practice.

The course to be followed in indoor instruction is laid down in Part II, Chapters I, II, III, IV; but in the discretion of the State authorities, any course embracing the elements given above may be adopted and followed. The recording rifle rod outfit or any other suitable device may be used in such a course. (346)

Gallery Practice

407. The principal objects of gallery practice are to continue in a different manner the instruction in aiming, positions, and trigger squeeze, and to determine, in certain cases, whether or not the individual shall be advanced to range practice.

(a) The following course in gallery practice is prescribed:

TABLE I

Range.	Targets.	Position.	Minimum number of shots.
Feet. 50	The iron gallery target issued by the Ordnance Department or one similar thereto, or paper targets.	Prone	10
50 50	...... do .... do	{ Sitting .... { Kneeling .. Standing ..	10 { 5 sitting. { 5 kneeling. 10

Where it is impracticable to use ranges of 50 feet, gallery practice may be conducted at a greater distance at a target whose dimensions and divisions have been proportionately increased. Firing will be by scores of five consecutive shots. Except in case of accident, a score once begun will be completed.

Qualification in Gallery Practice

(b) No officer or enlisted man who has failed to qualify as first class or better in a previous season shall be advanced to range practice until he has attained at least 90 points out of a possible 150 in the gallery practice course, by selecting his two best scores of five shots at each range. If a gallery range be not available, the recording rifle rod outfit or subtarget gun machine may be used to determine eligibility for range practice, under similar conditions, when specially authorized by the State authorities. (347)

Known Distance Firing

408. The qualification course is divided into Instruction Practice and Record Practice.

(a) Instruction practice embraces:

1. A prescribed course in which a certain proficiency must be attained in certain cases, before qualification practice is undertaken; this course may be shot through as many times as is necessary to insure proper instruction.

2. Such further preliminary practice at any range as is considered necessary to prepare the individual for the Record Practice.

3. Firing for recruits may be held at 100 yards in any position except standing, but does not count in determining proficiency in the instruction practice.

(b) The instruction practice and the number of shots at each range, upon which eligibility to advance to record practice is determined, are given in the following tables:

Instruction Practice

TABLE 2

Range.	Kind of fire.	Time	Shots.	Targets.	Position	Possible.
200	Slow fire.	No limit.	10	A	{ 5 sitting. 5 kneeling.	50
300	Slow fire.	No limit.	5	A	Prone.	25
500	Slow fire.	No limit.	10	B	Prone.	50

TABLE 3.—*Target D*
(Battle sight only will be used with this target.)

Range.	Kind of fire.	Time.	Shots.	Targets	Position.	Possible.
200	Slow.	No limit.	5	D	Kneeling.	25
200	Rapid.	1 minute	5	D	Kneeling from standing.	25
300	Slow.	No limit.	5	D	Prone.	25
300	Rapid.	1 minute.	5	D	Prone from standing.	25

Total, 225.

1. Each shot is marked in slow fire on target D. Rapid fire is conducted as prescribed in paragraph 110 except as to scores. At each range a total of 60 per cent of the possible must be attained before advancement to the next range.

[348]

2. After eligibility to fire-record practice has been determined according to the provisions of paragraphs (d) and (e), the record practice may be preceded by further preliminary practice in the discretion of the State authorities.

3. Firing in instruction practice will be by scores of five consecutive shots. A score once begun will be completed, unless accident or conditions of range weather interfere.

(c) The following grades of classification are obtained in instruction practice by selective scores of five consective shots each.

First-classman, 150; possible, 225.

Second-classman, 130; possible, 225.

Unqualified, below 130.

(d) No individual shall be advanced to record practice until he has attained the grade of first-classman, except as provided in paragraph (e).

(e) Instruction practice will be optional with the State authorities for all who have qualified as marksmen or better in the season immediately preceding. (348)

Record Practice

After completing the instruction practice, those who qualify as first-classmen are eligible to fire record practice, Organized Militia, or the qualification course, Regular Army.

The choice of courses shall be determined by the State authorities. Both courses may be pursued in the same state in the same season, but the course in regiments or in separate smaller tactical units shall be uniform; provided that an individual, who qualifies in record practice, Organized Militia course, as sharpshooter or expert rifleman, may be permitted to fire the Regular Army qualification course and qualify therein at the discretion of the State authorities. (349)

Qualification Course

(a) TABLE 4.—*Slow fire*

Range.	Time.	Shots.	Targets	Position
300	No limit.	10	A	Prone
500	No limit.	10	B	Prone.
600	No limit.	10	B	Prone, sandbag rest.[1]
		(2 s. s)		

[1] In firing with sandbag rest, either rifle or back of hand must rest on sandbag.

TABLE 5.—*Target D, rapid fire, battle sight*

Range.	Time.	Shots.	Targets.	Position.
200	1½ minutes.	10	D	Kneeling from standing.
300	2 minutes.	10	D	Prone from standing.

This course may be fired three times in any target season, the individual's classification being determined by the best of his three trials, but this provision shall not be constructed to permit the formation of a record based on scores selected from two or more trials—the basis of classification must be the result of one complete course in each case.

(b) Coaching is prohibited in record practice after the individual has taken his position at the firing point.

(c) Instruction and record practice may be fired on the same day, but record practice once begun must be completed without further instruction firing. (350)

Qualification Course, Regular Army.—The qualification course of the Regular Army shall be carried out by the Organized Militia with a strict adherence to all conditions and provisions required for the Regular Army, except that instruction and record practice may be fired on the same day, but record practice once begun must be completed without further instruction firing. (351)

Practice For Coast Artillery Reserves.—Special course "A," as described in paragraph 121, will be fired by coast artillery reserves of the Organized Militia, unless other courses be prescribed by the State authorities. (352)

Long-Distance Practice.—After the qualification course has been completed, those men who have qualified as experts and sharpshooters may be given long distance practice, at the discretion of the State authorities.

Practice for record may be preceded by instruction practice. Record practice will consist of any selective score of 10 consecutive shots each. The practice will be conducted as set forth for slow fire known distance.

TABLE 6.—*Target C*

Range.	Shots.	Position.
800	10	Prone.
1,000	10	Prone.

No one will be advanced to practice at 1,000 yards until he has attained a minimum total of 40 points at 800 yards in any score of 10 consecutive shots.

When an individual has attained a total of 85 points at 800 and 1,000 yards, including a minimum of 40 at 800 yards, by selective scores of 10 consecutive shots, he shall be considered to be qualified in long-distance practice but is not included in the table of classification nor in the computation of the figure of merit. He may be rewarded by suitable medals or other devices at the discretion of the State authorities. (353)

Practice with Telescopic Sights.—Practice with telescopic sights may be held at the discretion of the State authorities and, when held, will be conducted in accordance with paragraph 117. (354)

Miscellaneous.—(a) The conditions and requirements governing the conduct of target practice, except as modified in Part V, shall be the same for the Organized Militia as for the Regular Army.

(b) **Dress and Equipment.**—The dress and equipment of officers and men participating in target practice shall be prescribed by the State authorities. The cartridge belt will be worn at all times.

(c) **Marking.**—The provisions of these regulations governing marking which are impracticable for the Organized Militia may be modified by the State authorities except for the qualification course, Regular Army.

(d) **Scoring.**—Such provisions of these regulations as are impracticable for the Organized Militia may be modified by the State authorities except for the qualification course, Regular Army.

(e) **Estimating Distance.**—Estimating distance will not be required for the Organized Militia except when firing the Regular Army course, in which case paragraph 85 will be complied with. State authorities may require commissioned officers to qualify in estimating distance in the qualification course, Organized Militia.

(f) **Practice Season.**—The practice season will be determined by the State authorities.

(g) **Who Will Fire.**—

Required to Fire

1. Infantry, cavalry, and engineers:

Battalion and squadron staff officers, company and troop officers, inspectors small-arms practice, all enlisted men of companies and troops except cooks.

2. Coast artillery reserves:

Special course "A"—Company officers, all enlisted men of companies except cooks.

Authorized But Not Required to Fire

1. All other officers and enlisted men except those of the Medical Department and Chaplains.

2. All officers enumerated in the above table with more than 10 years commissioned or commissioned and enlisted service.

(h) **Amount of Fire.**—The amount of firing in one day for any individual is not restricted for the Organized Militia in any course.

(i) **Qualification.**—Qualification will be based on the results obtained in either one of the qualification courses or in the instruction practice, as set forth in the following table:

TABLE 7.—*Points required in qualification*

Courses.	Expert.	Sharpshooter.	Marksman.	1st class.	2nd class.	Unqualified[1]	Possible.	Insignia
1. Qualification course; instruction practice, Organized Militia.	...	...	...	150	130	{ Below 300	} 225	None
2. Qualification course; record practice, Organized Militia.	210	190	160	...	...	...	250	Bronze.
3. Qualification course; Regular Army.	253	238	202	177	152	...	300	Regular Army.
4. Special course "A," coast artillery reserves.	...	...	150	120	100	...	200	Bronze.

[1] All who fire the qualification course, instruction practice, Organized Militia; the qualification course Regular Army; the special course "A"; and who in any course fail to qualify as second class or better; and all who fail to complete a course, or fail to fire.

1. If in case authority is given to fire the qualification course, Regular Army, after a grade of qualification has been attained in the qualification course, Organized Militia, the higher qualification shall be used as the basis of record, and medals will be issued in each case.

2. If an individual fails to qualify as marksman or higher in firing the record practice, Organized Militia, his grading shall be first-class. In the Regular Army course and special course "A" the qualification shall be determined by the result of the firing in that course alone.

(J) **Holdover Qualifications.**—Holdover qualifications for the Organized Militia in the qualification courses (Organized Militia and Regular Army) will be limited to expert riflemen. An individual having attained this grade will retain that qualification for three consecutive target years, including the target season in which qualification was made. An individual entitled to a holdover qualification as an expert rifleman may be authorized by the State authorities to fire the qualification courses, in which case he forfeits no rights to such holdover qualification. If during such subsequent firing he again qualifies as an expert rifleman, his holdover privileges will begin from his latest qualification.

(k) **Insignia.**—For the qualification course, Organized Militia, bronze badges and pins.

For the qualification course, Regular Army, insignia similar to those issued to the Regular Army.

For special course "A," coast artillery reserves, bronze pins, marksmen only, when firing this course.

An individual qualifying as marksman, sharpshooter, or expert rifleman, will wear his badge or pin so long as he retains such a qualification.

(1) **Requalification Bars.**—No requalification bars are issued for marksmen.

Sharpshooters.—Requalification bars shall be issued at the rate of one bar for each three qualifications (not necessarily consecutive) as sharpshooter. The bar bears the last year of qualification.

Expert Riflemen.—Requalification bars are issued at the rate of one bar for each three qualifications as expert rifleman, holdover qualifications included, provided that when an individual requalifies as expert rifleman during a holdover period he is entitled to a bar for each three years of actual requalification. The bar bears the last year of qualification.

(m) **Combat Practice.**—Combat practice may be held by the Organized Militia whenever so directed by the State authorities. When held, it will be conducted in accordance with Part III, *Small Arms Firing Manual.*

(n) **Reports.**—A report of target firing shall be forwarded to the chief, Division of Militia Affairs, as soon after the close of the practice season as practicable, but not later than March 31 of the following year.

(o) **Figure of Merit.**—In each State there will be a company and regimental figure of merit to be calculated by the methods laid down by the Division of Militia Affairs. A proficiency test similar to that prescribed for the Regular Army is authorized at the discretion of the State authorities. (355)

Pistol Practice

409. **Preliminary Drills.**—(a) All officers and enlisted men belonging to organizations armed with the revolver or pistol shall be instructed in the care, preservation, and use of these arms, following the provisions of paragraphs 135-146, modified only as existing circumstances demand, in any case, in the discretion of the State authorities.

(b) The following course in pistol range practice is prescribed for officers and men armed with the Colt's revolver cal. .38, or the Smith & Wesson, cal. .38, as issued by the Ordnance Department:

TABLE 8

Range (yards).	Instruction practice					Record practice.[1]			
	Target L.				Target L, rapid fire.		Target L, rapid fire.		Target L, rapid fire.
	Slow fire.		Rapid fire.						
	Time limit.	Scores.	Time limit per score in seconds.	Scores.	Time limit per score in seconds.	Scores.	Time limit per score in seconds.	Scores.	Time limit per score in seconds.	Scores.
15	No limit.	Minimum of 1 at each range.	30	Minimum of 2 at each range	15	Minimum of 2 at each range.	....	...	15	2
25			30		15		30	2	15	2
50			30		....		30	2	....	...

[1] The record course will be fired but once.

A score consists of five consecutive shots.

(c) Rapid fire shall be conducted as prescribed in paragraph 172.

(d) The dismounted course prescribed in paragraphs 167-175 will be substituted for that prescribed above for any organization or individual of the Organized Militia armed with the Colt automatic pistol, cal. .45. The mounted course prescribed in paragraphs 176-199 is authorized but not required.

(f) **Who Will Fire.**

Arm or corps	To fire	Course.
Cavalry .... Field artillery .... Infantry .... Engineers .... Coast artillery	All officers and enlisted men armed with the pistol, except field officers for whom the course is authorized but not required.	Dismounted.
Staff departments except medical and chaplain.	Authorized but not required	Do.

(g) **Qualification and Insignia.**—Qualification and insignia for firing the revolver course shall be as given in the following table:

Table 9

Grade.	Points.	Possible.	Insignia
Expert pistol shot	320	400	Bronze badge.
First classman	300	400	Bronze pin
Second classman	250	400	Do.

The qualifications and insignia for firing the courses prescribed in paragraphs 167-175 or paragraphs 176-199 shall be as provided for in paragraphs 248 and 249.

(h) **Reports.**—The number of officers and enlisted men taking pistol practice will be reported on the report of small-arms firing. (356)

PART II

MISCELLANEOUS SUBJECTS PERTAINING TO COMPANY TRAINING AND INSTRUCTION

CHAPTER I

THE GOVERNMENT AND ADMINISTRATION OF A COMPANY

410. The proper performance of the duty of COMPANY COM-MANDER, like the proper performance of any other duty, requires work and attention to business.

The command of a company divides itself into two kinds of duty: government and administration.

The government includes the instruction, discipline, contentment, and harmony of the organization, involving, as it does, esprit de corps, rewards, privileges, and punishments.

The administration includes the providing of clothing, arms, ammunition, equipage, and subsistence; the keeping of records, including the rendition of reports and returns; and the care and accountability of Government and company property, and the disbursement of the company fund.

System and care are prerequisites of good administration.

The efficient administration of a company greatly facilitates its government.

THE CAPTAIN

411. With regard to his company the captain stands in the same light as a father to a large family of children. It is his duty to provide for their comfort, sustenance, and pleasure; enforce strict rules of obedience, punish the refractory and reward the deserving.

He should be considerate and just to his officers and men and should know every soldier personally and make him feel that he so knows him.

He should by word and act make every man in the company feel that the captain is his protector.

The captain should not be indifferent to the personal welfare of his men, and when solicited, being a man of greater experience, education, and information, he should aid and counsel them in such a way as to show he takes an interest in their joys and sorrows.

When any men are sick he should do everything possible for them until they can be taken care of by the surgeon. He can add much to the comfort and pleasure of men in the hospital by visiting them from time to time and otherwise showing an interest in their condition.

In fact, one of the officer's most important duties is to look after the welfare of his men—to see that they are well fed, well clothed and properly cared for in every other way—to see that they are happy and contented. The officer who does not look after the welfare of his men to the best of his ability, giving the matter his earnest personal attention, neglects one of the principal things that the Government pays him to do.

The soldier usually has a decided feeling for his captain, even though it be one of hatred. With regard to the higher grade of officers, he has respect for them according to regulations; otherwise, for the most

[357]

part, he is indifferent. At the very most, he knows whether his post or regimental commander keeps him long at drill, and particularly whether he has any peculiar habits. The average soldier looks upon his captain as by far the most important personage in the command.

There is no other position in the Army that will give as much satisfaction in return for an honest, capable and conscientious discharge of duty, as that of captain. There is a reward in having done his full duty to his company that no disappointment of distinction, no failure, can deprive him of; his seniors may overlook him in giving credits, unfortunate circumstances may defeat his fondest hopes, and the crown of laurel may never rest upon his brow, but the reward that follows upon the faithful discharge of his duty to his company he can not be deprived of by any disaster, neglect or injustice.

He is a small sovereign, powerful and great, within his little domain.

411a. Devolution of Work and Responsibility. The company commander should not attempt to do all the work—to look after all the details in person—he should not try to command directly every squad and every platoon. The successful company commander is the one who distributes work among his subordinates and organizes the help they are supposed to give him. By War Department orders, Army Regulations and customs of the service, the lieutenants and noncommissioned officers are charged with certain duties and responsibilities. Let every one of them carry the full load of their responsibility. The company commander should not usurp the functions of his subordinates—he should not relieve them of any of their prescribed or logical work and responsibility. On the contrary, he should give them more, and he should see that they "deliver the goods." Skill in distributing work among subordinates is one of the first essentials of leadership, as is the ability to get work out of them so that they will fill their functions to the full within the limits of their responsibility. Not only does devolution of work and responsibility cause subordinates to take more interest in their work (it makes them feel less like mere figure-heads), but it also teaches them initiative and gives them valuable experience in the art of training and handling men. Furthermore, it enables the company commander to devote more time to the larger and more important matters connected with the discipline, welfare, training, instruction and administration of the company.

The captain who allows his lieutenants to do practically nothing makes a mistake—he is doing something that will rob his lieutenants of all initiative, cause them to lose interest in the company, and make them feel like nonentities—like a kind of "fifth wheel"—it will make them feel that they are not, in reality, a part of the company—it will prevent them from getting a practical, working knowledge of the government and administration of a company.

By allowing his lieutenants to participate to the greatest extent possible in the government and administration of the company, and by not hampering and pestering them with unnecessary instructions about details, the captain will get out of his lieutenants the very best that there is in them.

The captain should require RESULTS from his lieutenants, and the mere fact that a lieutenant is considered inefficient and unable to do things properly, is no reason why he should not be required to do them. The captain is by Army Regulations responsible for the efficiency and instruction of his lieutenants regarding all matters pertaining to the company, and he should require them to perform all their duties properly, resorting to such disciplinary measures as may be considered necessary. The lieutenant who can not, or who will not, perform his duties properly is a drag on the company, and such a man has no business in the Army, or in the Organized Militia.

THE LIEUTENANT

412. To be able to perform well the duties of captain when the responsibility falls upon him, should be the constant study and ambition of the lieutenant.

He is the assistant of the captain and should be required by the captain to assist in the performance of all company duties including the keeping of records and the preparation of the necessary reports, returns, estimates and requisitions. The captain should give him lots to do, and should throw him on his own responsibility just as much as possible. He should be required to drill the company, attend the daily inspection of the company quarters, instruct the noncommissioned officers, brief communications, enter letters in the Correspondence Book, make out ration returns, reports, muster and pay rolls, etc., until he shows perfect familiarity therewith.

Whenever told to do a thing by your captain, do it yourself or see personally that it is done. Do not turn it over to some noncommissioned officer and let it go at that. If your captain wants some non commissioned officer to do the thing, he himself will tell him to do it—he will not ask you to do it.

It is customary in the Army to regard the company as the property of the captain. Should the lieutenant, therefore, be in temporary command of the company he should not make any changes, especially in the reduction or promotion of noncommissioned officers without first having consulted the captain's wishes in the matter.

It is somewhat difficult to explain definitely the authority a lieutenant exercises over the men in the company when the captain is present. In general terms, however, it may be stated the lieutenant can not make any changes around the barracks, inflict any punishment or put men on, or relieve them from, any duty without the consent of the captain. It is always better if there be a definite understanding between the captain and his lieutenants as to what he expects of them, how he wishes to have certain things done and to what extent he will sustain them.

If the lieutenant wants anything from the company in the way of working parties, the services of the company artificer or company clerk, the use of ordnance stores or quartermaster articles, he should always speak to the captain about the matter.

THE CAPTAIN AND THE LIEUTENANTS

410. The company officers should set an example to their men in dress, military bearing, system, punctuality and other soldierly qualities. It should be remembered that the negligence of superiors is the cue for juniors to be negligent.

If the men of a company are careless and indifferent about saluting and if they are shabby and lax in their dress, the company commander is to blame for it—company officers can always correct defects of this kind, if they will only try.

The character and efficiency of officers and the manner in which they perform their duties are reflected in the conduct and deportment of their men.

Of course, courage is a prerequisite quality for a good officer, and every officer should seek to impress his men that he would direct them to do nothing involving danger that he would not himself be willing to do under similar circumstances.

If a company officer be ignorant of his duties, his men will soon find it out, and when they do they will have neither respect for, nor confidence in, him.

Company officers should take an active interest in everything that affects the amusement, recreation, happiness and welfare of their men.

An officer just joining a company should learn without delay the names of all the men. A roll of the organization should be gotten and studied.

While an officer can gruffly order a soldier to do a thing and have his order obeyed, it should be remembered that, as a rule, human nature, especially American human nature, responds best to an appeal to pride, fairness, justice, reason, and the other nobler instincts of man. It is only in rare instances that the average man will give the best there is in him under coercion or pressure of authority.

There are but few men who have not some good in them, and this good can generally be gotten at, if one only goes about it in the right way. Study your men and try to arouse in them pride and interest in their work.

The soldier first learns to respect, then to honor and finally to love the officer who is strict but just; firm but kind—and this is the officer who will draw out of his men the very best there is in them.

413a. Treat your men like men, and remember there is nothing that will so completely take the spirit out of a man as to find fault with him when he is doing his best.

Young officers sometimes run to one of two extremes in the treatment of their men—they either, by undue familiarity, or otherwise, cultivate popularity with the men; or they do not treat them with sufficient consideration—the former course will forfeit their esteem; the latter, ensure their dislike, neither of which result is conducive to commanding their respect.

Treat your soldiers with proper consideration, dignity, and justice—remember they are members of your profession, the difference being one of education, rank, command, and pay—but they are men, like yourself, and should be treated as such.

Under no circumstances should you ever swear at a soldier—not only is this taking a mean, unfair advantage of your position, but it is also undignified, ungentlemanly and unmilitary. It is even more improper for you to swear at a soldier than it is for a superior to swear at you—in the latter case the insult can be properly resented; in the former, it must be borne in humiliating silence.

Remember, that if by harsh or unfair treatment you destroy a man's self-respect, you at the same time destroy his usefulness.

Familiarity is, of course, most subversive of discipline, but you can treat your men with sympathetic consideration without being familiar with them.

In dealing with enlisted men, do not use the same standard of intellect and morals that apply in the case of officers. And remember, too, that a thing that may appear small and trivial to an officer may mean a great deal to an enlisted man—study your men, learn their desires, their habits, their way of thinking, and then in your dealings with them try to look at things from their standpoint also. In other words in your treatment of your men be just as human as possible.

The treatment of soldiers should be uniform and just, and under no circumstances should a man be humiliated unnecessarily or abused. Reproof and punishment must be administered with discretion and judgment, and without passion; for the officer who loses his temper and flies into a tantrum has failed to obtain his first triumph in discipline. He who can not control himself can not control others.

Every officer should study himself carefully, he should analyze himself, he should place himself under a microscopic glass, so as to discover his weak points—and he should then try with his whole might and soul to make these weak points strong points. If, for instance, you realize that you are weak in applied minor tactics, or that you have no "bump of locality," or that you have a poor memory, or that you have a weak will, do what you can to correct these defects in your make-up. Remember "Stonewall" Jackson's motto: "A man can do anything he makes up his mind to do."

The Progress Company, Chicago, Ill., publishes "Mind Power," "Memory," "The Will," "The Art of Logical Thinking" (all by W. W. Atkinson), and several other books of a similar nature, that are both interesting and instructive. "The Power of the Will," by Haddock, for sale by Albert Lewis Pelton, Meriden, Conn., is an excellent book of its kind.

THE FIRST SERGEANT

414. It has been said the captain is the proprietor of the company and the first sergeant is the foreman.

Under supervision of the captain, he has immediate charge of all routine matters pertaining to the company.

In some companies in the Regular Army, it is customary for soldiers, except in cases of emergency, to get permission from the first sergeant to speak to the company commander at any time. In other organizations soldiers who wish to speak to the company commander away from the company quarters must first obtain the first sergeant's

415

permission, but it is not necessary to get this permission to speak to the company commander when he is at the barracks.

The first sergeant is sometimes authorized to place noncommissioned officers in arrest in quarters and privates in confinement in the guardhouse, assuming such action to be by order of the captain, to whom he at once reports the facts. However, with regard to the confinement of soldiers by noncommissioned officers, attention is invited to the Army Regulations on the subject.

THE NONCOMMISSIONED OFFICERS

(The status, duties, etc., of noncommissioned officers are covered in greater detail in Noncommissioned Officers' Manual, by the author. General agents: George Banta Publishing Co., Menasha, Wis.)

415. The efficiency and discipline of a company depend to such an extent on the noncommissioned officers that the greatest care and judgment should be exercised in their selection. They should be men possessing such soldierly qualities as a high sense of duty, cheerful obedience to orders, force of character, honesty, sobriety and steadiness, together with an intelligent knowledge of drills, regulations and orders.

They should exact prompt obedience from those to whom they give orders, and should see that all soldiers under them perform their military duties properly. They must not hesitate to reprove them when necessary, but such reproof must not be any more severe than the occasion demands.

The company officers must sustain the noncommissioned officers in the exercise of their authority, except, of course, when such authority is improperly or unjustly exercised. If they do wrong, they should be punished the same as the privates, but if it be simply an error of judgment they should merely be admonished. A noncommissioned officer should never be admonished in the presence of privates.

Judicious praising of noncommissioned officers in the presence of privates is not only gratifying to the noncommissioned officer, but it also tends to enhance the respect and esteem of the privates for him.

In addition to dividing the company into squads, each squad being under a noncommissioned officer as required by the Army Regulations, the company should also be divided into sections, each section being in charge of a sergeant. The squads and sections should, as far as possible, be quartered together in barracks, and the chiefs of squads and the chiefs of sections should be held strictly responsible for the conduct, dress, cleanliness, and the care of arms of the members of their respective squads and sections. Not only does this throw the corporals and the sergeants upon their own responsibility to a certain extent, but it also impresses upon them the importance of their position, and gets the privates in the habit of realizing and appreciating the authority exercised by noncommissioned officers.

When practicable, the noncommissioned officers should have separate rooms or tents, and should mess together at tables separate from the privates; for, everything that conduces to familiarity with inferiors tends to lower the dignity of the noncommissioned officers' position.

Throw your noncommissioned officers upon their own responsibility—throw them into deep water, so to speak, where they will either have to swim or sink. You can never tell what a man can really do until you have given him a chance to show you—until you have put him on his mettle—until you have tried him out. And very often men who seem to have nothing in them, men who have never before been thrown upon their own responsibility, will surprise you.

Do all you can to make your noncommissioned officers realize and appreciate the importance of their position. Consult them about different matters—get their opinions about various things. When going through the barracks at Saturday morning inspection, for instance, as you come to the different squads, have the squad leaders step to the front and follow you while you are inspecting their respective squads. If you find anything wrong with a man's bunk, speak to the squad leader about it. Also ask the squad leaders various questions about their squads.

Not only does such treatment of noncommissioned officers make them appreciate the importance, responsibility and dignity of their position, but it also gives them more confidence in themselves and raises them in the eyes of the privates.

Noncommissioned officers should always be addressed by their titles, by both officers and soldiers.

Noncommissioned officers are forbidden to act as barbers, or as agents for laundries, or in any other position of a similar character. (Cir. 34, '07.)

Everything possible should be done by the company officers to instruct the noncommissioned officers properly in their duties.[1]

So far as the company is concerned the noncommissioned officers are expected to assist the company commander in carrying out his own orders and those of his superiors—they should see that all company orders are obeyed and that the known wishes of the captain are carried out. If, for instance, the captain should tell the first sergeant that the men in the company may play cards among themselves, but that noncommissioned officers are not to play with privates and that men from other companies are not allowed to take part in, or to be present at the games, then it is the duty of the first sergeant to see that these instructions are carried out—it is his duty to make frequent inspections of the tables at which the men may be playing to see that no noncommissioned officers are playing and that no outsiders are present. The first sergeant who confined himself to publishing the order to the company and then doing nothing more, would be neglectful of his proper duty.

Noncommissioned officers clothed in the proper uniform of their grade are on duty at all times and places for the suppression of disorderly conduct on the part of members of the company in public places. Men creating disorder will be sent to their quarters in arrest and the facts reported to the company commander without delay.

[1] Silicate Roll Blackboards, which are perfectly flexible and can be rolled tightly, like a map, without injury, may be obtained from the New York Silicate Book Slate Co, 20 Vesey St, New York They are made in various sizes, but about the most convenient for use in noncommissioned officers' school is No. 3, three by four feet—price $2.

Noncommissioned officers can do much to prevent the commission of offenses by members of their commands, both when on and when off duty, and such prevention is as much their duty as reporting offenses after they are committed; in fact, it is much better to prevent the offense than to bring the offender to trial.

Company commanders should drill their noncommissioned officers thoroughly in the principles of discipline.

Noncommissioned Officers Authorized to Confine Enlisted Men. A company or detachment commander may delegate to his noncommissioned officers the authority to confine enlisted men in the guardhouse and to place them in arrest in quarters, provided the case is immediately reported to the company or detachment commander, who confirms the act of the noncommissioned officer and adopts it as his own.—W. D. decision, December, 1905.

Reduction and Resignation. A noncommissioned officer should never be reduced to ranks, except for grave and sufficient reasons. Nothing demoralizes the noncommissioned officers of a company so much and upsets discipline to such an extent as the feeling that upon the slightest pretext or fancy one is to be sent back to the ranks, to associate with the privates he has been required to discipline.

In some regiments noncommissioned officers are permitted to send in formal resignations, while in other regiments they are not, but, with the approval of the company commander, they may ask for reduction, giving proper, satisfactory and specific reasons. Of course, resignations submitted in a spirit of accepted insubordination or pique should not be considered, nor should they ever be in substitution for deserved disciplinary punishment. If a noncommissioned officer has good reasons for requesting reduction and the granting of the request would not result in detriment to the company, there is no reason why his application should not be favorably considered. However, in such a case, the noncommissioned officer should consult his company commander before submitting his request in writing. It is thought the preponderance of custom is against considering formal resignations.

Contentment and Harmony

416. The officers of the company should do everything possible to make the organization contented and harmonious. Contentment and harmony are not only conducive to good discipline and efficiency, but they also make the government of the company easy and reduce desertions to a minimum.

The showing of favoritism on the part of the captain is always a cause of great dissatisfaction amongst the soldiers in the company. Soldiers do not care how strict the captain is, just so he is fair and impartial, treating all men alike.

The Mess. The captain should give the mess his constant personal attention, making frequent visits to the kitchen and dining room while the soldiers are at meals so as to see for himself what they are getting, how it is served, etc.

It is not saying too much to state that, in time of peace, a good mess is the real basis of the contentment of a company.

Ascertain what the soldiers like to eat and then gratify their appetites as far as practicable.

Be careful that the cook or the mess sergeant doesn't fall into a rut and satiate the soldiers day after day with the same dishes.

Give the ration your personal attention—know yourself what the company is entitled to, how much it is actually getting, what the savings amount to, etc.

Library and Amusement Room. A library and an amusement room, supplied with good books, magazines, papers, a billiard or pool table, and a phonograph, are a source of much pleasure and contentment.

Athletic Apparatus. A judicious investment of the company fund in base balls, bats, dumb bells, Indian clubs, boxing gloves and other athletic goods, and the encouragement of baseball, basketball, quoits, etc., are in the interest of harmony and happiness.

Rewards and Privileges

417. 1. Deny all passes and requests for privileges of men whose conduct is not good, and on the other hand grant to men whose conduct is good, as many indulgences as is consistent with discipline.

2. Judicious praise in the presence of the first sergeant, a few noncommissioned officers, or the entire company, depending upon circumstances, very often accomplishes a great deal. After the according of such praise, let your action toward the man show that his good conduct is appreciated and that it has raised him in your estimation, and make him feel you are keeping your eye on him to see whether he will continue in his well doing.

3. Publication of commendatory orders, desirable special duty details, etc.

4. Promotion, and extra duty details which carry extra pay.

5. Meritorious conduct of importance should be noted in the soldier's military record and also on his discharge.

6. At the weekly company inspection, each chief of squad picks out the neatest and cleanest man in his squad—the captain then inspects the men so selected, the neatest and cleanest one being excused from one or two tours of kitchen police, or some other disagreeable duty; or given a two days' pass.

NOTE: Some officers do not think that good conduct should be especially rewarded, but that if all soldiers be held strictly accountable for their actions by a system of strict discipline, good conduct attains its own reward in the immunities it enjoys.

418. **Trials by Court-Martial.** As stated in the Army Regulations (Par. 953, '13), commanding officers should not bring every dereliction of duty before a court for trial, but in the case of minor offenses[1] the ends of discipline can often be served fully as well, if not even better, by requiring extra fatigue or by withholding privileges. If the soldier demands a trial, he cannot be given extra fatigue, although there is nothing in the Regulations preventing the giving of some other form of company punishment. This right to demand a trial must in every case be made known to the soldier before awarding extra fatigue.

Some Efficacious Forms of Company Punishment

419. 1. Extra fatigue under the Q. M. sergeant or the noncommissioned officer in charge of quarters, cleaning up around and in the company quarters, scrubbing pots, scouring tin pans, polishing stoves, cutting wood, policing the rears, cutting grass, pulling weeds, polishing the brass and nickel parts in the water closets and bath rooms, washing and greasing leather, cleaning guns, boiling greasy haversacks, and, in camp, digging drains and working around slop holes.

If the work be done well the offender may be let off sooner—if the work be not done well, he may be tried for it.

2. Men may not be allowed to leave the immediate vicinity of the barracks for periods ranging from one to ten days, during which time they are subject to all kinds of disagreeable fatigue, and required to report to the N. C. O. in charge of quarters at stated hours.

3. Breaking rocks for a given number of days. For every man so punished, a private of the same company is detailed as a sentinel and for every four men a corporal is detailed in addition—the idea being to cause every man in each organization to take an interest in preventing his own comrades from violating rules and regulations.

4. When two soldiers get into a row that is not of a serious nature, a good plan is to set them at work scrubbing the barrack windows—one on the outside and one on the inside, making them clean the same pane at the same time. They are thus constantly looking in each other's faces and before the second window is cleaned they will probably be laughing at each other and part friends rather than nursing their wrath.

5. Confinement to barracks, reporting to the noncommissioned officer in charge of quarters once every hour, from reveille to, say, 9 P. M.

NOTE: Some company commanders follow, for moral effect, the practice of publishing to their companies all summary court convictions of soldiers belonging to the organization.

Withholding of Privileges

1. Withholding of passes and of credit at the post exchange.

2. Withholding of furloughs.

420. Control of Drunken and Obscene Men. In order to control drunken and obscene men, they have been bucked and gagged until sufficiently sober to regain self-control and quiet down. The use of a cold water hose in such cases has been known to accomplish good results Great care and judgment, however, should be exercised and no more force used than is absolutely necessary.

It may also be said that persistently filthy men have been washed and scrubbed.

---

[1] For example, noisy or disorderly conduct in quarters, failure to salute officers, slovenly dressed at formations, rifle equipments not properly cleaned at inspection or other formations, overstaying pass, short absences without leave and absences from formations (especially for first offense)

421. Saturday morning and other company inspections are intended to show the condition of the organization regarding its equipment, military appearance and general fitness for service, and the condition of the quarters as regards cleanliness, order, etc. Usually everyone except the guard, one cook, and others whose presence elsewhere can not be spared, are required to attend inspections, appearing in their best clothes, their arms and accoutrements being shipshape and spick and span in every respect.

A man appearing at inspection with arms and equipments not in proper shape, especially if he be a recruit or if it be his first offense, may be turned out again several hours later, fully armed and equipped, for another inspection, instead of being tried by summary court.

Property Responsibility

422. Special attention should be given to the care and accountability of all company property.

1. All property (tents, axes, spades, chairs, hatchets, etc.), should be plainly marked with the letter of the company.

2. Keep a duplicate copy of every memorandum receipt given for property, and when such property is turned in or another officer's memorandum receipt is given covering the property, don't fail to get your original memorandum from the quartermaster.

3. See that the quartermaster gives you credit for all articles turned in, or property accounted for on statement of charges, proceedings of a surveying officer or otherwise.

4. Have a settlement with the quartermaster at the end of every quarter as required by Army Regulations, taking an inventory of all property held on memorandum receipt and submitting to the quartermaster a statement of charges and a certified list of the china and glassware unavoidably broken during the quarter.

5. Keep an account of all articles issued to the men, turned in to the quartermaster, condemned, expended, lost, stolen or destroyed.

6. Worn out and unserviceable property should be submitted to the action of an inspector as soon as practicable. If the time of the annual visit of the inspector be not near at hand and such property has accumulated to such an extent as to make the case one of emergency, application may be made to the department commander for the appointment of a special inspector, in which case a copy of the Inventory and Inspection Report, duly accomplished and signed, will be forwarded with the application.

7. Property that is to be submitted to the action of a surveying officer or an inspector should always first be carefully examined by the responsible officer in person, who should be prepared to give all necessary information in regard to it.

The property should be arranged in the order of enumeration in the survey or the inventory report, and should be arranged in rows of five, ten, or some other number, so that the numbers of the various articles can be counted at a glance.

The Army Regulations require that the responsible officer shall be present at the inspection of property by a regular inspector. He should also be present when property is acted on by a surveying officer.

423

Books and Records

123. The following books and records are required by Army Regulations to be kept in every company:

1. **Morning Report,** which shows the exact status of every member of the company. Changes that have occurred since the preceding reports are noted in figures and by name.

2. **Sick Report,** on which are entered the names of all enlisted men requiring medical attention and such officers as are excused from duty because of illness.

3. **Duty Roster.** A form on which is kept a record of all details for service in garrison and in the field, except the authorized special and extra duty details. For instructions regarding rosters, see "Rosters and Detachments," Manual of Interior Guard Duty.

4. **Order File,** consisting of a file of all orders received and issued.

5. **Company Fund Book,** in which are entered all receipts to, and expenditures from the company fund, together with the proceedings of the Company Council of Administration.

6. **The Company Target Records** consist partly of a series of sheets bound by the loose-leaf plan, one for each soldier, on which are entered his record practice and qualifications for each season of his three years' enlistment. Another part of the Company Target Record is given to a record of the collective fire and the figure of merit of the company.

7. **Correspondence Book,** with index, in which is entered with ink or indelible pencil a brief of each item of correspondence in respect to which a record is necessary and a notation of the action taken thereon.

8. **Document File,** containing the original documents or communications when these are retained, and carbon, letter press or other legible copies of all letters, indorsements or telegrams sent with regard to same. The file also contains similar copies of all letters, indorsements or telegrams originating in the company office.

9. **Descriptive List, Military Record and Clothing Account,** on which is kept a full description of every man, including the date of enlistment, personal description, a record of deposits, trials by court-martial, etc., also clothing allowances due soldier and amounts due U. S. for clothing drawn. The clothing account of every soldier is balanced June 30 and December 31 of each year, and when his service with an organization is terminated.

Every year the War Department publishes a general order giving the clothing allowance for the next fiscal year.

10. **A Record of Sizes of Clothing** for every man in the company as ascertained by measurements.

11. **Delinquency Book,** in which are noted disciplinary punishments awarded by the company commander.

12. **Property Book,** in which are entered all ordnance and quartermaster property in the possession of the members of the company.

CHAPTER II

DISCIPLINE

424. Definition. Discipline is not merely preservation of order, faithful performance of duty, and prevention of offenses—in other words, discipline is not merely compliance with a set of rules and regulations drawn up for the purpose of preserving order in an organization. This is only one phase of discipline. In its deeper and more important sense discipline may be defined as the habit of instantaneous and instinctive obedience under any and all circumstances—it is the habit whereby the very muscles of the soldier instinctively obey the word of command, so that under whatever circumstances of danger or death the soldier may hear that word of command, even though his mind be too confused to work, his muscles will obey. It is toward this ultimate object that all rules of discipline tend. In war, the value of this habit of instantaneous and instinctive obedience is invaluable, and during time of peace everything possible should be done to ingrain into the very blood of the soldier this spirit, this habit, of instantaneous, instinctive obedience to the word of command.

Methods of Attaining Good Discipline. Experience shows that drill, routine, military courtesy, attention to details, proper rewards for good conduct, and invariable admonition or punishment of all derelictions of duty, are the best methods of attaining good discipline—that they are the most effective means to that end.

Importance. History shows that the chief factor of success in war is discipline, and that without discipline no body of troops can hold their own against a well-directed, well-disciplined force.

Sound System. We must bear in mind that what may be considered a sound system of discipline at one epoch or for one nation, may be inapplicable at another epoch or for another nation. In other words, sound discipline depends upon the existing state of civilization and education, the political institutions of the country, the national trait and the national military system. For example, the system of discipline that existed in the days of Frederick the Great, and which, in modified form, exists today in certain European armies, whereby the soldier was so inured to a habit of subjection that he became a sort of machine—a kind of automaton. Such a system of discipline, while answering admirably well its purpose at that time and for those nations, would not do at all in this day and generation, and with a people like ours, in whom the spirit of personal freedom and individual initiative are born. Of course, the discipline that will insure obedience under any and all conditions—the discipline that will insure prompt and unhesitating obedience to march, to attack, to charge—is just as important today as it was a thousand years ago, but we can not attain it by the machine-making methods of former times. The system we use must be in keeping with the national characteristics of our people and the tactical necessities of the day, the latter requiring individual initiative. According to the old system, the company commander imposed his will upon a body of submissive units; under the new system the company commander, backed by authority and greater knowledge, leads obedient, willing units, exacting ready obedience and loyal co-operation. The company commander used to drive; now he leads.

What are the means of attaining and maintaining such discipline?

1. Explain to the men the importance of discipline and its value on the field of battle, and give the reasons that makes it necessary to subject soldiers to restrictions that they were not subjected to in civil life.

2. Do not impose unnecessary restrictions or hardships on your men, nor issue orders that have no bearing on their efficiency, health, cleanliness, orderliness, etc.

3. Demand a high standard of excellence in the performance of all duties whatsoever, and exact the utmost display of energy.

A system of discipline based on the above principles develops habits of self-control, self-reliance, neatness, order, and punctuality, and creates respect for authority and confidence in superiors.

Punishment. In maintaining discipline, it must be remembered the object of punishment should be two-fold: (a) To prevent the commission of offenses, and (b) to reform the offender. Punishment should, therefore, in degree and character depend upon the nature of the offense. Punishment should not be debasing or illegal, and the penalty should be proportionate to the nature of the offense. If too great, it tends to arouse sympathy, and foster friends for the offender, thus encouraging a repetition of the offense. A distinction, therefore, should be made between the deliberate disregard of orders and regulations, and offenses which are the result of ignorance or thoughtlessness. In the latter case the punishment should be for the purpose of instruction and should not go to the extent of inflicting unnecessary humiliation and discouragement upon the offender.

General Principles

In the administration of discipline the following principles should be observed:

1. Every one, officers and soldiers, should be required and made to perform their full duty. If the post commander, for instance, requires the company commanders to do their full duty, they will require their noncommissioned officers to do their full duty, and the noncommissioned officers will in turn require the men to do the same.

2. Subordinates should be held strictly responsible for the proper government and administration of their respective commands, and all changes or corrections should be made through them.

3. Subordinates should have exclusive control of their respective commands, and all orders, instructions and directions affecting their commands should be given through them.

4. If, in case of emergency, it be not practicable to make certain changes or corrections, or to give certain orders, instructions or directions, through the subordinates, they should be notified at once of what has been done.

5. After a subordinate has been placed in charge of a certain duty, all instructions pertaining thereto should be given through him,

and all meddling and interfering should be avoided. Interference by superiors relieves the subordinate of responsibility, and causes him to lose interest. become indifferent, and do no more than he is obliged to do.

6. The certainty of reward for, and appreciation of, meritorious conduct, should equal the certainty of punishment for dereliction of duty.

7. It is the duty of an officer or noncommissioned officer who gives an order to see that it is obeyed; carrying out orders received by him does not end with their perfunctory transmission to subordinates— this is only a small part of his duty. He must personally see that the orders so transmitted are made effective.

8. The treatment of soldiers should be uniform and just, and under no circumstances should a man be humiliated unnecessarily or abused. Reproof and punishment must be administered with discretion and judgment, and without passion; for an officer or noncommissioned officer who loses his temper and flies into a tantrum has failed to obtain his first triumph in discipline. He who can not control himself can not control others.

9. Punishment should invariably follow dereliction of duty, for the frequency of offenses depends, as a general rule, on the degree of certainty with which their commission is attended with punishment. When men know that their derelictions and neglects will be observed and reproved, they will be much more careful than they would be otherwise—that's human nature.

A strict adherence to the above general principles will instill into the minds of those concerned, respect for authority and a spirit of obedience.

CHAPTER III

GENERAL PRINCIPLES OF COMPANY TRAINING AND INSTRUCTION*

425. Object of Training and Instruction. The object of training and instructing a company is to thoroughly knit together its different parts, its various elements (individuals, squads and platoons), into a complete, homogeneous mass, a cohesive unit, that will under any and all conditions and circumstances respond to the will of the captain—a cohesive unit that knows how to march, that knows how to live properly in camp, that knows how to fight and that can be readily handled tactically on the field of battle. In short, the object of training and instruction is to make out of the company an efficient, wieldy fighting weapon, to be manipulated by the captain. There is but one way this object can be obtained. and that is by work, work, work—and then more work—by constant care, attention and pains—by co-operation, by team work, among the officers, the noncommissioned officers and the privates.

426. Method and Progression. Arrangement is an essential of sound teaching. Training and instruction in order to be easily understood and readily assimilated—in order to give the greatest results in the shortest time—must be carried on according to a methodical and progressive plan. Each subject or subjects upon a knowledge of which depend the proper understanding and mastering of another, should be studied and mastered before taking up the other subject, and the elementary and simpler aspects of a given subject must be mastered before taking up the higher and more difficult phases of the subject, which means that individual training and instruction must precede, and provide a sound foundation for, collective training and instruction—that is to say, for the higher tactical training and instruction of the company as a unit. These basic, fundamental principles of successful training and instruction apply to practical as well as theoretical training. For instance, in the subject of entrenchments we would first instruct the men individually in the use of the tools and in the construction and use of trenches, after which we would pass on to the tactical use of entrenchments by the company. Also, in training and instructing the company in fire discipline, we would first explain to the men the power and tactical value of the rifle, and instruct them in their duties on the firing line as regards adjustment of sights, attention to commands, economy of ammunition, etc.; we would explain to the platoon commanders and guides their duties as regards control of fire, enforcement of fire discipline, etc., after which we would practice the company as a unit in fire action, and fire control, ending up with an exercise showing the tactical application of the rules and principles explained. And again, in the training and instruction of the company in the attack, we would first train and

---

* This chapter is based on "Company Training", by General Haking, British Army, which is the best book the author has ever seen on the subject of company training "Field Training of a Company of Infantry", by Major Craufurd, British Army, an excellent little book, was also consulted.

instruct the company in all the formations and operations that naturally precede an attack (patrolling, outposts, advance guard, rear guard), and also in those that form an inherent part of an attack (extended order, field firing, use of cover, etc.).

427. Program. The training and instruction of a company, whether practical or theoretical, should be carried on in accordance with a fixed, definite program, in which the subjects are arranged in a natural, progressive order.

428. Simultaneous Instruction and Training. The next question that presents itself is: Should instruction and training in each branch be completed before proceeding to the next, or should instruction and training be carried on simultaneously in two or more different subjects, as one, for example, are taught mathematics, French and history at the same time, a different hour of the day being devoted to each subject? In other words, should we, for instance, devote one hour of the day to attack, one hour to defense, and one hour to the service of security, thus preventing the soldier from getting weary of doing the same thing that whole day? Our answer is:

1st. If the instruction and training is being given on the ground where the application of the principles of any given subject is varied so much by the type of the ground and the nature of the situation, each type of ground affording a different solution of the problem, it is thought the best results can be obtained by finishing each subject before proceeding to the next, thus not losing the "atmosphere" of one subject by switching to the next, and also confusing the minds of the men with different principles.

2nd. However, if the instruction and training be theoretical and the time available each day be several hours, better results can be obtained by studying two or more subjects simultaneously. This would also be the case if the work be practical, but if it be such that the type of the ground and the nature of the situation will not of themselves afford variety in the application of the same principles.

429. Responsibility. The Army Regulations and War Department orders hold the company commander responsible for the training and instruction of the company. The subject is a most important one and should receive serious thought and study. Before admonishing one of your men for not knowing a subject, always ask yourself, "Have I made an effort to teach it to him?"

430. Interest. Special effort should be made to make the training and instruction of the company interesting, so that the work will not become monotonous and irksome, and thus cause the men to lose interest and get stale. To accomplish this, these points should be borne in mind:

Variety. Inject variety into the work. Do not keep the men too long at one thing.

Clearness. Every exercise, lesson or lecture should have in view a well-defined object, the meaning and importance of which must be explained to, and understood by, the men at the beginning of the exercise, lesson or lecture. In other words, at the beginning, explain the main, governing idea of the subject, and then take pains to explain in a simple, conversational way each phase as you come to it. Give the reasons for everything. You can not expect men to take an interest in things the meaning of which they do not understand and the reason for which they do not see. Make sure by asking questions of different ones as you go along that your explanations are understood.

Thoroughness. Every lecture, talk, drill or exercise should be carefully planned and arranged before hand. Remember, that the men who are going to listen to your talk—the men who are going to go through the exercise—have the right to expect this of you, and you have no right to compel them to listen to lots of disconnected, half-baked statements, or make them go through a disjointed exercise or drill. In the case of tactical exercises always, if practicable, visit and examine the terrain beforehand. Of course, all this will mean work—additional work —but remember the government pays you to work.

Reality. Make all practical work as real as possible—do not permit the commission of absurdities—do not let men do things which manifestly they would not be able to do in actual practice—and you yourself be sure to make your exercises and tactical scheme as like real conditions of warfare as possible.

431. Individual Initiative. The effective range and great power of modern firearms cause troops in battle to be spread out over large areas, thus decentralizing control over men and operations, and consequently increasing the value and importance of individual initiative. The company commander should, therefore, practice, accustom and encourage the privates, noncommissioned officers and lieutenants in the development and exercise of individual initiative and responsibility. This should be borne in mind in all training and instruction.

Officers, noncommissioned officers and privates must not "lay down" just because they have no specific orders. Remember, the one thing above all others that counts in war, is **action, initiative.** Indeed, 'tis better to have acted and lost than never to have acted at all. Listen to what the Chief of Staff of the Army has to say about this in the preface to the Field Service Regulations: "Officers and men of all ranks and grades are given a certain independence in the execution of the tasks to which they are assigned and are expected to show initiative in meeting the different situations as they arise. Every individual, from the highest commander to the lowest private, must always remember that inaction and neglect of opportunities will warrant more severe censure than an error in the choice of means."

432. Determination and Individual Intelligence. While the value of discipline can hardly be overestimated, there are two other factors in battle that are fully as important, if not more so, and they are, **determination** to win, and **individual intelligence,** which, in war, as in all other human undertakings, almost invariably spell success. Therefore, make these two factors one of the basic principles of the instruction and training of the company, and do all you can to instill into your men a spirit of determination, and to develop in them individual intelligence. Every human being has in his soul a certain amount of determination, even though it be only enough to determine upon the small things of life. Some people are born with more determination than others, but it is a mistake to suppose that a man must remain through life with the same amount of determination that he brought into it. The attributes of the human mind, such as determination, bravery, ambition, energy, etc., are all capable of improvement and also of deterioration. It is essential, therefore, for us to endeavor by all means in our power to improve our strength of character—our determination. It is, of course, useless for us to learn the art of war if we have not sufficient determination, when we meet the enemy, to apply the principles we have studied. There is no reason, however, why every officer, noncommissioned officer and private should not improve his determination of character by careful training in peace. It can only be done by facing the difficulties, thoroughly understanding the dangers, and asking ourselves repeatedly whether we are prepared to face the ordeal in war. Let us not think, in a vague sort of a way, that in war we shall be all right and do as well as most people. We know that we are not gifted with tremendous personal courage, and we know that, whatever happens, we shall not run away. But that is not enough. We must train ourselves to understand that in the hour of trial we can harden our hearts, that we can assume the initiative, and retain it by constant advance and constant attack; unless we can fill our hearts with the determination to win, we can not hope to do our full duty on the field of battle and acquit ourselves with credit.

433. The Human Element. No system of training and instruction that does not take into account human nature, can be thoroughly effective. The human element probably enters into war more than it does into any other pursuit. The old idea of turning a human being into a machine, by means of discipline, and making him dread his captain more than the enemy, died long ago, especially with the American people. In modern war success depends to a great extent upon the initiative, the individual action of the soldier, and this action is greatly influenced by the soldier's state of mind at the moment, by the power that can be exercised over his mind by his comrades and those leading him. The company commander should, therefore, study the characteristics of the human mind with the object of ascertaining how he can influence the men under his command, so that in battle those human attributes which are favorable to success, may be strengthened and those which are favorable to defeat may be weakened. Of the former, courage, determination, initiative, respect, cheerfulness, comradeship, emulation and esprit de corps, are the principal ones; of the latter, fear, surprise, disrespect, and dejection, are the leading ones. By means of good, sound discipline, we can create, improve and foster the qualities mentioned that are favorable to success, and we can eliminate to a considerable extent, if not entirely, those that are detrimental to success.

434. Fear. The emotion of fear acts more powerfully upon the feelings of the individual soldier than any other emotion, and it is also probably the most infectious. Fear in a mild form is present in every human being. Nature wisely put it there, and society could not very well get along without it. For example, we stop and look up and down a crowded street before starting to cross, for fear of being run over; in going out in the cold we put on our overcoats, for fear of catching cold.

In fact, we hardly do anything in life without taking a precaution of some kind. These are all examples of reasonable fear, which, within bounds, is a perfectly legitimate attribute of a soldier in common with other human beings. For example, we teach the men to take advantage of cover when attacking, and we dig trenches when on the defense, in both cases for fear of being shot by the enemy. It is the unreasoning type of fear that plays havoc in war, and the most deadly and common form of it is a vague, indefinite, nameless dread of the enemy. If the average man was to analyze his feelings in war and was to ask himself if he were actually afraid of being killed, he would probably find that he was not. The ordinary soldier is prepared to take his chance, with a comfortable feeling inside him, that, although no doubt a number of people will be killed and wounded, he will escape. If, then, a man is not unreasonably afraid of being killed or wounded, is it not possible by proper training and instruction to overcome this vague fear of the enemy? Experience shows that it is. If a soldier is suffering from this vague fear of the enemy, it will at least be a consolation to him to know that a great many other soldiers, including those belonging to the enemy, are suffering in a similar manner, and that they are simply experiencing one of the ordinary characteristics of the human mind. If the soldier in battle will only realize that the enemy is just as much afraid of him as he is of the enemy, reason is likely to assert itself and to a great extent overcome the unpleasant feelings inside him. General Grant, in his Memoirs, relates a story to the effect that in one of his early campaigns he was seized with an unreasonable fear of his enemy, and was very much worried as to what the enemy was doing, when, all at once, it dawned upon him that his enemy was probably worrying equally as much about what he, Grant, was doing, and was probably as afraid as he was, if not even more so, and the realization of this promptly dispelled all of his, Grant's, fear. Confidence in one's ability to fight well will also do much to neutralize fear, and if a soldier knows that he can shoot better, march better, and attack better, than his opponent, the confidence of success that he will, as a result, feel will do much to dispel physical fear. By sound and careful training and instruction make your men efficient and this efficiency will give them confidence in themselves, confidence in their rifles, confidence in their bayonets, confidence in their comrades and confidence in their officers.

The physical methods of overcoming fear in battle are simply to direct the men's minds to other thoughts by giving them something for their bodies and limbs to do. It is a well-known saying that a man in battle frequently regains his lost courage by repeatedly firing off his rifle, which simply means that his thoughts are diverted by physical movements. This is no doubt one of the reasons why the attack is so much more successful in war than the defense, because in the attack the men are generally moving forward and having their minds diverted by physical motion from this vague dread of the enemy.

Courage. Courage, like all other human characteristics, is very infectious, and a brave leader who has no fear of the enemy will always get more out of his men than one who is not so well equipped in that respect. However, it is a well-known fact that a man may be brave far above his fellows in one calling or occupation, and extremely nervous in another. For example, a man may have greatly distinguished himself in the capture of a fort, who would not get on a horse for fear of being kicked off. Courage of this kind is induced chiefly by habit or experience—the man knows the dangers and how to overcome them, he has been through similar experiences before and he has come out of them with a whole skin. This type of courage can be developed by careful training during peace, and it can be increased by self-confidence—by so training the soldier that he knows and feels he will know what to do in any emergency which may arise, and how to do it; he will not be surprised by the unexpected event, which invariably occurs, and he will understand others besides himself are being troubled by unpleasant feelings, which it is his duty as a man and a soldier to overcome.

435. Surprise. Surprise may be said to be the mother of a panic, which is the worst form of fear. In such a case unreasoning fear sometimes turns into temporary insanity. Panic is most infectious, but, on the other hand, a panic can often be averted or stayed by the courageous action of one or more individuals, who can thus impose their will on the mass and bring the people to a reasonable state of mind. **Teach every man in the company that when surprised the only hope of success is to obey at once and implicitly the orders of his immediate commander.**

Surprises in war are not limited to the ordinary acceptance of the term, such as a sudden attack from an unexpected direction. The soldier who goes into battle, for instance, and hears the whiz of a bullet, or sees a shell burst in front of him, is surprised if he has not been taught in peace that these things have to be faced, and that for one bullet that hurts anyone thousands have to be fired. Similarly, a man sees a comrade knocked over: the horrors of war are immediately brought to him, and his courage begins to ebb—he has been surprised, because he has not realized in peace that men are bound to be killed in war. The whole atmosphere of the battle-field is a surprise to the average soldier with no previous experience—the enemy is everywhere, behind every bush, and lurking in every bit of cover, the air is full of bullets, and any advance towards the formidable-looking position held by the enemy is suicidal. However, if the soldier is properly trained and instructed in peace, he will not be greatly surprised at his novel surroundings; he will know that the enemy is not everywhere, and that one bullet sounds much more dangerous than it really is. A bullet sounds quite close when it is fifty yards away, and there is a popular saying that a man's weight in lead is fired for every man that is killed in war.

436. Respect. It is a mistake to imagine that all that is required from a soldier is respect to his officers and noncommissioned officers. Self-respect is fully as important. A soldier is a human being; if he possesses self-respect he will respect all that is good in his comrades, and they will respect all that is good in him. A man who respects himself knows how to respect other people. These are the men that form the backbone of the company, and are the best material on which

[377]

to work in order to raise the general standard of courage in battle. From a purely military point of view, it is absolutely necessary for an officer, noncommissioned officer, or private to possess some marked military qualifications in order to gain respect from others.

This respect engenders confidence in others. Self-respect in the individual can be encouraged, not by fulsome praise, but by a quiet appreciation of the good military qualities displayed by him, and by making use of those qualities whenever an opportunity occurs. For example, if a soldier is seen to do a good piece of scouting or patrolling, the first opportunity should be taken to give him a similar task, if possible in a more responsible position or on a more important occasion. Knowledge is a powerful factor in creating respect, and is probably second only to determination of character. It is essential, therefore, that all officers and noncommissioned officers should have a thorough knowledge of their duties—that they should be "on to their jobs."

437. Cheerfulness. Cheerfulness is a valuable military asset in war, and like all other characteristics of the human being, is very infectious, and in times of depression, such as during a long siege, or after the failure of an attack, it does more than anything else to restore the fighting power of the men.

438. Contentment. Contentment amongst troops in war is dependent upon these main factors: good leading, good food, and sufficient shelter and sleep. Of these, good leading is by far the most important, because it has been proved time and again that badly fed and badly quartered troops, who have suffered great hardships, will still be content and will fight in the most gallant and vigorous manner, provided they are well led. Although good leading emanates in the first instance from the highest military authorities, a great deal depends upon the company officers and noncommissioned officers. A good leader as a rule is careful of the comforts of his men; he obtains the best food and best shelter available, he does not wear out the men by unnecessary movements or unnecessary work, either in the field or in camp, and consequently when he does order them to do anything they know at once that it is necessary and they do it cheerfully.

439. Comradeship. Comradeship is a very valuable military characteristic. What a world of meaning there is in the words, "Me and my bunkie." A soldier may have many acquaintances and a number of friends, but he has but one "bunkie." In times of great danger two men who are "bunkies" will not shirk so easily as two independent men. The best in one man comes out to the surface and dominates any bad military points in the other. They can help each other in countless ways in war, and if one is unfortunately killed or wounded, the other will probably do his best to get even with the enemy at the earliest possible opportunity. This spirit may not be very Christian-like, but it is very human and practical, and helps to win battles, and to win battles is the only reason why soldiers go to war.

ART OF INSTRUCTION ON THE GROUND

440. **Advantages.** Whenever practicable, training and instruction should, in whole or in part, be imparted on the ground, as this gives the instruction a practical aspect that is most valuable, and enables the soldier to grasp and apply principles that he would not otherwise understand. Knowledge that a man can not apply has no value.

Different Methods. Instruction on the ground may be given according to one of these three methods:

1st Method. By means of a talk or lecture prepare the minds of the men for the reception and retention of the subject to be explained later on the ground. In other words, first explain the principles of the subject and then put a "clincher" on the information thus imparted by taking the men to some suitable ground, assuming certain situations and then by quizzing different men see how they would apply the principles just explained in the talk or lecture. For example, after a lecture on the selection of fire-positions take the men to some suitable near-by place and explain to them that the company is attacking toward that house and is being fired upon from that direction. Then continue:

Captain: Remember what I told you about the selection of good fire-positions during the advance. We want to use our rifles with effect, so we must be able to see the position of the enemy. On the other hand, we want to avoid being hit ourselves, if possible; so, we would like to get as much cover as possible. Now, Smith, do you think where we are at present standing is a good place for a fire-position?

Smith: No, sir.

Captain: Why not?

Smith: We can see the enemy from here, but he can see us better than we can see him, and can hit us easier than we can hit him.

Captain: Jones, can you choose a better place, either to the front or rear of where we are now standing?

Jones: I would choose a position along that row of bushes, about fifty yards to the front.

Captain: Why?

Jones: Because, etc., etc.

Twenty minutes' instruction in this manner, after a lecture, will firmly fix in the brains of the men the principles explained in the lecture.

It is a good plan to repeat the salient points of the lecture in the questions, as was done in the first question asked above, or to do so in some other way.

If a man can not give an answer, or choose a suitable place, explain the requirements again and help him to use his common sense.

2d Method. By practising the men on the ground in the subject about which the talk or lecture was delivered.

441. *3d Method.* This may be called the ocular demonstration method, which consists in having a part of the company go through the exercise or drill, while the rest of the company observes what is being done. This method is illustrated by the following example:

Attack. The company commander has just delivered a talk to the company on the second stage of the attack (See Par. 647), and has marched the company to a piece of ground suitable for practising this particular operation, and which the company commander has himself visited beforehand. (The ground should always be visited beforehand by the company commander, who should be thoroughly familiar with it. If possible, ground suitable for practising the operation in question should always be selected.) The operation should begin about 1200 yards from the enemy's position. After pointing out the enemy's position to the company, the particular part of his line it is intended to assault and the direction the company is to advance, the company commander would then proceed something like this: ''We are part of a battalion taking part in a battle, and there are companies to our right and left, with a support and reserve in our rear. So far we have been advancing over ground that is exposed to hostile artillery fire (or not exposed to hostile artillery fire, according to the actual country). We have just come under the enemy's infantry fire also, and consequently we must change our method of advancing. Our immediate object is to get forward, without expending more ammunition than is absolutely necessary, to a position close enough to the enemy to enable us to use our rifles with such deadly effect that we will be able to gain a superiority of fire. Now, is this place sufficiently close for the purpose? No, it is not—it's entirely too far away. Is that next ridge just in front of us close enough? No, it is not; it is at least 1,000 yards from the enemy's position. As a rule, we must get from eight to six hundred yards from the enemy's position before the real struggle for superiority of fire begins.

The following are the main points to which attention must be paid during this part of the advance:

1. We must halt in good fire-position from which we can see and fire at the enemy, and from which we can not be seen very clearly.

2. We must advance very rapidly over any open ground that is exposed to the enemy's artillery or rifle fire.

3. We must find halting places, if possible under cover, or under the best cover available, so as to avoid making our forward rushes so long that the men will get worn out, and begin to straggle long before they get close enough to the enemy to use their rifles with deadly effect.

4. Whenever possible, company scouts should be sent on ahead to select fire-positions.

Of course the above points will have been explained already in the lecture, but this short summary is given in order to focus the minds of the men upon the action that must be taken by the privates, and squad leaders and the platoon commanders.

We now take one platoon and the remainder of the company looks on. The platoon commander is reminded that he is under artillery and infantry fire, and is then directed to advance, in proper formation, to the first fire-position available.

We will suppose there is a gentle slope up to the next ridge or undulation of the ground, and that there are no obstructions to the view except those afforded by the ground itself. The platoon now advances, the captain remaining with the rest of the company, pointing out mistakes as well as good points, and asking the men questions, such as:

Captain: Corporal Smith, should the whole platoon have gone forward together, or would it have been better to advance by squads?

Corporal Smith: I think it should have advanced by squads.

Captain: No; it was all right to advance as they did. At this distance the enemy's infantry fire would not be very deadly, the platoon is well extended as skirmishers, it would take considerably longer to go forward to the next position by successive squads and we want to advance at this stage as rapidly as possible; for, the longer we took, the longer would the men be exposed to fire, and consequently the greater would be the number of casualties.

Captain: Sergeant Jones, why did the platoon advance at a run when moving down the slope, and begin to walk just before reaching the foot of the slope?

Sergeant Jones: Because the slope is exposed and it was necessary to get over it as quickly as possible. They began to walk just before reaching the foot of the slope, because they struck dead ground and were covered from the enemy's fire by the ridge in front.

Captain: Corporal Adams, shouldn't the platoon have halted when it reached cover, so as to give the men a rest?

Corporal Adams: No, sir; the men had not run very far and walking gave them sufficient rest. It would have been an unnecessary loss of time to halt.

Captain: Harris, why did that man run on ahead as soon as the platoon halted?

Pvt. Harris: So he could creep up the crest of the ridge and lie down in exactly the spot that is the best fire-position—that is, where he can just see to fire over the crest and where the enemy can not see him.

Captain: Yes, that's right. All the men in the platoon might not stop at the best fire-position and in the hurry and excitement of the moment the platoon commander might also fail to do so, but if a man goes forward and lies down, the whole platoon knows that they must not go beyond him. Individual men who, owing to slight undulations of ground, may not be able to fire when they halt in line with this man, can creep up until they can see. Others who, for the same reason as regards the ground, find that if they get up on a line with the man they will be unduly exposed, will halt before that time.

Captain: Sergeant Roberts, is it necessary for another platoon to provide covering fire during the advance of the platoon?

Sergeant Roberts: No, sir. At this range the enemy's infantry fire would not be very effective, and it is important to husband our ammunition for the later stages of the attack.

Having asked any other questions suggested by the situation or the ground, the captain will then take the rest of the company forward over the ground covered by the platoon, halting at the place where the platoon changed its pace from a rush to a walk, so that the men can see for themselves that cover from fire has been reached. He will then move

441 (contd.)

the rest of the company forward and tell them to halt and lie down in what each man considers to be the best fire-position, not necessarily adopting the same position as that chosen by the leading platoon. The platoon commanders will then go along their platoons and point out any mistakes.

The leading platoon will now join the company and another platoon will be deployed in the fire position, the platoon commander being directed to advance to the next fire-position.

As we are now about 1,000 yards from the enemy's position the question will again arise as to whether covering fire is necessary.

If the enemy's rifle fire were heavy and accurate it might be necessary, but it should be avoided if possible, on account of the expenditure of ammunition.

We will suppose that the ground falls gently towards the enemy and is very exposed to view for about 300 yards, and half this distance away there is a low bank running parallel to the front of the attack and with a small clump of three or four trees on the bank directly in front of the platoon. Four hundred yards away is the bottom of the valley covered with bushes and shrub. On the far side the ground rises with small undulations and low foot hills to the high ground occupied by the enemy.

There appears to be no marked fire-position which will afford any cover except the bank 150 yards away. The second platoon advances in the same manner as did the first and the captain with the commanders of the remaining platoons will continue to ask questions and point out what has been done right or wrong by the leading platoon. The first question which will arise is whether the platoon can reach the fire position offered by the bank in one rush, and secondly whether the bank is a good fire-position. A former question will again crop up as to whether the whole platoon should go forward at once or whether the advance should be made by squads.

A hundred and fifty yards is a long way to advance without a halt, and if a halt is made on such exposed ground fire must be opened. Probably three advances, each of about fifty yards, would be made, covering fire being provided by the other platoons, which will be occupying the fire-position which the leading platoon has just left. This covering fire would not endanger the leading platoon as it would be delivered from just behind the crest and the leading platoon would be over the crest and out of sight and therefore out of fire from the platoon in rear.

The selection of a fire-position during this advance would depend upon very minute folds of the ground, or very low bushes, grass, etc., which might give a certain amount of cover from view, and therefore make it difficult for the enemy to aim or range accurately. We will suppose that the leading platoon has halted to fire about fifty yards in front, the remaining platoons, in turn, should then be taken forward, examining the ground very carefully as they go, and each platoon commander asked to halt his platoon in what he considers to be the best place.

The possibility of using a scout to select a fire-position would be considered, and a fire-position selected by one platoon would be compared with that selected by another.

The third platoon would then lead during the advance to the next fire-position, and so on with the fourth platoon, if necessary, until the bank was reached. The bank will afford a good deal of material for discussion. Is it a good fire-position or is it not, should it be occupied as such or should it be avoided altogether?

If we ask an artillery officer his opinion about the matter, he will tell us that by means of the clump of trees the defenders' artillery will be able to range with absolute accuracy on that bank. The direction of the bank is parallel to their front, and therefore they can fire at any part of it for some distance right and left of the clump without materially altering their range, and if any infantry occupy the bank they can bring a very deadly fire to bear against them.

There appears to be no doubt, from an artillery point of view, that our platoon should avoid occupying it and get out of its neighborhood as rapidly as possible.

There is another drawback as regards the bank: it is some 850 yards from the enemy's position and may be expected to be under an effective rifle fire. It is no doubt a good mark for the enemy, and, now we come to the crux of the whole matter: his artillery and infantry fire might not do us much damage so long as we remain behind the bank, but they might make it very unpleasant for us directly we try to leave this cover and advance further.

Before finally deciding what to do we must consider human nature, which is entirely in favor of halting behind the bank, and if allowed to remain there long, will be opposed to leaving it. We cannot hope to gain superiority of fire over the enemy at a range of 850 yards, so that a long halt at the bank is out of the question. But it appears to be an extraordinary thing, when we are searching everywhere for cover, that we should be doubtful about occupying such good cover when we find it.

If we decide not to occupy it, the logical conclusion is that, when preparing a position for defense, we should construct a good fire position for the attack some 850 yards away, which is the last thing we should think of doing.

There is no doubt about it, that with badly-trained troops such a fire-position would be liable to become a snare, and that if they once occupied it, there would be great difficulty in getting them forward again, and probably the attack would be brought to a standstill at a critical time.

The answer appears to be found in the simple solution of good training. We must teach our men that when they get into such positions they must use the cover afforded, but for no longer than any other fire-position, and that they must get into the habit in peace of looking upon such localities with suspicion, and with the knowledge that they are not suitable for lengthy occupation in war, if the battle is to be won.

We now come to a still more difficult question of training, namely, how far can the company get forward from the bank without being compelled to stop in order to gain superiority of fire over the defence? In war we want to get as close as possible; the moral effect on the defence is greater, our fire is more effective, and we are likely to gain our object more rapidly. In peace there is no fire to stop us, and we move forward to ridiculous positions which we could not possibly reach in war without first gaining superiority of fire. The result of this is that we try to do the same thing when first we go to war, and we are stopped, probably much further back than we should have been if we had studied the question in peace.

Even on the most open ground we must get to within 600 yards of the enemy, and if the ground affords any cover in front, the exposed space must be rushed and the more forward position gained. Having pointed out this difficulty to the company during the previous lecture, and reminded them of it on the ground, we can now extend the whole company and move forward from the bank, using covering fire and letting each platoon commander decide how far he can get to the front after a series of rushes, the company acting as a whole.

The captain can then go down the line and discuss with each platoon the position it has reached. Whilst he is doing this, the remaining platoons can be trained in fire direction and control, which should be carefully watched and criticized by the platoon commanders. One platoon, owing to the nature of the ground in front of it, can get forward further than other platoons, and this should be brought home to each platoon, so as to avoid the possibility of playing the game of follow your leader, and one platoon halting merely because another has halted.

If there is still time available, and the ground is suitable, the company can be moved to a flank to choose a similar fire-position where the ground is more favorable to an advance, and where the company could get within 300 yards of the enemy, or even less, before it would be absolutely necessary to stop in order to gain superiority of fire.

If there is still time available, and the ground is suitable, the whole operation can be carried out in the opposite direction or in some other direction, and the platoons can thus be trained to appreciate that fire-positions which are good in one place are bad in another. A more thorough investigation of fire-positions which will suggest subjects for discussion on the ground will be found beginning with paragraph 649, dealing with this stage of the attack.

Defense. Demonstrations in defense can be carried out in a similar manner, the captain explaining to the company the general line of defense to be taken up, the portion allotted to the company, and the probable direction of the enemy's attack.

The co-operation of the artillery and infantry will have been pointed out in the previous lecture: how some part of the enemy's advance will be dealt with by artillery alone, some part by both artillery and infantry, and some part by infantry alone.

This can now be pointed out to the men on the ground. Having considered the assistance provided by the artillery, the next point to decide upon is the exact position of the fire trench. The best way to proceed is to allot a certain portion of the front occupied by the company to each platoon and to let the platoon commanders take charge of the operations. The platoon commander can direct one of his squads to select a position for the trench, and that squad can lie down there. The remaining squads will then select a position in turn. If two squads select the same they can lie down together. The platoon commander will then fall in his platoon, and make them lie down in the most retired position chosen; he will ask the squad leader why the squad chose that locality in preference to any other, why they did not go ten yards further forward or ten yards further back; and he will explain to the whole platoon the advantages and disadvantages of selecting this locality. He will then move the whole platoon forward to the next position chosen by another squad and deal with that locality. Finally, he will select the position he thinks the best, giving his reasons why he has decided upon it, and place the whole platoon on it. When all the platoons have decided upon their line of defense, the captain will move the whole company in turn from the ground occupied by one platoon to that occupied by another, asking the platoon commander in each case to explain why the position was chosen in preference to any other.

He will give his decision as regards each platoon, and he will finally arrange for the position to be occupied by the whole company. One platoon, for some good reason, may have chosen a place which it would not be safe to occupy, owing to the fire of another platoon on the flank. Another platoon may have chosen a place which was very good as regards the field of fire in a direction, which was already adequately defended by another platoon, but which had a bad field of fire over ground which no other platoon could fire upon. The company commander would adjust all these matters, and in the end one or more platoons might not be placed in the best position as regards their own particular front, but in the best as regards the whole company.

Having decided upon the exact site of the trenches and the general distribution of fire, the next matter to consider is the amount of clearing that is necessary, and the position and nature of any obstacles which may be required. Each platoon commander having been allotted a definite fire zone, can point out to his platoon what clearance is necessary; he can then ask each squad, as before, to choose the position for the obstacle. The company commander can then take the whole company to the position occupied by each platoon and tell the platoon commander to explain what ground they propose to clear, where they propose to place their obstacle, the material available for its construction, and in every case the reason why the decision has been arrived at. If digging is permitted, the trenches will now be constructed, and care will be taken that they are actually finished. It is far better to work overtime than to construct trenches which would be of little use in war and could not be properly defended. It is the exception rather than the rule to see trenches properly finished, fit for occupation, and capable of resisting a heavy attack. If the trenches cannot be dug the company can be taken to another part of the same position, where the ground in front is totally different, and the exercise can be repeated, the platoon and company commanders pointing out why a fire trench which was well sited in the first case would be badly sited if a similar position was selected in the second case.

Many other points which would come up for discussion, according to the nature of the ground in front, are dealt with in Chapters X and XI, Part III, "The Company in Defense."

Outpost. We can now turn to the method of training the company in outpost duty, making use of the same system of demonstration. Having pointed out to the company the locality where the main body is bivouacked, the fighting position which the main body will occupy in case a heavy attack is made against the outposts, and the general line of the outposts, the company commander will indicate on the ground the extent of front which is to be guarded by his company, stating whether imaginary companies continue the position on one or both flanks. He will point out the possible avenues of approach from the direction of the enemy to that portion of the position to be occupied by the company, and state from which direction the enemy is most likely to advance and why.

The first point to decide is the number of outguards and their exact position. In war this would always be done by the company commander, but if it is desired to give the junior officers of the company some instruction in this important detail, they should be sent out before the company arrives on the ground to reconnoiter the position and make their decisions. The exact siting of the trenches for the outguards, the construction of obstacles, and the clearance of the foreground having been decided upon and the positions selected for each outguard discussed, and a definite site selected, the next question to decide is the number and position of the sentries.

The platoon commander would then take each scheme in turn, visit with the whole platoon each position selected for the sentry, and decide finally what it would be best to do, giving, as usual, his reasons.

Having decided upon the positions of the sentries, and their line of retreat, so as not to mask the fire of the outguard, the next matter to consider would be the number of patrols that are required, and the particular areas of ground that must be examined by them periodically. The necessary trenches, obstacles, etc., would then be constructed.

Finally, the whole company should be assembled, marched to the position chosen for each outguard and the reasons for selecting the position explained by the company commander. The company should then be told off as an outpost company, and divided into outguards, supports, if any, and the necessary sentries over arms, patrols, etc., and marched to their respective posts.

If there is still time available each platoon commander can reconnoiter the ground for suitable positions for his outguards by night, take the outguards there, explain why the change of position is desirable, and direct the outguard commanders with their outguards to select positions for the sentries, following the same procedure as by day.

Although it is quite correct to select positions for night outposts during daylight, when possible, they should never be definitely occupied by the company before dark, when the forward movement could not be observed by the enemy. To practice night outposts by day is bad instruction, outguards and sentries are placed in positions which appear ridiculous to the ordinary mind, and the men get confused ideas on the subject. When it is desired to practice day and night outposts as an advanced exercise it is advisable to commence work in the afternoon, establish the day outposts, reconnoiter for the night outposts, make the change after dark and construct the necessary trenches, obstacles, etc., after dark.

It is, however, extremely important that the patrols should get to know their way about the country in front during daylight, when possible, so that they will have some practice in recognizing land marks by night.

It frequently occurs, when training the company in outpost duties, that long periods elapse during which the outguards are doing nothing. These opportunities should be taken to instruct the men in their duties when ordered to patrol to the front, the same system of demonstration being employed. For instance, the officer or noncommissioned officer commanding a piquet can select three men, point out certain ground in front which the sentries cannot see and which must be examined by a patrol, and proceed to instruct the whole picket in the best manner of carrying out this work. We will suppose that the patrol is working by day and that the ground to be visited is behind a small hill some 500 yards in front of the sentry. The commander of the picket will then explain to the men that the first object of the patrol is to reach the ground to be examined without being seen by any hostile patrols which may be moving about in front. Before proceeding further it is necessary for the patrol to decide upon the best line of advance. The various lines of advance will be discussed and the patrol asked to decide which they would select. Three other men can then be asked to give their opinion, and so on until all the men of the picket have expressed their views. The commander of the picket will then state which he considers the best line and give his reasons.

The next matter to decide is the method of advance to be adopted by the patrol. Are the three men to march past the sentry in one body and walk straight over the hill in front? If they do this there may be a hostile patrol hiding just behind the crest, watching the movements of our patrol, and directly the latter reach the hill they will be covered by the rifles of the hostile patrol at a few yards' range and will be captured or shot.

If the patrol is not to advance in one body how is it to act? There is plenty of time available, so that there are no objections to deliberate methods. The patrol should advance from cover to cover with one man always going forward protected by the rifles of the remaining two men who have halted in a good position to fire on any enemy that can fire on the leading man. The leading man having reached the cover in front will signal back all clear, and the two men in rear will join him. They will then make their next advance in a similar manner.

By looking at the hill the patrol can make a good guess at the locality which a hostile patrol would select if it was on the hill. It would be a place where it could get a good view towards our outpost line, and where the patrol could not be seen itself from the outpost line. If the hill was quite bare with nothing but grass on it and a flat round top, the best place for the enemy's patrol would be exactly on the top just behind the crest. In such a position he could not be seen by any sentry to the right or left of our picket. For example, if the hostile patrol chose a place on the side of the bare slope of the hill and looked over the crest line it would not be seen by our sentry, but it might be seen by another one on the flank.

The object of our patrol would be to approach the hill, not direct from the outguard, but either from the left or right of the hill and thus come on the flank of the enemy's patrol if he was there.

The whole picket can then be taken out to the front and follow the movements of the patrol from cover to cover until the hill is reached.

The next step will be to ascertain if there is any one on the top of the hill. If the hill is perfectly bare with a somewhat convex slope, it would be best for the three men to extend to about twenty yards interval and move forward together, prepared to drop on the first sign of the enemy, so that they can creep up and open fire on him without exposing themselves. Three men with magazine rifles, extended in this manner, opposed to a hostile patrol collected in one party, should be able to deal with the latter without much difficulty. Their fire would be converging, and coming from different directions would confuse the hostile patrol, especially if the advance was made from a flank. The men of the patrol when creeping up the hill should avoid exposing themselves in the direction of the ground behind the hill, if possible, because they want to examine that ground later on, and if seen by the enemy they might fall into an ambuscade. If it is impossible to avoid being seen from the ground beyond, it would be best for the patrol to retire as though they were going back to the outposts, and then move round the flank of the hill and advance to the ground beyond from an unexpected direction. All this would be considered by the officer or noncommissioned officer commanding the picket, together with many other points.

Sufficient has been said to explain how this system of demonstration can be worked in connection with any class of operation in the field. It is certainly slow, and takes a long time, but no one is ever idle and every one is constantly learning something fresh, for the simple reason that, although one may know every detail of the subject, the ground constantly differs and requires to be dealt with in a common-sense and skilful manner. The men are interested throughout, and one morning spent on this kind of work is worth several days of practice in the ordinary manner.

It should be remembered that this system of demonstration is only required to teach the men their work; when they have once learned it and thoroughly understand the necessary details they must be practiced in it, the company or platoon commander indicating what has been well done, what has been badly done, and what requires improvement.

OTHER EXAMPLES OF THE OCULAR DEMONSTRATION METHOD

442. The following illustrations will suggest other examples of the employment of the ocular demonstration method of instruction:

The advantages and disadvantages of close and extended order. Send a lieutenant or a noncommissioned officer with two or three squads of the older soldiers some distance to the front of the company, and have them advance toward the company, first in close order and then in extended order.

By ocular demonstration show the men who are watching the approach of the company how easy it would be even for the poorest shots to land bullets in the thick of a closed body, but how much of a less distinct target the extended order offers and how many spaces there are in the skirmish line for the bullets to pass through; also, how much more easily cover can be employed and the rifle used in the extended order. Let them see also how much more difficult it is for officers and noncommissioned officers to maintain control over the movements of troops in extended order, and the consequent necessity and duty of every soldier, when in extended order, doing all he can, by attention and exertion, to keep order and help his officers and noncommissioned officers to gain success.

The Use of Cover. Send a lieutenant or noncommissioned officer with a couple of squads of old soldiers a few hundred yards to the front and have them advance on the company as if attacking, first without taking advantage of cover and then taking advantage of all available cover, the part of the company that is supposed to be attacked lying down and aiming and snapping at the approaching soldiers. Then reverse the operation—send the defenders out and have them advance on the former attackers. Explain that the requisites of good cover are: Ability to see the enemy; concealment of your own body; ability to use the rifle readily. Then have a number of men take cover and snap at an enemy in position, represented by a few old soldiers. Point out the defects and the good points in each case.

443. Practice in Commanding Mixed Squads. In order to practice noncommissioned officers in commanding mixed firing squads, and in order to drill the privates in banding themselves together and obeying the orders of anyone who may assume command, it is good training for two or more companies to practice re-enforcing each other by one company assuming a given fire-position and the other sending up re-enforcements by squads, the men being instructed to take positions anywhere on the firing line where they may find an opening. However, explain to the men that whenever possible units should take their positions on the firing line as a whole, but that in practice it is very often impossible to do this, and that the drill is being given so as to practice the noncommissioned officers in commanding mixed units on the firing line and also to give the privates practice in banding themselves into groups and obeying the command of any noncommissioned officer who may be over them.

444. Operating Against Other Troops. There is no better way of arousing interest, enthusiasm, and pride in training troops than by creating a feeling of friendly rivalry and competition amongst the men, and the best way to do this is to have one part of the company operate against the other in all such practical work as scouting, patrolling, attacking, etc. Whenever practicable, blank ammunition should be used. One of the sides should wear a white handkerchief around the hat or some other distinguishing mark. The troops that are sent out must be given full and explicit instructions as to just exactly what they are to do, so that the principles it is intended to illustrate may be properly brought out.

CHAPTER IV

GENERAL COMMON SENSE PRINCIPLES OF APPLIED MINOR TACTICS[1]

445. To begin with, you want to bear in mind that there is nothing difficult, complicated or mysterious about applied minor tactics—it is just simply the application of plain, every-day, common horse sense—the whole thing consists in familiarizing yourself with certain general principles based on common sense and then applying them with common sense. Whatever you do, don't make the mistake of following blindly rules that you have read in books.

446. One of the ablest officers in the Army has recently given this definition of the Art of War:

> One-fifth is learned from books;
> One-fifth is common sense;
> Three-fifths is knowing men and how to lead them.

The man who would be successful in business must understand men and apply certain general business principles with common sense; the man who would be a successful hunter must understand game and apply certain general hunting principles with common sense, and even the man who would be a successful fisherman must understand fish and apply certain general fishing principles with common sense. And so likewise the man who would lead other men successfully in battle must understand men and apply certain general tactical principles with common sense.

Of course, the only reason for the existence of an army is the possibility of war some day, and everything the soldier does—his drills, parades, target practice, guard duty, schools of instruction, etc.—has in view only one end: The preparation of the soldier for the field of battle.

446a. While the responsibilities of officers and noncommissioned officers in time of peace are important, in time of battle they are much more so; for then their mistakes are paid for in human blood.

What would you think of a pilot who was not capable of piloting a boat trying to pilot a boat loaded with passengers; or, of an engineer who was not capable of running a locomotive trying to run a passenger train? You would, of course, think him criminal—but do you think he would be more criminal than the noncommissioned officer who is not capable of leading a squad in battle but who tries to do so, thereby sacrificing the lives of those under him?

You can, therefore, appreciate the importance, the necessity, of every officer and noncommissioned officer doing everything that he possibly can during times of peace to qualify himself for his duties and responsibilities during times of war.

---

[1] In the preparation of the first part of this chapter, extracts of words and of ideas, were made from a paper on Applied Minor Tactics read before the St. Louis Convention of the National Guard of the United States, in 1910, by Major J. F. Morrison, General Staff, U. S. Army.

If we are going to have a good army we must have good regiments; to have good regiments we must have good battalions; to have good battalions we must have good companies—but to have good companies we must have efficient company officers and noncommissioned officers.

As stated before, everything in the life of the soldier leads to the field of battle. And so is it that in the subject of minor tactics all instruction leads to the battle. First we have map problems; then terrain exercises; next the war game; after that maneuvers, and finally the battle.

447. Map Problems and Terrain Exercises. In the case of map problems you are given tactical problems to solve on a map; in the case of terrain exercises you are given problems to solve on the ground. (The word "Terrain," means earth, ground.) These are the simplest forms of tactical problems, as you have only one phase of the action, your information is always reliable and your imaginary soldiers always do just exactly what you want them to do.

448. War Game. Next comes the war game, which consists of problems solved on maps, but you have an opponent who commands the enemy—the phases follow one another rapidly and the conditions change —your information is not so complete and reliable. However, your men being slips of cardboard or beads, they will, as in the case of your imaginary soldiers in the map problems and terrain exercises, go where you wish them to and do what you tell them to do—they can't misunderstand your instructions and go wrong—they don't straggle and get careless as real soldiers sometimes do.

Map problems, terrain exercises and war games are but aids to maneuvers—their practice makes the maneuvers better; for you thus learn the principles of tactics and in the simplest and quickest way.

449. Maneuvers. In the case of the maneuver the problem is the same as in the war game, except that you are dealing with real, live men whom you can not control perfectly, and there is, therefore, much greater chance for mistakes.

The Battle. A battle is only a maneuver to which is added great physical danger and excitement.

General rules and principles that must be applied in map problems, terrain exercises, the war game and maneuvers

450. Everything that is done must conform in principle to what should be done in battle—otherwise your work is wasted—your time is thrown away.

In solving map problems and in the war game, always form in your mind a picture of the ground where the action is supposed to be taking place—imagine that you see the enemy, the various hills, streams, roads, etc., that he is firing at you, etc.—and don't do anything that you would not be able to do if you were really on the ground and really in a fight.

Whether it be a corporal in command of a squad or a general in command of an army, in the solution of a tactical problem, whether it be a map problem, a terrain exercise, a war game, maneuver or battle, he will have to go through the same operation:

1st. Estimate the situation;
2d. Decide what he will do;
3d. Give the necessary orders to carry out his decision.

At first these three steps of the operation may appear difficult and laborious, but after a little practice the mind, which always works with rapidity in accustomed channels, performs them with astonishing quickness.

The child beginning the study of arithmetic, for example, is very slow in determining the sum of 7 and 8, but later the answer is announced almost at sight. The same is true in tactical problems—the process may be slow at first, but with a little practice it becomes quick and easy.

451. Estimating the Situation. This is simply "sizing up the situation," finding out what you're "up against," and is always the first thing to be done. It is most important, and in doing it the first step is to determine your MISSION—what you are to do, what you are to accomplish—the most important consideration in any military situation.

Consider next your own forces and that of the enemy—that is, his probable strength and how it compares with yours.

Consider the enemy's probable MISSION[1] and what he will probably do to accomplish it.

Consider the geography of the country so far as it affects the problem—the valleys to cross, defiles to pass through, shortest road to follow, etc.

Now, consider the different courses open to you with the advantages and disadvantages of each.

You must, of course, in every case know what you're up against before you can decide intelligently what you're going to do.

In making your plan always bear in mind not only your own MISSION, but also the general mission of the command of which you form a part, and this is what nine men out of ten forget to do.

You are now ready to come to a decision, which is nothing more or less than a clear, concise determination of what you're going to do and how you're going to do it.

452. The Decision. It is important that you should come to a clear and correct decision—that you do so promptly and then execute it vigorously.

The new Japanese Field Service Regulations tell us that there are two things above all that should be avoided—inaction and hesitation. "To act resolutely even in an erroneous manner is better than to remain inactive and irresolute"—that is to say do something. Frederick the Great, expressed the same idea in fewer words: "Don't haggle."

Having settled on a plan, push it through—don't vacillate, don't waver. Make your plan simple. No other has much show. Complicated plans look well on paper, but in war they seldom work out. They require several people to do the right thing at the right time and this under conditions of excitement, danger and confusion, and, as a result, they generally fail.

---

[1] The word *"mission"* is used a great deal in this text. By your *"mission"* is meant your business, what you have been told to do, what you are trying to accomplish.

The Order. Having completed your estimate of the situation and formed your plan, you are now ready to give the orders necessary to carry it out.

You must first give your subordinates sufficient information of the situation and your plan, so that they may clearly understand their mission.

The better everyone understands the whole situation the better he can play his part. Unexpected things are always happening in war—a subordinate can act intelligently only if he knows and understands what his superior wants to do.

Always make your instructions definite and positive—vague instructions are sometimes worse than none.

Your order, your instructions, must be clear, concise and definite —everyone should know just exactly what he is to do.

A Few General Principles

453. The man who hunts deer, moose, tigers and lions, is hunting big game, but the soldier operating in the enemy's territory is hunting bigger game—he's hunting for human beings—but you want to remember that the other fellow is out hunting for you, too; he's out "gunning" for you. So, don't fail to be on the alert, on the look out, all the time, if you do he'll "get the drop" on you. Remember what Frederick the Great said: "It is pardonable to be defeated, but never to be taken by surprise."

Do not separate your force too much; if you do, you weaken yourself—you take the chance of being "defeated in detail"—that is, of one part being defeated after another. Remember the old saying: "In union there is strength." Undue extension of your line (a mistake, by the way, very often made) is only a form of separation and is equally as bad.

While too much importance can not be attached to the proper use of cover, you must not forget that sometimes there are other considerations that outweigh the advantages of cover. Good sense alone can determine. A certain direction of attack, for instance, may afford excellent cover but it may be so situated as to mean ruin if defeated, as where it puts an impassable obstacle directly in your rear. And don't forget that you should always think in advance of what you would do in case of defeat.

What is it, after all, that gives victory, whether it be armies or only squads engaged? It's just simply inflicting on the enemy a loss which he will not stand before he can do the same to you. Now, what is this loss that he will not stand? What is the loss that will cause him to break? Well, it varies; it is subject to many conditions—different bodies of troops, like different timbers, have different breaking points. However, whatever it may be in any particular case it would soon come if we could shoot on the battlefield as we do on the target range, but we can not approximate it.

There are many causes tending to drag down our score on the battlefield, one of the most potent being the effect of the enemy's fire. It is cited as a physiological fact that fear and great excitement cause the pupil of the eye to dilate and impair accuracy in vision and hence of shooting. It is well established that the effectiveness of the fire of one side reduced proportionately to the effectiveness of that of the other.

Bear in mind then these two points—we must get the enemy's breaking point before he gets ours, and the more effective we make our fire the less effective will be his.

Expressed in another way—to win you must gain and keep a fire superiority.

This generally means more rifles in action, yet a fire badly controlled and directed, though great in volume, may be less effective than a smaller volume better handled.

The firing line, barring a few exceptional cases, then, should be as heavy as practicable consistent with the men's free use of their rifles.

This has been found to be about one man to the yard. In this way you get volume of fire and the companies do not cover so much ground that their commanders lose their power to direct and control.

If it becomes necessary to hold a line too long for the force available, it is then better to keep the men close together and leave gaps in the line. The men are so much better controlled, the fire better directed, the volume the same, and the gaps are closed by the cross fire of parties adjacent.

CHAPTER V

GENERAL PLAN OF INSTRUCTION IN MAP PROBLEMS FOR NONCOMMISSIONED OFFICERS AND PRIVATES—INSTRUCTION IN DELIVERING MESSAGES

(The large wall map to be used for this instruction can be obtained from the George Banta Publishing Co., Menasha, Wis., at a cost of $1.50.)

454. The noncommissioned officers and the privates of the squad, section, platoon or company are seated in front of the instructor, who, with pointer in hand, is standing near the map on the wall.

The instructor assumes certain situations and designates various noncommissioned officers to take charge of squads for the purpose of accomplishing certain missions; he places them in different situations, and then asks them what they would do. He, or the noncommissioned officer designated to perform certain missions, designates certain privates to carry messages, watch for signals, take the place of wounded noncommissioned officers, etc. For example, the instructor says: ''The battalion is marching to Watertown (see Elementary Map in pocket at back of book) along this road (indicating road); our company forms the advance guard; we are now at this point (indicating point). Corporal Smith, take your squad and reconnoiter the woods on the right to see if you can find any trace of the enemy there, and rejoin the company as soon as you can. Corporal Jones, be on the lookout for any signals that Corporal Smith may make.''

Corporal Smith then gives the command, ''1. Forward, 2. MARCH,'' and such other commands as may be necessary.

Instructor: Now, when you reach this point (indicating point), what do you see?

(Corporal Smith holds his rifle horizontally above his head.)

Corporal Jones: Captain, Corporal Smith signals that he sees a small body of the enemy.

Corporal Smith: Lie down. Range, 700. 1. Ready; 2. AIM; 3. Squad; 4. FIRE. 1. Forward; Double time; 2. MARCH, etc.

The noncommissioned officers and the privates who are thus designated to do certain things must use their imagination as much as possible. They must look at the map and imagine that they are right on the ground, in the hostile territory; they must imagine that they see the streams, hills, woods, roads, etc., represented on the map, and they must not do anything that they could not do if in the hostile territory, with the assumed conditions actually existing.

The general idea of this system of instruction is to make the noncommissioned officers and the privates think, to make them use common sense and initiative in handling men in various situations, in getting out of difficulties. By thus putting men on their mettle in the presence of their comrades and making them bring into play their common sense and their powers of resourcefulness, it is comparatively easy to hold the attention of a whole squad, section, platoon or company, for those who are not actually taking part in the solution of a particular problem are curious to see how those who are taking part will answer different questions and do different things—how they will "pan out."

Everything that is said, everything that is done, should, as far as practicable, be said and done just as it would be said and done in the field. The commands should be actually given, the messages actually delivered, the reports actually made, the orders and instructions actually given, the signals actually made, etc., just the same as they would be if the operations were real. Of course, sometimes it is not practicable to do this, and again at other times it would be advisable not to do so. If, for instance, in the solution of a problem there were a great many opportunities to give commands to fire, to make signals, to deliver messages, etc., and if these things were actually done every time, it would not only become tiresome but it would also delay the real work and instruction. Common sense must be used. Just bear this in mind: In the solution of map problems the noncommissioned officers and the privates are to be given proper and sufficient instruction in giving commands, making signals, sending and delivering messages, making reports, etc., the instructor using his common sense in deciding what is proper and sufficient instruction. In carrying out this feature of the instruction it would be done thus, for instance:

Instead of a platoon leader saying, "I would give the order for the platoon (two, three or four squads) to fire on them," he would say, for instance, "I would then give the command, 'AT LINE OF MEN. RANGE, 600. FIRE AT WILL,'" and would continue the firing as long as necessary." Should the instructor then say, for instance, "Very well; the enemy's fire has slackened; what will you do now?" The platoon leader would answer, for instance, "I would signal: 1. By squads from the right; 2. RUSH."

Instead of saying for instance, "I would advance my squad to the top of this hill at double time," the squad leader should say, "I would give the command: '1. Forward, double time; 2. MARCH,'" and upon reaching the top of this hill, I would command, '1. Squad; 2. HALT,' cautioning the men to take advantage of cover."

Instead of saying, "I would signal back that we see the enemy in force," the squad leader should take a rifle and make the signal, and if a man has been designated to watch for signals, the man would say to the captain (or other person for whom he was watching for signals): "Captain, Corporal Smith has signaled that he sees the enemy in force."

Instead of saying, "I would send a message back that there are about twenty mounted men just in rear of the Jones' house; they are dismounted and their horses are being held by horseholders," say, "Smith, go back and tell the captain (or other person) there are about twenty mounted men just in rear of the Jones' house. They are dismounted and their horses are being held by horseholders." Private Smith would then say to the captain (or other person), "Captain, Corporal Harris sends word there are about twenty men just in the rear of the Jones' house. They are dismounted and their horses are being held by horseholders."

For problems exemplifying this system of instruction, see Par. 509.

The instruction may be varied a little by testing the squad leaders in their knowledge of map reading by asking, from time to time during the solution of the problem, such questions as these:

Captain: Corporal Smith, you are standing on Lone Hill (See Elementary Map), facing north. Tell me what you see?

Corporal: The hill slopes off steeply in front of me, about eighty feet down to the bottom land. A spur of the hill runs off on my right three-fourths of a mile to the north. Another runs off on my left the same distance to the west. Between these two spurs, down in front of me, is an almost level valley, extending about a mile to my right front, where a hill cuts off my view. To my left front it is level as far as I can see. A quarter of a mile in front of me is a big pond, down in the valley, and I can trace the course of a stream that drains the pond off to the northwest, by the trees along its bank. Just beyond the stream a railroad runs northwest along a fill and crosses the stream a mile and a half to the northwest, where I can see the roofs of a group of houses. A wagon road runs north across the valley, crossing the western spur of this hill 600 yards from Lone Hill. It is bordered by trees as far as the creek. Another road parallels the railroad, the two roads crossing near a large orchard a mile straight to my front.

Captain: Can you see the Chester Pike where the railroad crosses it?

Corporal: No, sir.

Captain: Why?

Corporal: Because the hill "62," about 800 yards from Lone Hill, is so high that it cuts off my view in that direction of everything closer to the spur "62" than the point in the Salem-Boling road, where the private lane runs off east to the Gray house.

Captain: Sergeant Jones, in which direction does the stream run that you see just south of the Twin Hills?

Sergeant: It runs south through York, because I can see that the northern end starts near the head of a valley and goes down into the open plain. Also it is indicated by a very narrow line near the Twin Hills which becomes gradually wider or heavier the further south it goes. Furthermore, the fact that three short branch streams are shown joining together and forming one, must naturally mean that the direction of flow is towards the one formed by the three.

Captain: Sergeant Harris does the road from the Mason farm to the Welsh farm run up or down hill?

Sergeant: It does both, sir. It is almost level for the first half mile west of the Mason farm; then, as it crosses the contour marked 20 and a second marked 40, it runs up hill, rising to forty feet above the valley, 900 yards east of the Mason farm. Then, as it again crosses a contour marked 40 and a second marked 20, it goes down hill to the Welsh farm. That portion of the road between the points where it crosses the two contours marked 40, is the highest part of the road. It crosses this hill in a "saddle," for both north and south of this summit on the road are contours marked 60 and even higher.

Captain: Corporal Wallace, you are in Salem with a patrol with orders to go to Oxford. There is no one to tell you anything about this section of the country and you have never been there before. You have this map and a compass. What would you do?

Corporal: I would see from my map and by looking around me that Salem is situated at the crossing of two main roads. From the map I would see that one leads to Boling and the other was the one to take for Oxford. Also, I would see that the one to Boling started due north out of Salem and the other, the one I must follow, started due west out of Salem. Taking out my compass, I would see in what direction the north end of the needle pointed; the road running off in that direction would be the one to Boling, so I would start off west on the other.

Captain: Suppose you had no compass?

Corporal: I would look and see on which side of the base of the trees the moss grew. That side would be north. Or, in this case, I would probably not use a compass even if I had it; for, from the map, I know that the road I wish to start off on crosses a railroad track within sight of the crossroads and on the opposite side of the crossroads from the church shown on the map; also, that the Boling road is level as far as I could see on the ground, while the Chester Pike crosses the spur of Sandy Ridge, about a half mile out of the village.

Captain: Go ahead, corporal, and explain how you would follow the proper route to Oxford.

Corporal: I would proceed west on the Chester Pike, knowing I would cross a good sized stream, on a stone bridge, about a mile and a half out of Salem; then I would pass a crossroad and find a swamp on my right, between the road and the stream. About a mile and a half from the crossroad I just mentioned, I would cross a railroad track and then I would know that at the fork of the roads one-quarter of a mile further on I must take the left fork. This road would take me straight into Oxford, about a mile and three-quarters beyond the fork.

Captain: Sergeant Washington, do the contours about a half mile north of the Maxey farm, on the Salem-Boling road, represent a hill or a depression?

Sergeant: They represent a hill, because the inner contour has a higher number, 42, than the outer, marked 20. They represent sort of a leg-of-mutton shaped hill about 42 feet higher than the surrounding low ground.

Variety and interest may be added to the instruction by assuming that the squad leader has been killed or wounded and then designate some private to command the squad; or that a man has been wounded in a certain part of the body and have a soldier actually apply his first aid packet; or that a soldier has fainted or been bitten by a rattlesnake and have a man actually render him first aid.

454a. The privates may be given practical instruction in delivering messages by giving them messages in one room and having them deliver them to someone else in another room. It is a good plan to write out a number of messages in advance on slips of paper or on cards, placing them in unsealed envelopes. An officer or a noncommissioned officer in one room reads one of the messages to a soldier, then seals it in an envelope and gives it to the soldier to hand to the person in another room to whom he is to deliver the message. The latter checks the accuracy of the message by means of the written message. Of course, this form of instruction should not be given during the solution of map problems by the men. (For model messages, see page 51.)

The same slips or cards may be used any number of times with different soldiers. A soldier should never start on his way to deliver a message unless he understands thoroughly the message he is to deliver.

CHAPTER VI

THE SERVICE OF INFORMATION

(Based on the Field Service Regulations.)

PATROLLING

455. Patrols are small bodies of infantry or cavalry, from two men up to a company or troop, sent out from a command at any time to gain information of the enemy and of the country, to drive off small hostile bodies, to prevent them from observing the command or for other stated objects, such as to blow up a bridge, destroy a railroad track, communicate or keep in touch with friendly troops, etc. Patrols are named according to their objects, reconnoitering, visiting, connecting, exploring, flanking patrols, etc. These names are of no importance, however, because the patrol's orders in each case determine its duties.

456. The size of a patrol depends upon the mission it is to accomplish; if it is to gain information only, it should be as small as possible, allowing two men for each probable message to be sent (this permits you to send messages and still have a working patrol remaining); if it is to fight, it should be strong enough to defeat the probable enemy against it. For instance, a patrol of two men might be ordered to examine some high ground a few hundred yards off the road. On the other hand, during the recent war in Manchuria a Japanese patrol of 50 mounted men, to accomplish its mission, marched 1,160 miles in the enemy's country and was out for 62 days.

457. Patrol Leaders. (a) Patrol leaders, usually noncommissioned officers, are selected for their endurance, keen eyesight, ability to think quickly and good military judgment. They should be able to read a map, make a sketch and send messages that are easily understood. Very important patrols are sometimes lead by officers. The leader should have a map, watch, field glass, compass, message blank and pencils.

(b) The ability to lead a patrol correctly without a number of detailed orders or instructions, is one of the highest and most valuable qualifications of a noncommissioned officer. Since a commander order ing out a patrol can only give general instructions as to what he desires, because he can not possibly foresee just what situations may arise, the patrol leader will be forced to use his own judgment to decide on the proper course to pursue when something of importance suddenly occurs. He is in sole command on the spot and must make his decisions entirely on his own judgment and make them instantly. He has to bear in mind first of all his mission—what his commander wants him to do.

Possibly something may occur that should cause the patrol leader to undertake an entirely new mission and he must view the new situation from the standpoint of a higher commander.

(c) More battles are lost through lack of information about the enemy than from any other cause, and it is the patrols led by noncommissioned officers who must gather almost all of this information. A battalion or squadron stands a very good chance for defeating a regiment if the battalion commander knows all about the size, position and movements of the regiment and the regimental commander knows but a little about the battalion; and this will all depend on how efficiently the patrols of the two forces are led by the noncommissioned officers.

458. Patrols are usually sent out from the advance party of an advance guard, the rear party of a rear guard, the outguards of an outpost, and the flank (extreme right or left) sections, companies or troops of a force in a fight, but they may be sent out from any part of a command.

The commander usually states how strong a patrol shall be.

459. Orders or Instructions—(a) The orders or instructions for a patrol must state clearly whenever possible:

1. Where the enemy is or is supposed to be.

2. Where friendly patrols or detachments are apt to be seen or encountered and what the plans are for the body from which the patrol is sent out.

3. What object the patrol is sent out to accomplish; what information is desired; what features are of especial importance; the general direction to be followed and how long to stay out in case the enemy is not met.

4. Where reports are to be sent.

(b) It often happens that, in the hurry and excitement of a sudden encounter or other situation, there is no time or opportunity to give a patrol leader anything but the briefest instructions, such as "Take three men, corporal, and locate their (the enemy's) right flank." In such a case the patrol leader through his knowledge of the general principles of patrolling, combined with the exercise of his common sense, must determine for himself just what his commander wishes him to do.

460. Inspection of a Patrol Before Departure. Whenever there is time and conditions permit, which most frequently is not the case, a patrol leader carefully inspects his men to see that they are in good physical condition; that they have the proper equipment, ammunition and rations; that their canteens are full, their horses (if mounted) are in good condition, not of a conspicuous color and not given to neighing, and that there is nothing about the equipment to rattle or glisten. The patrol leader should also see that the men have nothing with them (maps, orders, letters, newspapers, etc.) that, if captured, would give the enemy valuable information. This is a more important inspection than that regarding the condition of the equipment.

Whenever possible the men for a patrol should be selected for their trustworthiness, experience and knack of finding their way in a strange country.

461. Preparing a Patrol for the Start. The patrol leader having received his orders and having asked questions about anything he does not fully understand, makes his estimate of the situation (See Par. 451). He then selects the number of men he needs, if this has been left to him, inspects them and carefully explains to them the orders he has received and how he intends to carry out these orders, making sure the men understand the mission of the patrol. He names some prominent

place along the route they are going to follow where every one will hasten if the patrol should become scattered.

For example: An infantry company has arrived at the town of York (See Elementary Map). Captain A, at 2 P. M., calls up Corporal B and three men of his squad.

Captain A: Corporal, hostile infantry is reported to be at Oxford. Nothing else has been heard of the enemy. The company remains here tonight. You will take these three men and reconnoiter about two miles north along this road (indicates the Valley Pike) for signs of a hostile advance in this direction.

Stay out until dusk.

Corporal C has been sent out that road (points east along the county road).

Send messages here. Do you understand?

Corporal B: Yes, sir; I am to—(here he practically repeats Captain A's orders, the three men listening). Is Corporal C to cover that hill (points toward Twin Hills)?

Captain A: No; you must cover that ground. Move out at once, corporal. (Corporal B quickly glances at the men and sees that they have their proper equipment.)

Corporal B (to his men): You heard the captain's orders. We will make for that hill (points to Twin Hills). Jones, I want you to go 150 yards in advance of me; Williams, follow me at 100 yards; Smith, you'll stay with me. Jones, you'll leave this road after crossing the creek and march on that clump of trees. I want both you and Williams to be on the alert and watch me every minute for signals. In case we become scattered, make for that hill (points to Twin Hills).

Private Jones: Corporal, shall I keep 150 yards from you or will you keep the correct distance?

Corporal B: You keep the correct distance from me. Forward, Jones.

Of course, the patrol leader makes all these preparations if he has time; but, as we have said before, there will be a great many occasions when he is required to start out so promptly that he will not have any time for the inspection described and he will have to make an estimate of the situation and give his detailed orders to the members of his patrol as they start off.

461a. Co-ordination Before Departure. Every member of a patrol should notice for himself the direction taken and all landmarks that are passed, and every man should keep his eyes and ears open all the time. Before leaving an outpost position or other place to which it is to return, the patrol commander should "co-ordinate" himself—he should see where he is with respect to certain mountains, high buildings and other prominent objects, and after the patrol has left, he should frequently turn his head around and see what the starting point looks like from where he is. This will help him to find his way back without difficulty.

THE PRINCIPLES OF PATROLLING

462. Paragraphs 462 to 507 describe the methods of leading a patrol—the points a patrol leader should fully understand. In other words, they state the principles of patrolling. When you first study this chapter, simply read over these principles without trying to memorize any of them. Whenever one of the principles is applied in the solution of any of the problems on patrolling given in this book you will generally find the number of the paragraph which states that principle enclosed in brackets. Turn back and study the paragraph referred to until you thoroughly understand its meaning and you feel sure that you know how to apply that principle whenever the occasion might arise in actual patrolling. Try to impress its common sense meaning (never the mere words) on your mind, so that when a situation arises requiring the sort of action indicated in the principle, YOU WILL NOT FAIL TO RECOGNIZE IT.

463. **Formation of Patrols.** (a) Figure 1 gives some examples of various ways of forming patrols. These are merely examples for the purpose of giving a general idea of the arrangement of the men. In practice common sense must dictate to the patrol leader the best formation in each case.

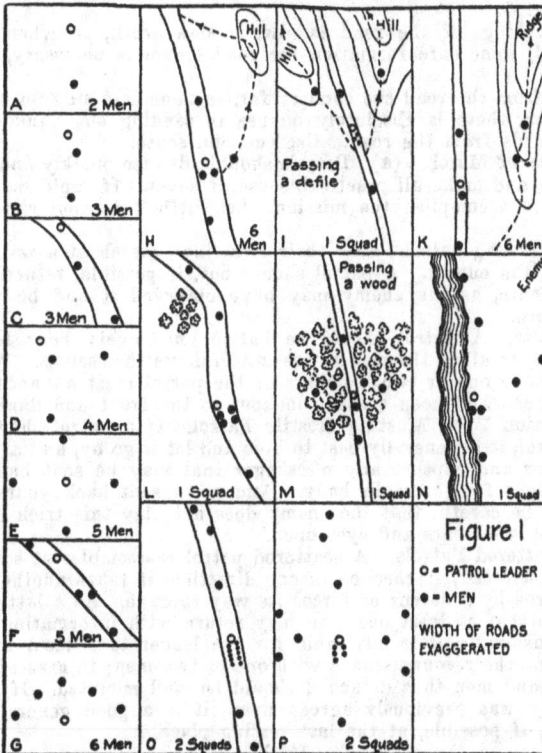

Figure 1

(b) In very small patrols the leader is usually in advance where he can easily lead the patrol through not always (See E, Figure 1. The distance between men depends upon the character of the country and the situation. In L, Figure 1, it might be anywhere from 150 to 400 yards from the leading man to the last, the distance being greater in level or open country. Some such formation as G, Figure 1, could be used in going through high brush, woods, or over very open country.

(c) The men must be so arranged that each man will be within signaling distance of some member of the patrol and the escape of at least one man, in case of surprise, is certain.

It must be remembered that the patrol may have to march a long distance before it is expected that the enemy will be encountered, or it may have a mission that requires it to hurry to some distant point through very dangerous country. In such cases the patrol will probably have to follow the road in order to make the necessary speed, and it will not be possible for flankers to keep up this rate marching off the road. The formation in such cases would be something like those shown in F, H and O.

Marching off the road is always slow work, so when rapidity is essential, some safe formation for road travel is necessary, as in F, L and O.

If, from the road the country for, say one-half mile on each side, can be seen, there is absolutely no use in sending out flankers a few hundred yards from the road. Use common sense.

464. Rate of March. (a) Patrols should advance quickly and quietly; be vigilant and make all practicable use of cover. If rapid marching is necessary to accomplish the mission, then little attention can be paid to cover.

(b) Returning patrols, near their own lines, march at a walk, unless pressed by the enemy. A patrol should not, if possible, return over its outgoing route, as the enemy may have observed it and be watching for its return.

465. Halts. A patrol should be halted once every hour for about ten minutes, to allow the men to rest and relieve themselves. Whenever a halt is made one or two members of the patrol must advance a short distance ahead and keep a sharp lookout to the front and flanks.

466. Action Upon Meeting Hostile Patrol. If a patrol should see a hostile patrol, it is generally best to hide and let it go by, and afterwards look out for and capture any messenger that may be sent back for it with messages for the main body. And when sent back yourself with a message, be careful that the enemy does not play this trick on you—always keep your ears and eyes open.

467. Scattered Patrols. A scattered patrol reassembles at some point previously selected; if checked in one direction, it takes another; if cut off, it returns by a detour or forces its way through. As a last resort it scatters, so that at least one man may return with information

Occasionally it is advisable for the leader to conceal his patrol and continue the reconnaissance with one or two men; in case of cavalry the leader and men thus detached should be well mounted. If no point of assembly was previously agreed upon, it is a good general rule to reassemble, if possible, at the last resting place.

468. Return by Different Route. A patrol should always make it a rule to return by a different route, as this may avoid its being captured by some of the enemy who saw it going out and are laying in wait for it.

469. Guard Against Being Cut Off. When out patrolling always guard against being cut off. Always assume that any place that affords good cover is held by the enemy until you know that it is not, and be careful not to advance beyond it without first reconnoitering it; for, if you do, you may find yourself cut off when you try to return.

470. Night Work. Patrols far from their commands or in contact with the enemy, often remain out over night. In such cases they seek a place of concealment unknown to the inhabitants, proceeding thereto after nightfall or under cover. Opportunities for watering, feeding and rest must not be neglected, for there is no assurance that further opportunities will present themselves. When necessary the leader provides for subsistence by demand or purchase.

471. Civilians: In questioning civilians care must be taken not to disclose information that may be of value to the enemy. Strangers must not be allowed to go ahead of the patrol, as they might give the enemy notice of its approach. Patrol leaders are authorized to seize telegrams and mail matter, and to arrest individuals, reporting the facts as soon as possible.

472. Patrol Fighting. (a) A patrol sent out for information never fights unless it can only get its information by fighting or is forced to fight in order to escape. This principle is the one most frequently violated by patrol leaders, particularly in peace maneuvers. They forget their mission—the thing their commander sent them out to do—and begin fighting, thus doing harm and accomplishing no important results.

(b) A patrol sent out to drive off hostile detachments has to fight to accomplish its mission. Sometimes a patrol has orders both to gain information and to drive back hostile patrols. In this case it may be proper to avoid a fight at one moment and to seek a fight at another. The patrol leader must always think of his mission when deciding on the proper course to follow, and then use common sense.

473. Signals. The following should be clearly understood by members of a patrol:

Enemy in sight in small numbers: Hold the rifle above the head horizontally.

Enemy in force: Same as preceding, raising and lowering the rifle several times.

Take cover: A downward motion of the hand.

Other signals may be agreed upon before starting, but they must be simple and familiar to the men; complicated signals must be avoided. Signals must be used cautiously, so as not to convey information to the enemy.

The patrol leader should see that all his men thoroughly understand that whenever they are away from the center of the patrol they must look to the nearest man for signals at least once every minute. It should never be necessary for the patrol leader to call to a man in order to get his attention. All movements of men at a distance should be

474 (contd.)

regulated by signals and the men should constantly be on the lookout for these signals.

474. Messages. (a) The most skillful patrol leading is useless unless the leader fully understands when to send a message and how to write it.

(b) A message, whether written or verbal, should be short and clear, resembling a telegram. If it is a long account it will take too much time to write, be easily misunderstood, and if verbal, the messenger will usually forget parts of it and confuse the remainder.

(c) Always state when and where things are seen or reported. If haste is required, do not use up valuable moments writing down the day of the month, the hour the message is written, etc. These data are essential as a matter of future record for formal telegrams and should be put in patrol messages only when time is abundant, but never slight the essential points of information that will give valuable help to your chief. Always try to put yourself in his place—not seeing what you see and read your message—and then ask yourself, What will he want to know?

(d) The exact location of the enemy should be stated; whether deployed, marching or in camp, his strength, arm of the service (cavalry, infantry or artillery), and any other detail that you think would be valuable information for your chief. In giving your location do not refer to houses, streets, etc., that your chief in the rear has no knowledge of. Give your direction and distance from some point he knows of or, if you have a map like his, you can give your map location.

(e) Be sure your message is accurate. This does not mean that something told you should not be reported, but it should be reported, not as a fact, but as it is—a statement by somebody else. It is well to add any information about your informant, such as his apparent honesty, the probability of his having correct information, etc.—this may help your chief.

(f) A message should always end with a short statement of what you are going to do next. For example: "Will remain in observation," "Will continue north," "Will work around to their rear," etc. Time permitting, the bearer of a verbal message should always be required to repeat it before leaving.

(g) The following is a reproduction of a message blank used in field service. The instructions on the envelope are also given. A patrol leader will usually be furnished with a pad of these blanks:

U. S. ARMY FIELD MESSAGE	No.	Sent by	Time	Rec'd by	Time	Check
	(These spaces for Signal Operators only)					

Communicated by	(Name of sending detachment)
Buzzer, Phone, Telegraph, Wireless, Lantern, Helio, Flag, Cyclist, Foot Messenger, Mounted Messenger. Underscore means used	*From* ....
	(Location of sending detachment)
	At . ...
	Date . . . *Hour* . ..*No*. ....

To	..		. .		. . .	.. ..	
				. ...			. ... .	
.	...	.	. .. .	....	.	....	... .	
...	... ...	.	. ...	. ...	..	. ...	. ...	
.		. ...		. ...	... .	... .	
Received	.	.. ..	......	.. ..	... .	

The heading "From" is filled in with the *name* of the detachment sending the information; as "Officer's Patrol, 7th Cav." Messages sent on the same day from the same source to the same person are numbered consecutively The address is written briefly; thus, "Commanding Officer, Outpost, 1st Brigade." In the signature the writers surname only and rank are given.

This blank is four and a half by eight inches, including the margin on the left for binding. The back is ruled in squares and provided with scales for use in making simple sketches explanatory of the message. It is issued by the Signal Corps in blocks of forty with duplicating sheets. The regulation envelope is three by five and one fourth inches and is printed as follows:

U. S. ARMY FIELD MESSAGE

To............................... *No*...........

(For Signal operators only)

When sent ....*No*. ... .

Rate of speed ... .

Name of Messenger ... • .. . .

When and by whom rec'd ....

THIS ENVELOPE WILL BE RETURNED TO BEARER

MODEL MESSAGES

1. **Verbal.** "Four hostile infantrymen one mile north of our camp, moving south. I will continue north."

2. **Verbal.** "About one hundred hostile infantrymen two miles north of our camp at two o'clock, marching south. Will observe them."

474 (contd.)

3. **Verbal.** "Long column of troops marching west in Sandy Creek Valley at two o'clock. Will report details later."

4. **Verbal.** "Just fired on by cavalry patrol near Baker's Pond. Will work to their rear."

5. **Written.**   Patrol from Support No. 2,
                                                               Lone Hill,
                                          26 Mch. 11, 8-15 A. M., No. 1.

C. O.,
    Support No. 2.
      See hostile troop of cavalry halted at x-roads, one mile S. of our outguards. Nothing else in sight. Will remain here in observation.

James,
Corporal.

6. **Written** (very hurriedly).   Lone Hill, 8-30, No. 2.

C. O.,
    Support No. 2.
      Column of about 300 hostile cavalry trotting north towards hostile troop of cavalry now halted at x-roads one mile south of our outguards. Will remain here.

James,
Cpl.

7. **Written.**   Patrol from 5th Inf.,
                                          S. E. corner Boling Woods,
                                          3 Apl. 11, 2-10 P. M., No. 2.

Adjutant,
    5th Inf., near Baker House.
      Extreme right of hostile line ends at R. R. cut N. E. of BAKER'S POND. Entrenchments run S. from cut along crest of ridge. Line appears to be strongly held. Can see no troops in rear of line. Will reconnoiter their rear.

Smith,
Sergeant.

8. **Written** (from cavalry patrol far to front).
                                          Patrol from Tr. B, 7th Cav.,
                                                               Boling,
                                          14 June, 12, 10 A. M., No. 3.

To C. O.,
    Tr. B, 7th Cav.,
      S. on Chester Pike.
      No traces of enemy up to this point. Telegraph operator here reports wires running north from Boling were cut somewhere at 8-30 A. M. Inhabitants appear friendly. Will proceed north.

Jones,
Sergeant.

9. **Written** (from cavalry patrol far to front).

Patrol from Tr. B, 7th Cav.,
Oxford,
8 July, 12, 10-15 A. M., No. 2.

To C. O.,
1st Sq. 7th Cav.,
On Valley Pike, S. of York.
Bearer has canteen found in road here, marked "85 CAV.—III CORPS." Inhabitants say no enemy seen here. They appear hostile and unreliable. No telegraph operator or records remain here. Roads good macadam. Water and haystacks plentiful. Will move rapidly on towards CHESTER.

Lewis,
Sergeant.

Patrol from Support No. 3,
On Ry. ¾ mi. N. of County Road,
2 Aug. 12, 9-15 P. M., No. 1.

C. O.,
Support No. 2,
Near Maxey House.
R. R. crosses creek here on 80-foot steel trestle. Hostile detachment is posted at N. end. Strength unknown. Creek 5 ft. deep by 60 ft. wide, with steep banks, 5 ft. high. Flows through meadow land. Scattered trees along banks. R. R. approaches each end of trestle on 10-foot fill. R. R. switch to N. E. 700 yds. S. of bridge. (See sketch on back.) I will cross creek to N. of bridge.

Brown,
Corporal.

474a. A message should be sent as soon as the enemy is first seen or reported. Of course, if the enemy is actually known to be in the vicinity and his patrols have been seen, etc., you must by all means avoid wasting your men by sending them back with information about small hostile patrols or other things you know your chief is already aware of and did not specifically tell you to hunt for.

If you have properly determined in your own mind what your mission is then you will have no trouble in deciding when to send messages. For example, suppose your orders are "To reconnoiter along that ridge and determine if the enemy is present in strength," and you sight a patrol of eight men. You would waste no time or men sending back any message about the patrol, for your mission is to find out if strong bodies of the enemy are about. But suppose that while working under the above orders you located a hostile battalion of infantry—a large body of troops. In this case you would surely send a detailed message, as your mission is to determine if the enemy was present in strength.

Again, suppose that while moving towards the ridge indicated by your chief in his orders, you saw his force suddenly and heavily fired on from a new and apparently unexpected quarter, not a great distance from you, but not on the ridge referred to. You know or believe none of your patrols are out in that neighborhood. In this case you should realize instantly, without any order, that your mission had changed and you should hasten to discover the size and position of this new enemy and send the information back to your chief, first notifying him of your intended change of direction.

Never forget your mission in the excitement of leading your own little force.

Absence of the Enemy. It is frequently just as important to send a message to your chief that the enemy is not in a certain locality as it is to report his actual whereabouts. You must determine from your mission when this is the case. For example, if you were ordered "To patrol beyond that woods and see if any hostile columns are moving in that direction," and on reaching the far side of the woods you had a good view of the country for some distance beyond, it would be very important to send a message back telling your chief that you could see, say, one-half mile beyond the woods and there was no enemy in sight. This information would be of the greatest importance to him. He might feel free to move troops immediately from that vicinity to some more dangerous place. You would then continue your reconnaissance further to the front.

Suggestions for Gaining Information About the Enemy

475. Enemy on the March. (a) The patrol should observe the march of the column from a concealed position that hostile patrols or flankers are not apt to search (avoid conspicuous places). Always try to discover if one hostile detachment is followed by another—if what can be seen appears to be an advance guard of a larger body not yet in view. The distance between the detachments, their relative size, etc., is always important.

(b) **Estimating Strength of Column.** The strength of a column may be estimated from the length of time it takes to pass a selected point. As infantry in column of squads occupies half a yard per man, cavalry one yard per horse and artillery in single file twenty yards per gun or caisson (ammunition wagon), a selected point would be passed in one minute by 175 infantry; 110 cavalry (at a walk); 200 cavalry at a trot and 5 guns or caissons. If marching in columns of twos, take one-half of the above figures.

(c) **Dust.** The direction of march, strength and composition (infantry, cavalry or artillery) of a column can be closely estimated from the length and character of the cloud of dust that it makes. Dust from infantry hangs low; from cavalry it is higher, disperses more quickly, and, if the cavalry moves rapidly, the upper part of the cloud is thinner; from artillery and wagons, it is of unequal height and disconnected. The effect of the wind blowing the dust must be considered.

(d) **Trail of Column.** Evenly trodden ground indicates infantry; prints of horseshoes mean cavalry and deep and wide wheel tracks indicate artillery. If the trail is fresh, the column passed recently; if narrow, the troops felt secure and were marching in column of route; if broad they expected an action and were prepared to deploy. A retreating army makes a broad trail across fields, especially at the start.

Always remember that the smallest or most insignificant things, such as the number of a regiment or a discarded canteen or collar ornament, may give the most valuable information to a higher commander. For example, the markings on a discarded canteen or knapsack might prove to a general commanding an army that a certain hostile division, corps, or other force was in front of him when he thought it had not been sent into the field. The markings on the canteen would convey little or no meaning to the patrol leader, but if he realized his duty he would take care to report the facts. Cavalry patrols working far ahead of the foot troops should be most careful to observe and report on such details.

(e) **Reflection of Weapons.** If brilliant, the troops are marching toward you, otherwise they are probably marching away from you.

Enemy in Position. (a) If an outpost line, the patrol locates the line of sentinels, their positions, the location and strength of the out-guards and, as far as possible, all troops in rear. The location of the flanks of the line, whether in a strong or weak position, is of the utmost importance. Places where the line may be most easily penetrated should be searched for and the strength and routes of the hostile patrols observed.

As outposts are usually changed at dawn this is the best time to reconnoiter their positions.

(b) A hostile line of battle is usually hard to approach, but its extent, where the flanks rest and whether or not other troops are in rear of these flanks, should be most carefully determined.

Information as to the flanks of any force, the character of the country on each flank, etc., is always of the greatest importance, because the flanks are the weakest portions of a line. In attacking an enemy an effort is almost always made to bring the heaviest fire or blow to bear on one of his flanks. Naturally all information about this most vulnerable part of an enemy is of great importance.

476. Prisoners. When a patrol is ordered to secure prisoners they should be questioned as soon as captured, while still excited and their replies can in a way be verified. Their answers should be written down (unknown to them) and sent back with them as a check on what they may say on second thought.

Prisoners should always be questioned as to the following points: What regiment, brigade, division, etc., they belong to; how long they have been in position, on the march, etc.; how much sickness in their organization; whether their rations are satisfactory; who commands their troops, etc. Always try to make the prisoners think the questions are asked out of mere curiosity.

477. Camp Noises. The rumble of vehicles, cracking of whips, neighing of horses, braying of mules and barking of dogs often indicate the arrival or departure of troops. If the noise remains in the same place and new fires are lighted, it is probable that reënforcements have arrived. If the noise grows more indistinct, the troops are probably withdrawing. If, added to this, the fires appear to be dying out, and the enemy seems to redouble the vigilance of the outposts, the indications of retreat are strong.

478. Abandoned Camps. (a) Indications are found in the remains of camp fires. They will show, by their degree of freshness, whether much or little time elapsed since the enemy left the place, and the quantity of cinders will give an indication of the length of time he occupied it. They will also furnish a means of estimating his force approximately, ten men being allowed to each fire.

(b) Other valuable indications in regard to the length of time the position was occupied and the time when it was abandoned may be found in the evidence of care or haste in the construction of huts or shelters, and in the freshness of straw, grain, dung or the entrails of slaughtered animals. Abandoned clothing, equipments or harness will give a clue to the arms and regiments composing a retreating force. Dead horses lying about, broken weapons, discarded knapsacks, abandoned and broken-down wagons, etc., are indications of the fatigue and demoralization of the command. Bloody bandages lying about, and many fresh graves, are evidences that the enemy is heavily burdened with wounded or sick.

479. Flames or Smoke. If at night the flames of an enemy's camp fires disappear and reappear, something is moving between the observer and the fires. If smoke as well as flame is visible, the fires are very near. If the fires are very numerous and lighted successively, and if soon after being lighted they go out it is probable the enemy is preparing a retreat and trying to deceive us. If the fires burn brightly and clearly at a late hour, the enemy has probably gone, and has left a detachment to keep the fires burning. If, at an unusual time, much smoke is seen ascending from an enemy's camp, it is probable that he is engaged in cooking preparatory to moving off.

If lines of smoke are seen rising at several points along a railway line in the enemy's rear, it may be surmised that the railroad is being destroyed by burning the crossties, and that a retreat is planned.

480. Limits of vision. (a) On a clear day a man with good vision can see:

At a distance of 9 to 12 miles, church spires and towers;

At a distance of 5 to 7 miles, windmills;

At a distance of 2 to 2½ miles, chimneys of light color;

At a distance of 2,000 yards, trunks of large trees;

At a distance of 1,000 yards, single posts;

At 500 yards the panes of glass may be distinguished in a window.

(b) Troops are visible at 2,000 yards, at which distance a mounted man looks like a mere speck; at 1,200 yards infantry can be distinguished from cavalry; at 1,000 yards a line of men looks like a broad belt; at 600 yards the files of a squad can be counted, and at 400 yards the movements of the arms and legs can be plainly seen.

(c) The larger, brighter or better lighted an object is, the nearer it seems. An object seems nearer when it has a dark background than when it has a light one, and closer to the observer when the air is clear than when it is raining, snowing, foggy or the atmosphere is filled with smoke. An object looks farther off when the observer is facing the sun than when he has his back to it. A smooth expanse of snow, grain fields or water makes distances seem shorter than they really are.

Suggestions for the Reconnaissance of Varous Positions and Localities

481. Cross roads should be reconnoitered in each direction for a dis-tance depending on how rapidly the patrol must continue on, how far from the main road the first turn or high point is, etc. The main body of the patrol usually remains halted near the crossroads, while flankers do the reconnoitering.

482. Heights. In reconnoitering a height, if the patrol is large enough to admit of detaching them, one or two men climb the slope on either flank, keeping in sight of the patrol, if possible. In any case, one man moves cautiously up the hill, followed by the others in the file at such distance that each keeps his predecessor in view.

483. Defiles. On approaching a defile, if time permits, the heights on either side are reconnoitered by flankers before the patrol passes through, in single file at double time, the distance being the same as in ascending a hill. The same method is adopted in reconnoitering a railroad cut or sunken road.

484. Bridges and Fords. At a bridge or ford, the front of the patrol is contracted so as to bring all the men to the passage. The leading patrolers cross first and reconnoiter the far side to prevent the possibility of the enemy surprising the main body of the patrol as it is crossing the bridge. The patrol then crosses rapidly, and takes up a proper formation. A bridge is first examined to see that it is safe and has not been tampered with by the enemy.

485. Woods. The patrol enters a wood in skirmishing order, the intervals being as great as may be consistent with mutual observation and support on the part of the members of the patrol. On arriving at the farther edge of the wood, the patrol remains concealed and carefully looks about before passing out to open ground. When there is such a growth of underbrush as to make this method impracticable, and it is necessary to enter a wood by a road, the road is reconnoitered as in case of defile, though not usually at double time.

486. Enclosures. In reconnoitering an enclosure, such as a garden, park or cemetery, the leading patrolers first examine the exterior, to make sure that the enemy is not concealed behind one of the faces of the enclosure. They then proceed to examine the interior. Great care is taken in reconnoitering and entering an enclosure to avoid being caught in a confined or restricted space by the enemy.

487. Positions. In approaching a position, but one man advances (one is less liable to be detected than two or more), and he crawls cautiously toward the crest of the hill or edge of the wood or opening of the defile, while the others remain concealed in the rear until he signals them to advance.

488. Houses. When a house is approached by a patrol, it is first reconnoitered from a distance, and if nothing suspicious is seen, it is then approached by one or two men, the rest of the party remaining concealed in observation. If the patrol is large enough to admit of it, four men approach the house, so as to examine the front and back entrances at the same time. Only one man enters the door, the others remaining outside to give the alarm, should a party of the enemy be concealed in the house. The patrol does not remain in the vicinity of

[413]

the house any longer than necessary, as information relative to its numbers and movements might be given to the enemy, if a hostile party should subsequently visit the place. Farmhouses are searched for newspapers and the inhabitants questioned. If necessary to go up to a building, wood or hill, where an enemy is likely to be concealed, run for the last couple of hundred yards, having your rifle ready for instant use, and make for some point that will afford you cover when you get close up. In the case of a building, for instance, you would make for one of the corners. Such a maneuver would probably be disconcerting to anyone who might be lying in wait for you, and would be quite likely to cause them to show themselves sooner than they intended, and thus give you a chance to turn around and get away. If they fired on you while you were approaching at a run, they would not be very likely to hit you.

489. Villages. (a) In approaching a small village one or two men are sent in to reconnoiter and one around each flank, but the main body does not enter until the scouts have reported. In small patrols of three to six men so much dispersion is not safe and only one section of the village can be reconnoitered at a time.

(b) If the presence of the enemy is not apparent, the patrol enters the village. A suitable formation would be in single file at proper distance, each man being on the opposite side of the street from his predecessor, thus presenting a more difficult target for hostile fire and enabling the men to watch all windows.

(c) If the patrol is strong enough, it seizes the postoffice, telegraph office and railroad stations, and secures all important papers, such as files of telegrams sent and received, instructions to postmasters, orders of town mayor, etc., that may be there. If the patrol is part of the advance guard, it seizes the mayor and postmaster of the place and turns them over to the commander of the vanguard with the papers seized.

(d) While searching a village sentinels are placed at points of departure to prevent any of the inhabitants from leaving. Tall buildings and steeples are ascended and an extensive view of the surrounding country obtained.

(e) At night a village is more cautiously approached by a small party than by day. The patrol glides through back alleys, across gardens, etc., rather than along the main street. If there are no signs of the enemy, it makes inquiry. If no light is seen, and it seems imprudent to rouse any of the people, the patrol watches and captures one of the inhabitants, and gets from him such information as he may possess.

(f) The best time for the patrol to approach a village it at early dawn, when it is light enough to see, but before the inhabitants are up. It is dangerous in the extreme for a small patrol to enter a village unless it is certain that it is not occupied by the enemy, for the men could be shot down by fire from the windows, cellarways, etc., or entrapped and captured. As a rule large towns and cities are not entered by small patrols, but are watched from the outside, as a small force can not effectively reconnoiter and protect itself in such a place.

Facts Which Should Be Obtained by Patrols to Certain Objects

490. Roads. Their direction, their nature (macadamized, corduroy, plank, dirt, etc.), their condition of repair, their grade, the nature of crossroads, and the points where they leave the main roads; their borders (woods, hedges, fences or ditches), the places at which they pass through defiles, cross heights or rivers, and where they intersect railroads, their breadth (whether suitable for column of fours or platoons, etc.).

491. Railroads. Their direction, gauge, the number of tracks, stations and junctions, their grade, the length and height of the cuts, embankments and tunnels.

492. Bridges. Their position, their width and length, their construction (trestle, girder, etc.), material (wood, brick, stone or iron), the roads and approaches on each bank.

493. Rivers and Other Streams. Their direction, width and depth, the rapidity of the current, liability to sudden rises and the highest and lowest points reached by the water, as indicated by drift wood, etc., fords, the nature of the banks, kinds, position and number of islands at suitable points of passage, heights in the vicinity and their command over the the banks.

494. Woods. Their situation, extent and shape; whether clear or containing underbrush; the number and extent of "clearings" (open spaces); whether cut up by ravines or containing marshes, etc.; nature of roads passing through them.

495. Canals. Their direction, width and depth; condition of towpaths; locks and means of protecting or destroying them.

496. Telegraphs. Whether they follow railroads or common roads; stations, number of wires.

497. Villages. Their situation (on a height, in a valley or on a plain); nature of the surrounding country; construction of the houses, nature (straight or crooked) and width of streets; means of defense.

498. Defiles. Their direction; whether straight or crooked; whether heights on either side are accessible or inaccessible; nature of ground at each extremity; width (frontage of column that can pass through).

499. Ponds and Marshes. Means of crossing; defensive use that might be made of them as obstacles against enemy; whether the marshy grounds are practicable for any or all arms.

500. Springs and Rivulets. Nature of approaches; whether water is drinkable and abundant.

501. Valleys. Extent and nature; towns, villages, hamlets, streams, roads and paths therein; obstacles offered by or in the valley, to the movement of troops.

502. Heights. Whether slopes are easy or steep; whether good defensive positions are offered; whether plateau is wide or narrow; whether passages are easy or difficult; whether the ground is broken or smooth, wooded or clear.

Suggestions for Patrols Employed in Executing Demolition

(Destruction or blocking of bridges, railroads, etc.)

503. Patrols never execute any demolition unless specifically ordered to do so. Demolition may be of two different characters: Temporary demolition, such as cutting telegraph wires in but a few places or merely burning the flooring of bridges, removing a few rails from a track, etc., and permanent demolition, such as cutting down an entire telegraph line, completely destroying bridges, blowing in tunnels, etc. Only temporary demolition will be dealt within this book.

504. Telegraph Line. To temporarily disable telegraph lines, connect up different wires close to the glass insulators, wrap a wire around all the wires and bury its ends in the ground (this grounds or short circuits the wire), or cut all the wires in one or two places.

505. Railroads. To temporarily disable railroads remove the fish plates (the plates that join the rails together at the ends) at each end of a short section of track, preferable upon an embankment, then have as many men as available raise the track on one side until the ties stand on end and turn the section of track so that it will fall down the embankment; or, cut out rails by a charge of dynamite or gun cotton placed against the web and covered up with mud or damp clay. Eight to twelve ounces of explosive is sufficient. Or blow in the sides of deep cuts or blow down embankments. Bridges, culverts, tunnels, etc., are never destroyed except on a written order of the commander-in-chief.

506. Wagon Road.—(a) Bridges can be rendered temporarily useless by removing the flooring, or, in the case of steel bridges, by burning the flooring (if obtainable, pour tar or kerosene on flooring), particularly if there is not time to remove it.

Short culverts may sometimes be blown in.

A hastily constructed barricade across a bridge or in a cut of trees, wagons, etc., may be sufficient in some cases where only the temporary check of hostile cavalry or artillery is desired.

(b) The road bed may be blocked by digging trenches not less than thirty feet wide and six feet deep, but as this would take a great deal of time patrols would rarely be charged with such work.

507. Report on Return of Patrol. On returning the patrol leaders should make a short verbal or written report, almost always the former, briefly recounting the movements of the patrol, the information obtained of the enemy, a description of the country passed over and of friendly troops encountered. Of course, this is not practicable when the situation is changing rapidly and a returning patrol is immediately engaged in some new and pressing duty.

508.   **Model Reports of Patrol Leaders**

1. **Verbal.**

    Patrol Leader (Corporal B): Sir, Corporal B reports back with his patrol.

    Captain A: I received two messages from you, corporal. What else did you discover?

    Corporal B: That was a regiment of infantry, sir, with one battalion thrown out as advance guard. The main body of two battalions went into bivouac at the crossroads and the advance guard formed an outpost line along the big creek two miles south of here.

Captain A: Give me an account of your movements.

Corporal B: We followed this main road south to the creek, where we avoided a mounted patrol moving north on the road at 1-45 P. M., and then reconnoitered the valley from a ridge west of the road. We followed the ridge south for half a mile to a point where we could see a road crossing the valley and the main road at right angles, three miles south of here. There we halted, and at 2:20 what seemed to be the point and advance party (about forty men) of an infantry advance guard appeared, marching north up this road, the head at the crossroad. I then sent you message No. 1 by Private Brown.

In fifteen minutes three companies had appeared 600 yards in rear of the advance party, and I could see a heavy, low column of dust about one-half mile further to the rear. Message No. 2 was then sent in by Privates Baker and Johnson, and to avoid several hostile patrols, I drew off further to the northwest.

The advance guard then halted and established an outpost line along the south of the creek, two miles from here. The cloud of dust proved to be two more battalions and a wagon train. These two battalions went into bivouac on opposite sides of this road at the crossroads and sent out strong patrols east and west on the crossroad. Five wagons went forward to the outpost battalion and the reserve built cook fires.

As Private Rush, here, was the only man I had left, we started back, sketching the valley, ridge and positions of the main body and outpost. Here is the sketch, sir. The fields are all cut crops or meadow.

We sighted two foot patrols from the outpost, moving north about a mile from here, one following the road and one further east.

I did not see any of our patrols.

That is all, sir.

2. **Written.**

Report of Sergeant Wm. James' Patrol of Five Men

Support No. 1,
Outpost of 6th Inf., Near Dixon,
22 Aug. 12, 2-30 to 5 P. M.

The patrol followed the timber along the creek for one mile S. from our outguards and leaving the creek bottom moved ½ mile S. E. to the wooded hill (about 800 ft. high), visible from our lines.

From this hill top the valley to the east (about one mile wide) could be fairly well observed. No signs of the enemy were seen and a message, No. 1 was sent back by Private Russell.

A wagon road runs N. and S. through the valley, bordered by four or five farms with numerous orchards and cleared fields. Both slopes of the valley are heavily wooded.

The patrol then moved S. W., until it struck the macadam pike which runs N. and S., through our lines. Proceeding S. 400 yds. on this pike to a low hill a farmer, on foot, was met. Said he lived one mile further S.; was looking for some loose horses; that four hostile cavalrymen, from the east, stopped at his farm at noon, drank some milk, took oats for their horses, inquired the way to Dixon and rode off in that

509

direction within fifteen minutes. He said they were the first hostiles he had seen; that they told nothing about themselves, and they and their horses looked in good condition. Farmer appeared friendly and honest.

The patrol then returned to our lines following the pike about two miles. Road is in good condition, low hedges and barbed wire fences, stone culverts and no bridges in the two miles. Bordering country is open and gently rolling farming country and all crops are in. A sketch is attached to this report. None of our patrols was seen.

Respectfully submitted,

Wm. James,
Sergeant, Co. A, 6th Infy.

509.   Problem in Patrol Leading and Patrolling

In studying or solving tactical problems on a map you must remember that unless you carefully work out your own solution to the problem before looking at the given solution, you will practically make no progress.

It is best, if your time permits, to write out your solutions, and when you read over the given solutions, compare the solution of each point with what you thought of that same point when you were solving the problem, and consider why you did just what you did. Without this comparison much of the lasting benefit of the work is lost.

In some of these problems both the problem and solutions are presented in dialogue form so as to give company officers examples of the best method of conducting the indoor instruction of their men in minor tactics. It also gives an example of how to conduct a tactical walk out in the country, simply looking at the ground itself, instead of a map hanging on the wall. The enlarged Elementary Map referred to in Par. 454, is supposed to be used in this instruction as well as in the war games.

Problem No. 1. (Infantry)

The Elementary Map (scale 12 inches to the mile) being hung on the wall, about two sergeants and two squads of the company are seated in a semicircle facing it, and the captain is standing beside the map with a pointer (a barrack cleaning rod makes an excellent pointer).

Captain. We will suppose that our company has just reached the village of York. The enemy is reported to be in the vicinity of Boling and Oxford (he points out on the map all places as they are mentioned). We are in the enemy's country.

Corporal James, I call you up at 3 P. M. and give you these orders: "Nothing has been seen of the enemy yet. Our nearest troops are three miles south of here. Take four men from your squad and reconnoiter along this road (County Road) into the valley on the other side of that ridge over there (points to the ridge just beyond the cemetery), and see if you can discover anything about the enemy. Report back here by 5 o'clock. I am sending a patrol out the Valley Pike." Now, Corporal, state just what you would do.

Corporal James: I would go to my squad, fall in Privates Amos, Barlow, Sharp and Brown; see that they had full canteens; that their

[418]

arms were all right; that they were not lame or sick and I would have them leave their blanket rolls, haversacks and entrenching tools with the company. (Par. 460.)

I would then give these orders (Par. 459): "We are ordered out on patrol duty. Nothing has been seen of the enemy yet. Our nearest troops are three miles south of here. We are ordered to reconnoiter along this road into the valley on the other side of that ridge, and see if we can discover anything about the enemy. Another patrol is going up the Valley Pike. Reports are to be sent here. In case we are scattered we will meet at that woods on the hill over there (indicates the clump of trees just west of Mills' farm).

I will go ahead. Amos, follow about fifty yards behind me. Barlow, you and Sharp keep about 100 yards behind Amos, and Brown will follow you at half that distance. All keep on the opposite side of the road from the man ahead of you." (Par. 463.)

Captain: All right, Corporal, now describe what route you will follow.

Corporal James: The patrol will keep to the County Road until the crest of the ridge near the stone wall is reached, when what I see in the valley beyond will decide my route for me.

Captain: How about the woods west of the stone walls?

Corporal James: If I did not see anyone from our patrol on the Valley Pike reconnoitering there, I would give Barlow these orders just after we had examined the cemetery, when the patrol would have temporarily closed up somewhat: "Barlow, take Sharp and examine that little woods over there. Join us at the top of this hill." I would then wave to Brown to close up and would proceed to the hill top.

Captain: Barlow, what do you do?

Private Barlow: I would say, "Sharp, out straight across for that woods. I will follow you." I would follow about 100 yards behind him. When he reached the edge of the woods I would signal him to halt by holding up my left hand. After I had closed up to about fifty yards I would say to him, "Go into the woods and keep me in sight." I would walk along the edge of the woods where I could see Sharp and the corporal's patrol on the road at the same time.

Captain: That is all right, Barlow. Corporal, you should have instructed Amos or Brown to keep a close watch on Barlow for signals.

Corporal James: I intended to watch him myself.

Captain: No, you would have enough to do keeping on the alert for what was ahead of you. Now describe how you lead the patrol to the top of the hill, by the stone wall.

Corporal James: When I reached the crest I would hold up my hand for the patrol to halt and would cautiously advance and look ahead into the valley. If I saw nothing suspicious I would wave to the men to close up and say, "Amos, go to that high ground about 250 yards over there (indicates the end of the nose made by the 60-foot contour just north of the east end of the stone wall), and look around the country." I would keep Brown behind the crest, watching Barlow's movements.

Captain: Now, Corporal, Amos reaches the point you indicated and Barlow and Sharp join you. What do you do?

Corporal James: Can I see the Steel Bridge over Sandy Creek?

Captain: No, it is three-fourths of a mile away and the trees along the road by Smith's hide it. You can see the cut in the road east of the bridge and the Smith house, but the cross roads are hidden by the trees bordering the roads. You see nothing suspicious. It is a clear, sunny afternoon. The roads are dusty and the trees in full foliage. The valley is principally made up of fields of cut hay, corn stubble and meadow land.

Corporal James: Does Private Amos give me any information?

Captain: No, he makes you no signals. You see him sitting behind a bush looking northwest, down the valley.

Corporal James: I would say, "Barlow, head straight across to where that line of trees meets the road (indicates the point where the lane from Mills' farm joins the Chester Pike). Sharp. keep about fifty yards to my right rear." I would follow Barlow at 150 yards and when I had reached the bottom land I would wave to Amos to follow us.

Captain: How about Brown?

Corporal James: I had already given him his orders to follow as rear guard and he should do so without my telling him.

Captain: Amos, what do you do when you see the corporal wave to you?

Private Amos: I would go down the hill and join him.

Captain: No, you could do better than that. You are too far from the corporal for him to signal you to do much of anything except stay there or join him. You should join him, but you should not go straight down to him. You should head so as to strike the Mills' Lane about 100 yards east of the house and then go down the lane, first looking along the stone wall. In this way you save time in reconnoitering the ground near the Mills' farm and protect the patrol against being surprised by an enemy hidden by the line of trees, or the wall along the lane. You are not disobeying your orders but just using common sense in following them out and thinking about what the corporal is trying to do.

Now, Corporal, why didn't you go to the Smith house and find out if the people there had seen anything of the enemy?

Corporal James: You said we were in the enemy's country, sir, so I thought it best to avoid the inhabitants until I found I could not get information in any other way. I intended first to see if I could locate any enemy around here, and if not, to stop at houses on my return. In this way I would be gone before the people could send any information to the enemy about my patrol.

Captain: Barlow reaches the Chester Pike where the Mills' lane leaves it. You are about 150 yards in his rear. Sharp is 50 yards off to your right rear, Amos 100 yards to your left rear and Brown 50 yards behind you. Just as Barlow starts to climb over the barbed wire fence into the Chester Pike you see him drop down on the ground. He signals, "Enemy in sight." Tell me quickly what would you do?

[420]

Corporal James: I would wave my hand for all to lie down, and I would hasten forward, stooping over as I ran, until I was about twenty yards from him, when I would crawl forward to the fence, close by him. Just before I reached him I would ask him what he saw.

Captain: He replies, "There are some hostile foot soldiers coming up this road."

Corporal James: I would crawl forward and look.

Captain: You see three or four men, about 500 yards north of you, coming up the Chester Pike. They are scattered out.

Corporal James: I would say, "Crawl into the lane, keep behind the stone wall, watch those fellows, and work your way to that farm" (indicates the Mills' farm). I would start towards the Mills' farm myself, under cover of the trees along the lane and would wave to the other men to move rapidly west, towards the hills.

Captain: Why didn't you try to hide near where you were and allow the hostile men to pass?

Corporal James: There does not seem to be any place to hide near there that a patrol would not probably examine.

Captain: What is your plan now?

Corporal James: I want to get my patrol up to that small woods near the Mills' farm, but I hardly expect to be able to get them up to that point without their being seen. In any event, I want them well back from the road where they can lie down and not be seen by the enemy when he passes.

Captain: You succeed in collecting your patrol in the woods without their being seen, and you see four foot soldiers in the road at the entrance to the land. One man starts up the lane, the others remaining on the road.

Corporal James: I say, "Brown, go through these woods and hurry straight across to York. You should be able to see the village from the other side of the woods. Report to the captain that a hostile patrol of four foot men is working south up the valley, two miles northeast of York. We will go further north. Repeat what I have told you." (Par. 474.)

Captain: Why didn't you send this message before?

Corporal James: Because we were moving in the same direction that the messenger would have had to go, and, by waiting a very few minutes, I was able to tell whether it was a mere patrol or the point of an advance guard.

Captain: Do you think it correct to send a messenger back with news about a small patrol?

Corporal James: Ordinarily it would be wrong, but as nothing has been seen of the enemy until now, this first news is important because it proves to the Captain that the enemy really is in this neighborhood, which it seems to me is a very important thing for him to know and what my mission required me to do. (Par. 474a.)

Captain: What are you going to do now, Corporal?

Corporal James: We have traveled about two miles and stopped frequently, so it must be about 4 o'clock. It is one and one-third miles back to York, where I should arrive about 5 o'clock. It would take me twenty-five minutes to go from here to York, so I have about thirty-five minutes left before 5 o'clock. This will permit me to go forward another mile and still be able to reach York on time. It is two-thirds of a mile to the Mason farm, and if the hostile patrol appears to be going on, I will start for that point. Did anyone at the Mills' farm see us?

Captain: No, but tell me first why you do not go along this high ground that overlooks the valley?

Corporal James: Because our patrol that started out the Valley Pike is probably near Twin Hills and I want to cover other country. The orchard at Mason's would obstruct my view from the hills.

Captain: The hostile patrol goes on south. Describe briefly your next movements.

Corporal James: I lead my patrol over to Mason's and, concealing two of the men so that both roads and the house can be watched, I take one man and reconnoiter around the farm yard and go up to the house to question the inhabitants. (Par. 488.)

Captain: You find one woman there who says some other soldiers, on foot, passed there a few minutes ago, marching south. She gives you no other information about the enemy or country.

Corporal James: I would send Amos over to see how deep and wide Sandy Creek is (Par. 493.) When he returned I would take the patrol over to Twin Hills, follow the ridge south to the stone wall on the County Road, watching the valley for signs of the hostile patrol, and follow the road back to York; then make my report to the Captain, telling him where I had gone, all I had seen, including a description of the country. If I had not been hurried, I would have made a sketch of the valley. I can make a rough one after I get in. (Par. 507.)

Captain: Suppose on your way back you saw hostile troops appearing on the County Road, marching west over Sandy Ridge. Would you stay out longer or would you consider that you should reach Oxford by 5 o'clock?

Corporal James: I would send a message back at once, and remain out long enough to find out the strength and probable intention of the new enemy.

Captain (to one platoon of his troop of cavalry): We will suppose that this troop has just (9 A. M.) arrived in Boling (Elementary Map) on a clear, dry, summer day. The enemy is supposed to be near Salem and we have seen several of his patrols this morning on our march south to Boling. Sergeant Allen, I call you up and give you these instructions: ''Take Corporal Burt's squad (eight men) and reconnoiter south by this road (indicates the Boling-Morey house road) to Salem. I will take the troop straight south to Salem and you will join it there about 10:15. It is four and one-half miles to Salem. Start at once '' (You have no map.)

Sergeant Allen: I would like to know just what the Captain wishes my patrol to do. (Par. 461.)

Captain: We will suppose that this is one of the many occasions in actual campaign where things must be done quickly. Where there is no time for detailed orders. You know that the troop has been marching south towards Salem where the enemy is supposed to be. You also know we have seen several of his patrols. I have told you what the troop is going to do, and from all this you should be able to decide what your mission is in this case. We will, therefore, consider that there is no time to give you more detailed orders, and you have to decide for yourself. Of course, if you had failed to hear just what I said, then, in spite of the necessity for haste, I would repeat my instructions to you. [Par. 459(b).]

Sergeant Allen: I would ride over to Corporal Burt's squad and lead it out of the column to the road leading to the Morey house, and say, "The troop is going on straight south to Salem, four and one-half miles away. This squad will reconnoiter south to Salem by this road, joining the troop there about 10:15. In case we become separated, make for Salem. Corporal, take Brown and form the point. I will follow with the squad about 300 yards in rear. Regulate your gait on me after you get your distance. Move out now at a trot." [Par. 459(a).]

After Corporal Burt had gotten 150 yards out I would say, "Carter, move out as connecting file." I would then say, "Downs, you will follow about 150 yards behind us as rear guard." When Carter had gone 150 yards down the road I would order, "1. Forward; 2. trot; 3. MARCH," and ride off at the head of the four remaining men (in column of twos). (Par. 463.)

Captain: Sergeant, tell me briefly what is your estimate of the situation—that is, what sort of a proposition you have before you and how you have decided to handle it.

Sergeant Allen: As the enemy is supposed to be near Salem and we have already seen his patrols, I expect to encounter more patrols and may meet a strong body of the enemy, on my way to Salem. As I have no map, I cannot tell anything about the road, except that it is about four and one-half miles by the direct road the troop will follow, therefore my route will be somewhat longer. I have been given an hour and fifteen minutes in which to make the trip, so, if I move at a trot along the safer portions of the road, I will have time to proceed very slowly and cautiously along the dangerous portions. My patrol will be stretched out about 500 yards on the road, which should make it difficult for the enemy to surprise us and yet should permit my controlling the movements of the men. (Par. 463.)

I consider that my mission is to start out on this road and find my way around to Salem in about an hour and, particularly, to get word across to the Captain on the other road of anything of importance about the enemy that I may learn.

Captain: Very well. When you reach the cut in the road across the south nose of Hill 38, your point has almost reached the Morey house. Do you make any change in your patrol?

Sergeant Allen: I order, "1. Walk, 2. MARCH," and watch to see if the connecting file observes the change of gait and comes to a walk.

Captain: Suppose he does not come to a walk?

Sergeant Allen: I would say, "Smith, gallop ahead and tell Carter to walk and to keep more on the alert."

Captain: Corporal Burt, you reach the road fork at Morey's. What do you do?

Corporal Burt: I say, "Brown, wait here until Carter is close enough to see which way you go and then trot up to me." I would walk on down the road.

Captain: Wouldn't you make any inspection of the Morey house?

Corporal Burt: Not unless I saw something suspicious from the road. I would expect the main body of the patrol to do that.

Captain: Don't you make any change on account of the woods you are passing?

Corporal Burt: No, sir. It has very heavy underbrush and we would lose valuable time trying to search through it. A large force of the enemy would hardly hide in such a place.

Captain: Sergeant Allen, you reach the road fork. What do you do?

Sergeant Allen: I would have two men go into the Morey house to question anyone they found there. I would order one of the other two men to trot up (north) that road 200 yards and wait until I signaled to him to return. With the other man I would await the result of the inspection of the Morey house. Corporal Burt should have gone ahead without orders to the cut in the road across Long Ridge, leaving Brown half way between us. (Pars. 481 and 488.)

Captain: You find no one at the Morey house.

Sergeant Allen: I would signal the man to the north to come in. I would then order two men to "find a gate in the fence and trot up on that hill (indicating Long Ridge), and look around the country and join me down this road." (Par. 463.) I would then start south at a walk, halting at the cut to await the result of the inspection on the country from the hill.

Captain: Foster, you and Lacey are the two men sent up on Long R'dge. When you reach the hill top you see four hostile cavalrymen trotting north on the Valley Pike, across the railroad track.

Pr'vate Foster: I signal like this (enemy in sight), and wait to see if they go on north. (Par. 473.) Do I see anything else behind or ahead of them?

Captain: You see no other signs of the enemy on any road. Everything looks quiet. The hostile cavalrymen pass the Baker house and cont'nue north.

Private Foster: I would then take Lacey, trot down the ridge to Sergeant Allen, keeping below the crest and report, "Sergeant, we saw four hostile mounted men trotting north on the road about three-quarters of a mile over there (pointing), and they kept on north, across that road (pointing to the Brown-Baker-Oxford road). There was nothing else in sight." I would then tell him what the country to the south looked like, if he wanted to know.

Captain: Sergeant Allen, what do you do now?

Sergeant Allen: I would continue toward the Brown house at a trot. I would send no message to you as you already know there are hostile patrols about and therefore this information would be of little or no importance to you. (Par. 474a.)

Captain: You arrive at Brown's house.

Sergeant Allen: I would send two men in to question the people and I would continue on at a walk. I would not send any one up the road towards Oxford as Foster has already seen that road.

Captain: You should have sent a man several hundred yards out the Farm Lane. (Par. 481.) If he moved at a trot it would only have taken a very short time. Continue to describe your movements.

Sergeant Allen: I would halt at the railroad track until I saw my two men coming on from the Brown house. I would then direct the other two men who were with me to go through the first opening in the fence to the west and ride south along that ridge (62—Lone Hill—Twin Hills' ridge) until I signaled them to rejoin. I would tell them to look out for our troop over to the east. If there were a great many fences I would not send them out until we were opposite the southern edge of that woods ahead of us. There I would send them to the high ground to look over the country, and return at once.

Captain: There are a great many fences west of the road and practically none east of the road to Sandy Creek. Just as you arrive opposite the southern edge of those woods and are giving orders for the two men to ride up the hill, you hear firing in the direction of Bald Knob. In the road at the foot of the south slope of Bald Knob, where the trail to the quarry starts off, you can see quite a clump of horses. You see nothing to the west of your position or towards Mason's. What do you do?

Sergeant Allen: I signal "RALLY" to Carter and Downs. If there is a gate nearby I lead my men through it. If not, I have them cut or break an opening in the fence and ride towards the railroad fill at a fast trot, having one man gallop ahead as point.

When we reach the fill, the point having first looked beyond it, I order, "DISMOUNT. Lacey, hold the horses. 1. As skirmishers along that fill, 2. MARCH." When Corporal Burt, Brown, Carter and Downs come up Lacey takes their horses and they join the line of skirmishers. Captain, what do I see from the fill?

Captain: There appear to be about twenty or thirty horses in the group. The firing seems to come from the cut in the road just north of the horses and from the clump of trees by the Quarry. You can also hear firing from a point further north on the road, apparently your troop replying to the fire from Bald Knob. You see nothing in the road south of the horses as far as Hill 42, which obstructs your view. What action do you take?

Sergeant Allen: I order, "AT THE FEET OF THOSE HORSES. RANGE, 850. CLIP FIRING."

Captain: What is your object in doing as you have done?

Sergeant Allen: I know the captain intended to go to Salem with the troop. From the fact that he is replying to the hostile fire I judge he still wishes to push south. I was ordered to reconnoiter along this road, but now a situation has arisen where the troop is being prevented or delayed in doing what was desired and I am in what appears to be a very favorable position from which to give assistance to the troop and enable them to push ahead. I am practically in rear of the enemy

and within effective range of their lead horses. I therefore think my
mission has at last temporarily changed and I should try and cause
the twenty or thirty hostile troopers to draw off [Par. 457(b)']. Besides, I think it is my business to find out what the strength of this
enemy is and whether or not he has reinforcements coming up from
Salem, and send this information to the captain. From my position I
can still watch the Chester Pike.

Captain: After you have emptied your clips you see the enemy
running down out of the cut and from among the trees mount their
horses and gallop south. What do you do?

Sergeant Allen: I would send Foster across the creek above the
trestle (south of trestle), to ride across to that road (pointing towards
the cut on Bald Hill) and tell the captain, who is near there, that
about thirty men were on the hill and they have galloped south, and
that I am continuing towards Salem. I would have Foster repeat the
message that I gave him. I would then trot back to the Chester Pike
and south to Mason's, taking up our old formation.

Captain: You see nothing unusual at Mason's and continue
south until you reach the cross roads by the Smith farm. Corporal
Burt and Private Brown are near the stone bridge south of Smith's;
Private Carter is half way between you and Corporal Burt; and Private
Downs is 100 yards north of Smith's. You have three men with you.
What do you do?

Sergeant Allen: What time is it now?

Captain: It is now 10:45 a. m.

Sergeant Allen: I would say, "Lacey, take Jackson and gallop as far as that cut in the road (points east) and see if you can
locate the enemy or our troop in the valley beyond. I will wave my
hat over my head when I want you to return." I would then say to
Private Moore, "Gallop down to Corporal Burt and tell him to fall back
in this direction 100 yards, and then you return here bringing the other
two men with you." I would then await the result of Private Lacey's
reconnaissance, sending Carter to the turn in the road 200 yards west
of the cross roads.

Captain: Lacey, what do you do?

Private Lacey: I order Jackson, "Follow 75 yards behind me
and watch for signals from Sergeant Allen," and I then gallop across
the steel bridge and half way up the hill. I then move cautiously up to
the cut and, if the fences permit, I ride up on the side of the cut, dismounting just before reaching the crest of the ridge, and walk forward
until I can see into the valley beyond.

Captain: You see no signs of the enemy in the valley, but you
see your own troop on the road by the Gibbs farm with a squad in advance in the road on Hill 42.

Private Lacey: I look towards Sergeant Allen to see if he is
signaling. I make no signals.

Captain: What do you do, Sergeant?

Sergeant Allen: I wave my hat for Private Lacey to return.
I wave to Private Downs to join me and when Private Lacey arrives
I signal "ASSEMBLE" to Corporal Burt and then say, "Lacey, join

Corporal Burt and tell him to follow me as rear guard. Martin, join Carter and tell him to trot west. We will follow. You stay with him." After he got started I would order, "Follow me. 1. Trot; 2. **MARCH.**"

Captain: When Private Carter reaches the crest of the ridge about one-half mile west of Smith's he signals, "Enemy in sight in large numbers," and he remains in the road with Martin fifty yards in rear. (Par. 473.)

Sergeant Allen: I order, "1. **Walk;** 2. **MARCH.** 1. **Squad;** 2. **HALT,**" and gallop up to Private Carter, dismount just before reaching the crest, give my horse to Private Martin, and run forward.

Captain: Carter points out what appears to be a troop of cavalry standing in the road leading north out of York, just on the edge of the town. You see about four mounted men 200 yards out of York on your road, halted, and about the same number on the Valley Pike near where it crosses the first stream north of York. What do you do?

Sergeant Allen: I wait about three minutes to see if they are going to move.

Captain: They remain halted, the men at York appear to be dismounted.

Sergeant Allen: I write the following message:

> Hill ½ mile N. E. of York,
> 10 A. M.

Captain X:

A hostile troop of cavalry is standing in road at YORK (west of SALEM) with squads halted on N. and N. E. roads from YORK. Nothing else seen. Will remain in observation for the present.

> Allen,
> Sgt. (Pars. 474 and 474a.)

I would give the message to Martin, who had previously brought my horse up close in rear of the crest, and would say to him, "Take this message to the captain, straight across to the road the troop is on, and turn south towards Salem if you do not see them at first. Take Lacey with you. Tell him what you have seen. He knows where the troop is." I would have Carter hold my horse, and watch the remainder of the patrol for signals, while I observed the enemy.

Captain: At the end of five minutes the hostile troop trots north on the Valley Pike, the patrol on your road rides across to the Valley Pike and follows the troop.

Sergeant Allen: I would wait until the troops had crossed the creek north of York and would then face my patrol east and trot to the cross roads at Smith's, turn south and continue to Salem, sending one man to ride up on Sandy Ridge, keeping the patrol in sight.

Captain: We have carried out the problem far enough. It furnishes a good example of the varying situations a patrol leader has to meet. Good judgment or common sense must be used in deciding on the proper course to follow. You must always think of what your chief is trying to do and then act in the way you think will best help him to accomplish his object. If you have carefully decided just what mission you have been given to accomplish, you cannot easily go wrong. In handling a mounted patrol you must remember that if the men become widely separated in strange country, or even in country they are fairly familiar with, they are most apt to lose all contact with each other or become lost themselves.

Problem No. 3. (Infantry)

Captain (to one platoon of his company): We will suppose it is about half an hour before dawn. One platoon of the company is deployed as skirmishers, facing north, in the cut where the County Road crosses Sandy Ridge. It is the extreme right of a line of battle extending west along the line of the County Road. The fight has not commenced. This platoon is resting in a wheat field between the railroad and the foot of the slope of Sandy Ridge, 200 yards south of the County Road. Sergeant Allen, I call you up and give you these instructions: "The enemy's line is off in that direction (pointing northwest). Take six men and work north along the railroad until it is light enough to see; then locate the hostile line and keep me informed of their movements. I will be in this vicinity. You have a compass. Start at once." Describe briefly the formation of your patrol while it is moving in the dark.

Sergeant Allen: One man will lead. A second man will follow about fifteen yards in rear of him. I will follow the second man at the same distance with three more men, and the last man will be about twenty yards in rear of me. All will have bayonets fixed, loaded and pieces locked. One short, low whistle will mean, **Halt,** two short whistles will mean, **Forward,** and the word "Sandy" will be the countersign by which we can identify each other.

Captain: Very well. We will suppose that you reach the steel trestle over Sandy Creek just at dawn and have met no opposition and heard nothing of the enemy. On either side of Sandy Creek are fields of standing corn about six feet tall. In the present dim light you can only see a few hundred yards off.

Sergeant Allen: The patrol being halted I would walk forward to the leading man (Brown) and say, "Brown, take Carter and form the point for the patrol, continuing along this railroad. We will follow about 150 yards in rear." I would then rejoin the main body of the patrol and order the man in rear to follow about 75 yards in rear of us. When the point had gained its distance I would move forward with the main body, ordering one man to move along the creek bank (west bank), keeping abreast of us until I signaled to him to come in.

Captain: Just as you reach the northern end of the railroad fill your point halts and you detect some movement in the road to the west of you. It is rapidly growing lighter.

Sergeant Allen: I would move the main body by the left flank into the corn, signaling to the man following the creek to rejoin, and for the rear guard to move off the track also. I would expect Brown to do the same, even before he saw what we had done. I would then close up on the point until I could see it and, halting all the patrol, I would order Foster to take Lacey and work over towards the road to see what is there and to report back to me immediately.

Captain: In a few minutes Foster returns and reports, "The enemy is moving south in the road and in the field beyond, in line of squads or sections. A hostile patrol is moving southeast across the field behind us. We were not seen.''

(Note: This situation could well have been led up to by requiring Private Foster to explain how he conducted his reconnaissance and having him formulate his report on the situation as given.)

Sergeant Allen: I would then work my patrol closer to the road, keeping Foster out on that flank, and prepare to follow south in rear of the hostile movement.

Captain: The information you have gained is so important that you should have sent a man back to me with a verbal message, particularly as you are in a very dangerous position, and may not be able to send a message later. While you have not definitely located the left of the enemy's line, you have apparently discovered what appears to be a movement of troops forward to form the left of the attacking line. Your action in turning south to follow the troops just reported, is proper, as you now know you are partly in rear of the hostile movement and must go south to locate the hostile flank that your mission requires you to report on.

You men must picture in your minds the appearance of the country the sergeant is operating through. His patrol is now in a field of high standing corn. Unless you are looking down between the regular rows of corn you can only see a few yards ahead of you. The road has a wire fence and is bordered by a fairly heavy growth of high weeds and bushes. The ground is dry and dusty. Sergeant, how do you conduct your movement south?

Sergeant Allen: As my patrol is now in a very dangerous neighborhood and very liable to be caught between two hostile lines, with a deep creek between our present position and our platoon, I think it best to move cautiously southeast until I reach the creek bank (I cannot see it from where I now am), and then follow the creek south. I think I am very apt to find the enemy's left resting on this creek. Besides, if I do not soon locate the enemy, I can hold the main body of my patrol close to the creek and send scouts in towards the road to search for the enemy. It will also be much easier to send information back to the platoon from the creek bank, as a messenger can ford it and head southeast until he strikes the railroad and then follow that straight back to our starting point. It would thus be very difficult for him to get lost.

Captain: You move southeast and strike the creek bank just south of the railroad trestle. You now hear artillery fire off to the west and rifle fire to the southwest which gradually increases in volume. You see a high cloud of dust hanging over the road on the hill west of Mason's and south of this road on the north slope of the northernmost knoll of the Twin Hills, you can occasionally see the flash of a gun, artillery, being discharged. There seems to be no rifle firing directly in your front.

Sergeant Allen: I hurriedly write the following message:

At Ry. trestle 1 mi. N. of Platoon,

5:15 A. M.

Captain X:

Can see arty. firing from position on N. slope of knoll on high ridge to W. of me. and ¼ mi. S. of E. and W. road. Hostile line is S. of me. Have not located it. Will move S.

Allen,

Sgt. (Par. 474.)

I hand this to Private Smith and say to him, ''Carry this quickly to the captain. Follow the railroad back until you cross a wagon road. Our platoon should be to the west of the track just beyond the road.'' I also read the message to Smith and point out the hostile artillery. I have considered that I sent a message before telling about the hostile advance.

I then continue south, ·moving slowing and with great caution. I instruct the remaining four men that in case we are surprised to try to cross the creek and follow the railroad back to the platoon.

Captain: Your information about the hostile artillery position was important and should have been sent in, provided you think your description of the hostile position was sufficiently clear to be understood by an observer within your own lines.

There is some question as to the advisability of your remaining on the west bank of the creek. Still you would not be able to tell from where you were what direction the creek took, so you probably would remain on the west bank for the present.

You continue south for about 150 yards and your leading man halts, comes back to you, and reports that the corn ahead is broken and trampled, showing it has been passed over by foot troops. About the same time you hear rifle fire to your immediate front. It sounds very close.

Sergeant Allen: I say, ''Cross this creek at once,'' and when we reach the other bank and the patrol forms again, we move slowly south, all the men keeping away from the creek bank, except myself, and I march opposite the two men constituting the main body.

Captain: About this time you detect a movement in the corn across the creek in rear of the place you have just left. You think it is a body of troops moving south. The firing in front seems to be delivered from a point about two or three hundred yards south of you and you can hear heavy firing from off in the direction of your company, a few bullets passing overhead. There are scattered trees along the creek and some bushes close to the edge.

Sergeant Allen: I would conceal myself close to the bank, the patrol being back, out of sight from the opposite bank, and await developments.

Captain: Sergeant, your patrol is in a dangerous position. The enemy will very likely have a patrol or detachment in rear and beyond his flank. This patrol would probably cross the railroad trestle and take you in rear. You should have given the last man in your patrol particular instructions to watch the railroad to the north. It would have been better if you had sent one man over to the railroad, which is only a short distance away, and had him look up and down the track and also make a hurried survey of the country from an elevated position on the fill.

I also think it would be better not to await developments where you now are, but to push south and make sure of the position of the left of the enemy's firing line. Later you can devote more time to the movements in rear of the first line. You are taking too many chances in remaining where you are. I do not mean that you should leave merely because you might have some of your men killed or captured, but because if this did occur you would probably not be able to accomplish your mission. Later you may have to run a big chance of sacrificing several of your men, in order to get the desired information, which would be entirely justifiable. Tell me how your men are arranged and what your next movement would be.

Sergeant Allen: I have four men left. I am close to the stream's bank, under cover; two men are about 25 yards further away from the stream; Private Brown is up stream as far off as he can get and still see the other two men, and Private Foster is down stream the same distance. Both Brown and Foster are well back from the stream. The two men in the middle, the main body of the patrol, make their movements conform to mine, and Brown and Foster regulate their movements on the main body. I will move south until I can locate the enemy's advance line.

Captain: When you are about opposite the Mason house, Brown comes back to you, having signaled halt, and reports he can see the enemy's firing line about 100 yards ahead on the other side of the stream, and that a small detachment is crossing the stream just beyond where he was. What do you do?

Sergeant Allen: I creep forward with Brown to verify his report. The remainder of the patrol remains in place.

Captain: You find everything as Brown reported. You see that the firing line extends along the southern edge of the cornfield, facing an uncultivated field covered with grass and frequent patches of weeds two or three feet high. You cannot determine how strong the line is, but a heavy fire is being delivered. You cannot see the detachment that crossed the creek south of you because of the standing corn.

Sergeant Allen: I crawl back to the main body, leaving Brown, and write the following message:

5/6 mi. N. of Platoon,

Captain X: 5:32 A. M.

Enemy's left rests on creek ¾ mile to your front, along S. edge of cornfield. Creek is 5 ft. deep by 60 ft. wide. Hostile patrols have crossed the creek. Will watch their rear.

Allen,

Sgt.

I give this to Private James and say, "Go over to the railroad (pointing), then turn to your right and follow the track until you cross a wagon road. Our platoon is just beyond that, on this side of the track. Give this message to the captain. Hurry."

509 (contd.)

Captain: You should have either read the message to James or had him read it. You should also have cautioned him to watch out for that hostile detachment. It might be better to send another man off with a duplicate of the message, as there is quite a chance that James may not get through and the message is all-important. James, you get back to the wagon road here (pointing) and find yourself in the right of your battle line, but cannot locate me or the company right away.

Private James: I would show the note to the first officer I saw in any event, and in this case, I would turn it over to the officer who appeared to be in command of the battalion or regiment on the right of the line, telling him what company the patrol belonged to, when we went out, etc.

Captain: What do you do, sergeant?

Sergeant Allen: I start to move north a short distance in order to find out what reënforcements are in rear of the hostile line.

Captain: After you have moved about 75 yards you are suddenly fired into from across the creek, and at the same time from the direction of the railroad trestle. Your men break and run east through the corn and you follow, but lose sight of them. When you cross the railroad fill you are fired on from the direction of the bridge. You finally stop behind the railroad fill on the quarry switch, where two of your men join you.

Sergeant Allen: I would start south to rejoin the company and report.

Captain: That would be a mistake. It would require a long time for a second patrol to make its way out over unknown ground, filled with hostile patrols, to a point where they could observe anything in rear of the hostile flank. You are now fairly familiar with the ground, you also know about where the hostile patrols are and you have two men remaining. After a brief rest in some concealed place nearby, you should start out again to make an effort to determine the strength of the troops in rear of the hostile flank near you, or at least remain out where you could keep a sharp lookout for any attempted turning movement by the enemy. Should anything important be observed you can send back a message and two of you remain to observe the next developments before returning. The information you might send back and the additional information you might carry back, would possibly enable your own force to avoid a serious reverse or obtain a decided victory.

Your work would be very hazardous, but it is necessary, and while possibly resulting in loss of one or two of your men, it might prevent the loss of hundreds in your main force.

CHAPTER VII

THE SERVICE OF SECURITY

(Based on the Field Service Regulations.)

General Principles

510. **The Service of Security** embraces all those measures taken by a military force to protect itself against surprise, annoyance or observation by the enemy. On the march, that portion of a command thrown out to provide this security is called an advance, flank or rear guard, depending on whether it is in front, to the flank or in rear of the main command; in camp or bivouac, it is called the outpost.

The principal duties of these bodies being much the same, their general formations are also very similar. There is (1) the cavalry covering the front; next (2) a group (4 men to a platoon) or line of groups in observation; then (3) the support, or line of supports, whose duty is to furnish the men for the observation groups and check an enemy's attempt to advance until reinforcements can arrive; still farther in rear is (4) the reserve.

In small commands of an infantry regiment or less there usually will not be any cavalry to cover the front, and the reserve is generally omitted. Even the support may be omitted and the observation group or line of groups be charged with checking the enemy, in addition to its regular duties of observation. But whatever the technical designation of these subdivisions, the rearmost one is always in fact a reserve. For example, if the command is so small that the subdivision formally designated at the reserve is omitted, the rear element (squad or platoon or company, etc.,) is used as a reserve. As this text deals principally with small commands and only those larger than a regiment usually have the subdivision termed the reserve, this distinction between the element in the Field Service Regulations called the reserve and the actual reserve, must be thoroughly understood.

The arrangements or formations of all detachments thrown out from the main force to provide security against the enemy, are very flexible, varying with every military situation and every different kind of country. The commander of such a detachment must, therefore, avoid blindly arranging his men according to some fixed plan and at certain fixed distances. Acquire a general understanding of the principles of the service of security and then with these principles as a foundation use common sense in disposing troops for this duty.

ADVANCE GUARD

511. **Definition and Duties.** An advance guard is a detachment of a marching column thrown out in advance to protect the main column from being surprised and to prevent its march from being delayed or interrupted. (The latter duty is generally forgotten and many irritating, short halts result, which wear out or greatly fatigue the main body, the strength of which the advance guard is supposed to conserve.)

511 (contd.)

In detail the duties of the advance guard are:

1. To guard against surprise and furnish information by reconnoitering to the front and flanks.

2. To push back small parties of the enemy and prevent their observing, firing upon or delaying the main body.

3. To check the enemy's advance in force long enough to permit the main body to prepare for action.

4. When the enemy is met on the defensive, to seize a good position and locate his lines, care being taken not to bring on a general engagement unless the advance guard commander is authorized to do so.

5. To remove obstacles, repair the road, and favor in every way possible the steady march of the column.

Strength. The strength of the advance guard varies from one-ninth to one-third of the total command. The larger the force the larger in proportion is the advance guard, for a larger command takes relatively longer to prepare for action than a small one. For example, a company of 100 men would ordinarily have an advance guard of from one to two squads, as the company could deploy as skirmishers in a few seconds. On the other hand, a division of 20,000 men would ordinarily have an advance guard of about 4,500 men, all told, as it would require several hours for a division to deploy and the advance guard must be strong enough to make a stubborn fight.

Composition. The advance guard is principally composed of infantry, preceded if possible, by cavalry well to the front. When there is only infantry, much more patrolling is required of the front troops than when cavalry (called "Advance cavalry") is out in advance. This book does not deal with large advance guards containing artillery and engineers. Machine guns, however, will be frequently used in small advance guards to hold bridges, defiles, etc.

Distance From Main Body. The distance at which the advance guard precedes the main body or the main body follows the advance guard depends on the military situation and the ground. It should always be great enough to allow the main body time to deploy before it can be seriously engaged. For instance, the advance guard of a company, say 1 squad, should be 350 to 500 yards in advance of the company. The distance from the leading man back to the principal group of the squad should generally be at least 150 yards. This, added to the distance back to the main body or company, makes a distance of from 500 to 650 yards from the leading man to the head of the main body.

Examples:

Command.	Advance Guard.	Distance (yds.)
Patrol of 1 squad	2 men	100 to 300
Section of 3 squads	4 men	200 to 400
Inf. platoon of 50 men	1 squad	300 to 450
Cav. platoon of 20 men	4 men	300 to 450
Inf. company of 108 men	1 to 2 squads	350 to 500
Cav. troop of 86 men	½ platoon	450 to 600
Inf. battalion	½ to 1 company	500 to 700
Cav. squadron	½ to 1 troop	600 to 800

These are not furnished as fixed numbers and distances, but are merely to give the student an approximate, concrete idea.

511a. Connecting Files. It should be remembered that between the advance guard and the main body, and between the several groups into which the advance guard is subdivided, connecting files are placed so as to furnish a means of communicating, generally by signals, between the elements (groups) of the column. There should be a connecting file for at least every 300 yards. For example, suppose the advance guard of a platoon is 300 yards in front of the main body. In ordinary rolling country, not heavily wooded, a connecting file would be placed half way between the two elements—150 yards from each one.

It is generally wiser to use two men together instead of one, because this leaves one man free to watch for signals from the front while the other watches the main body. However, in very small commands like a company, this is not practicable, as the extra man could not be spared.

FORMATION OF ADVANCE GUARDS

Subdivisions. The advance guard of a large force like a brigade or division is subdivided into a number of groups or elements, gradually increasing in size from front to rear. The reason for this is that, as has already been explained, a larger group or force requires longer to deploy or prepare to fight than a smaller one, therefore the small subdivisions are placed in front where they can quickly deploy and hold the enemy temporarily in check while the larger elements in rear are deploying. The number of these subdivisions decreases as the strength of the advance guard decreases, until we find the advance guard of a company consists of one or two squads, which naturally cannot be subdivided into more than two groups; and the advance guard of a squad composed of two men, which admits of no subdivision.

		Distance to next element in rear.
Advance Cavalry	....	1 to 5 miles
Support { (furnishes patrols) {	Advance party.... { Point	150 to 300 yds.
	Advance party proper....	300 to 600 yds.
	Support proper	400 to 800 yds.
Reserve (usually omitted in small commands)	....	500 yds. to 1 mile

The distances vary principally with the size of the command—slightly with the character of the country.

The advance cavalry is that part of the advance guard going in front of all the foot troops. It is generally one to five miles in advance of the infantry of the advance guard, reconnoitering at least far enough to the front and flanks to guard the column against surprise by artillery fire—4,500 yards.

511b. Support. (a) The support constitutes the principal element or group of all advance guards. It follows the advance cavalry, when there is any, and leads the advance guard when there is no cavalry. The support of a large command is subdivided within itself in much the same manner as the advance guard as a whole is subdivided. It varies in strength from one-fourth to one-half of the advance guard.

(b) **Advance Party.** As the support moves out it sends forward an advance party several hundred yards, the distance varying with the nature of the country and size of the command. For example, the advance party of a support of one company of 108 men, would ordinarily be composed of one section of three squads, and would march about 300 yards in advance of the company in open country, and about 200 yards in wooded country.

The advance party sends out the patrols to the front and flanks to guard the main body of the support from surprise by effective rifle fire. Patrols are only sent out to the flanks to examine points that cannot be observed from the road. As a rule they will have to rejoin some portion of the column in rear of the advance party. As the advance party becomes depleted in strength in this manner, fresh men are sent forward from the main body of the support to replace those who have fallen behind while patrolling. When there is advance cavalry, much less patrolling is required of the infantry.

(c) The point is a patrol sent forward by the advance party 150 to 300 yards. When the advance party is large enough the point should ordinarily consist of a complete squad, commanded by an officer or experienced noncommissioned officer. It is merely a patrol in front of the column and takes the formation described for patrols.

(d) The commander of the support ordinarily marches with the advance party. He should have a map and control of the guide, if any is present. He sees that the proper road is followed; that guides are left in towns and at crossroads; that bridges, roads, etc., are repaired promptly so as not to delay the march of the column and that information of the enemy is promptly sent back to the advance guard commander; he verifies the correctness of this information, if possible.

511c. (a) A thorough understanding of the arrangement of the support and the duties of the leaders of its subdivisions—point, flank patrols, advance party and main body (of the support)—is of the greatest importance to a noncommissioned officer. For example, the ignorance of one noncommissioned officer leading the advance party of a column of troops six miles long can cause the entire column to be delayed. If he halts because a few shots are fired at his men, and conducts a careful reconnaissance before attacking (instead of pushing right in on the enemy, forcing him to fall back quickly, if a weak detachment; or, to disclose his strength, if strong), the entire column, six miles long, is halted, the march interrupted, valuable time lost, and what is more important, the men irritated and tired out.

(b) The leader of the point must understand that as the principal duty of an advance guard is to secure the safe and uninterrupted march of the main body, he is the first man to discharge this duty. If, for example, his squad receives a volley of shots from some point to the front, he cannot take the time and precautions the commander of a larger body would take to reconnoiter the enemy's position, determine something about his strength, etc., before risking an attack. If he did he would not be securing the uninterrupted march of the main body. He has to deploy instantly and press the enemy hard until the hostile opposition disappears or the advance party comes up and its commander takes charge. The point will lose men in this way, but it is necessary, for otherwise one small combat patrol could delay the march time after time.

(c) The same problem must be met in much the same manner by the leader of the advance party. In this case there is more time to think, as the point, being in advance, will have begun the fight before the advance party arrives; but the leader of the advance party must use his men freely and quickly to force the enemy to "show his hand," thus preventing small harassing or combat detachments from delaying the march.

(d) As the subdivisions of the advance guard become larger their leaders act with increasing caution, for as soon as it develops that the enemy in front is really present in some strength, then a halt becomes obligatory and a careful reconnaissance necessary.

(e) The leader of every subdivision must always start a reconnaissance the instant the enemy develops. He may, as in the case of the point, only send one man around to discover the enemy's strength; or, if the leader of the main body of the support, he may send an entire squad. In almost every case the instant he has given his orders for deploying and firing at or rushing the enemy, he sends out his man or men to work around to a position permitting a view of the hostile force. Every noncommissioned officer should impress this on his memory so that he will not forget it in the excitement of a sudden engagement.

(f) No attempt should be made to subdivide the advance guard of a small force into all the elements previously described. For example, the advance guard of a squad is simply a point of one or two men; the advance guard of a company is usually no more than a squad acting as a point, the squad actually having several men from 100 to 150 yards in advance, who really constitute a point for the squad; the advance guard of a battalion would usually consist of a company or less distributed as an advance party proper and a point. The advance guard of a regiment would have no reserve—if. for example, a battalion were used as the advance guard of a regiment, there would be only a support, which would be distributed about as follows: A support proper of about three companies and an advance party (point included) of about one company.

(BATTALION ACTING AS·ADVANCE GUARD. NO RESERVE)

SUPPORT

1. Support proper
(3 Cos.)

2. Advance party
(1 Co.)

1. Advance party proper
(3 Squads)

2. Point
(1 Squad)

400 TO 500 YDS.   150 TO 200 YDS.

Reserve. An advance guard large enough to have a reserve would be distributed as follows:

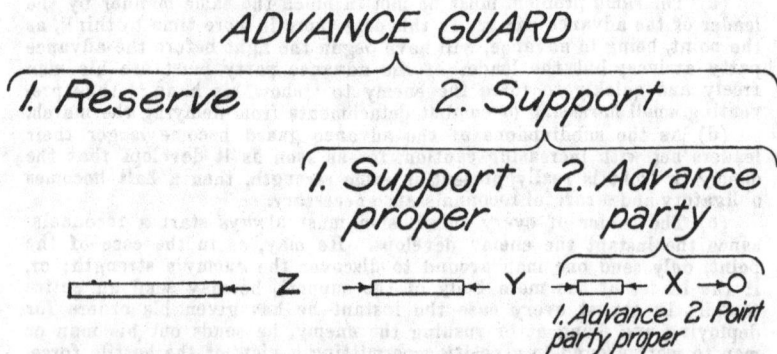

ADVANCE GUARD

1. Reserve   *2. Support*

1. Support   *2. Advance*
proper   *party*

```
┌──────────┐ ← Z →  ┌────┐ ← Y → ┌──┐← X →  O
└──────────┘        └────┘       └──┘
                              1. Advance 2. Point
                              party proper
```

The distance Z would be greater than Y and Y would be greater than X. For example, a regiment acting as the advance guard of a brigade would, under ordinary conditions, be distributed about as follows:

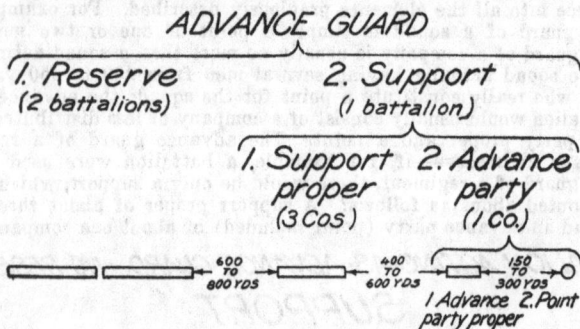

ADVANCE GUARD

1. Reserve   *2. Support*
(2 battalions)   *(1 battalion)*

1. Support   *2. Advance*
proper   *party*
(3 Cos)   *(1 CO.)*

```
┌──────────┐  ┌────┐  600   ┌────┐ 400  ┌──┐ 150  O
└──────────┘  └────┘  TO    └────┘ TO    └──┘ TO
                      800 YDS       600 YDS    300 YDS
                                       1. Advance 2. Point
                                       party proper
```

As only large commands have a reserve, which would always be commanded by an officer, noncommissioned officers need not give this much consideration, but it must be understood that while this fourth subdivision of the advance guard is the only one officially termed reserve, the last subdivision of any advance guard actually is a reserve, no matter what its official designation.

The advance guard of a cavalry command adopts formations similar to those described above, except that the distances are increased because of the rapidity with which the command can close up or deploy. An advance party with a few patrols is usually enough for a squadron, and precedes it from 600 to 1,000 yards.

Reconnaissance. In reconnaissance the patrols are, as a rule, small (from two to six men).

The flanking patrols, whether of the advance cavalry or of the advance party, are sent out to examine the country wherever the enemy might be concealed. If the nature of the ground permits, these patrols march across country or along roads and trails parallel to the march of the column. For cavalry patrols this is often possible; but with infantry patrols and even with those that are mounted, reconnaissance is best done by sending the patrols to high places along the line of march to overlook the country and examine the danger points. These patrols signal the results of their observations and, unless they have other instructions, join the columns by the nearest routes, other patrols being sent out as the march proceeds and as the nature of the country requires.

Deserters, suspicious characters and bearers of flags of truce (the latter blindfolded), are taken to the advance guard commander.

Advance Guard Order. On receipt of the order for a march designating the troops for the advance guard, the commander of the latter makes his estimate of the situation; that is, he looks at the map or makes inquiries to determine what sort of a country he must march through and the nature of the roads; he considers what the chances are of encountering the enemy, etc., and then how he should best arrange his advance guard to meet these conditions, and what time the different elements of his advance guard must start in order to take their proper place in the column. He then issues his order at the proper time—the evening before if possible and he deems it best, or the morning of the march.

The order for a large advance guard would ordinarily be written; for a small command it would almost invariably be verbal, except that the commander or leader of each element should always make written notes of the principal points, such as the road to be followed, time to start, distances, etc.

511d. Written advance guard order:

(See Fort Leavenworth map in pocket at back of book.)

Field Orders
No. 1, (x)
Troops

Advance Guard, Det. 1st Div.
Leavenworth, Kansas,
10 Aug. '08, 5·30 A. M.

(a) Advance Cavalry
Captain B
Tr. A, 1st Cav.
(less 1 squad)

1 A Red force of all arms is reported to have camped near ATCHISON[1] last night Its cavalry patrols were seen near KICKAPOO yesterday.

(b) Support·
Major C
1st Bn. 1st Inf.
1 Squad Tr. A,
1st Cav.

Our main body will follow the advance guard at one-half mile.

Det. Co. A, Engrs.

2. This advance guard will march on KICKAPOO.

(c) Reserve—in
order of march:

3. (a) The advance cavalry will leave camp at once and march via ATCHISON CROSS to KICKAPOO, SHERIDAN'S DRIVE and the country west of the line of march will be carefully observed

Hq. and 2nd Bn
1st Inf.
Btry. B, 5th
F. A.
3d Bn. 1st Inf.
Det. Amb. Co. No 1

(b) The point of the support will start at 5:45 A. M. and march by the ATCHISON CROSS-FRENCHMAN-KICKAPOO road

(c) The reserve will follow the support at 800 yards.

(x) This order is issued pursuant to a previous "march order," and assumes that the troops designated for the advance guard have been notified when and where to assemble.

[1] About fourteen miles northwest of Ft. Leavenworth.

4. The field train will assemble near 70 at 7 A. M. under Captain X, Quarter-master, 1st Inf., and join the field train of the main body as that train passes.

5. I shall be at the head of the reserve.

Y,
Colonel,
Commanding.

Delivered verbally to assembled troop, battalion and battery commanders and staff; copy to det. commander by Lt. N.

Note: The paragraphs on the left lettered (a), (b), (c), etc., are called the distribution, and those on the right numbered 1, 2, 3, etc., are called body.

In issuing his order to an advance guard a noncommissioned officer should follow the form above, except that it should be verbal, and the troops in each part of the advance guard should be named in the body of the order. For example, in giving an advance guard order for one platoon, the noncommissioned officer would say, for instance, to his platoon:

"A Red battalion is reported to have camped near Atchison last night. Our battalion will march towards Kickapoo this morning.

"This platoon will form the advance guard and will march out the Atchison Pike, followed by the main body at 500 yards.

"The point will start at 5:45 A. M. and move by the Atchison Cross-Frenchman-Kickapoo road. The remainder of the advance guard will follow at 300 yards.

"I will march with the point."

ADVANCE GUARD PROBLEMS
Problem No. 1. (Infantry)

512. Captain (to one platoon of his company): We will assume that our battalion camped last night at Oxford (Elementary Map) in the enemy's country. It is now sunrise, 5:30 A. M.; camp has been broken and we are ready to march. The officers have returned from reporting to the major for orders and I fall in the company and give the following orders:

"A regiment of the enemy's cavalry is thought to be marching towards Salem from the south. Our battalion will march at once towards Salem to guard the railroad trestle over Sandy Creek, following this road (pointing southeast along the road out of Oxford) and the Chester Pike, which is one and three-quarters miles from here.

"This company will form the advance guard.

"Sergeant Adams, you will take Corporal Baker's squad and form the point, followed by the remainder of the company at about 400 yards. Patrols and connecting files will be furnished by the company.

"The company wagon will join the wagons of the battalion.

"I will be with the company.

"Move out at once."

The weather is fine and the roads are good and free from dust. It is August and nearly all the crops are harvested. Bushes and weeds form a considerable growth along the fences bordering the road.

Sergeant, give your orders.

Sergeant Adams: 1st squad, 1. **Right**, 2. **FACE**, 1. **Forward**, 2. **MARCH.** Corporal Baker, take Carter (Baker's rear rank man) and go ahead of the squad about 200 yards. Move out rapidly until you get your distance and then keep us in sight.

I would then have the two leading men of the rest of the squad follow on opposite sides of the road, as close to the fence as possible for good walking. This would put the squad in two columns of files of three men each, leaving the main roadway clear and making the squad as inconspicuous as possible, without interfering with ease of marching or separating the men. [Par. 511b (c).] What sort of crops are in the fields on either side of the road?

Captain: The field on the right (south) is meadow land; that on the left, as far as the railroad, is cut hay; beyond the railroad there is more meadow land.

Sergeant Adams: I would have told Corporal Baker to wait at the cross roads by the Baker house for orders and—

Captain: If you were actually on the ground you probably could not see the cross roads from Oxford. In solving map problems like these do not take advantage of seeing on the map all the country that you are supposed to go over, and then give orders about doing things at places concerning which you would not probably have any knowledge if actually on the ground without the map.

Besides, in this particular case, it was a mistake to have your point wait at the cross roads. If there was any danger of their taking the wrong road it would be a different matter, but here your mission requires you to push ahead. (Par. 511c.) The major is trying to get south of the trestle towards Salem before the cavalry can arrive and destroy it.

Sergeant Adams: I would march steadily along the road, ordering the last man to keep a lookout to the rear for signals from the connecting file (Par. 511a), and I would direct one of the leading men to watch for signals from Corporal Baker.

Captain: You should have given the direction about watching for signals earlier, as this is very important. You also should have ordered two men to follow along the timber by the creek to your south until you signaled for them to come in. The trees along the creek would obstruct your view over the country beyond the creek.

Sergeant Adams: But I thought, Captain, that the patrolling was to be done by the company.

Captain: Yes, the patrolling is to be done by the company, but the creek is only a quarter of a mile, about 400 yards, from the road you are following and the men sent there are merely flankers, not a patrol. You have eight men under your command and you are responsible for the ground within several hundred yards on either side of your route of march. Long Ridge is almost too far for you to send your men, because they would fall far behind in climbing and descending its slopes but it would not be a great mistake if you sent two men there. As Long Ridge affords an extended view of the valley through which the Chester Pike runs, a patrol should go up on it and remain there until the battalion passes, and this would be more than the leading squad could be expected

to attend to. The creek is almost too far from the road in places, but
as it is open meadow land you can keep the men within easy touch of you
and recall them by signal at any moment you desire. In this work you
can see how much depends on good judgment and a proper understanding of one's mission.

Corporal Baker, explain how you would move out with Carter.

Corporal Baker: We would alternate the walk and double time
until we had gotten about 200 yards ahead of the squad. I would then
say, "Carter, walk along this side of the road (indicates side), keeping
on the lookout for signals from the squad. I will go about fifty yards
ahead of you." I would keep to the opposite side of the road from Carter,
trying to march steadily at the regular marching gait, and keeping a
keen watch on everything in front and to the flanks.

Captain: Very good. When you arrive at the cross roads you see
a man standing in the yard of the Baker house.

Corporal Baker: I would not stop, but would continue on by the
cross roads, as I have no time to question the man and the Sergeant will
want to do that. I would call to him and ask him if he had seen any of
the enemy about and how far it was to the Chester Pike. If anything
looked suspicious around the house or barnyard, I would investigate.

Captain: Sergeant, you arrive at the cross roads, and see the
Corporal and Carter going on ahead of you.

Sergeant Adams: I would have already signaled to the two men
following the creek to come in and would send a man to meet them with
the following order: "Tell Davis to move along the railroad fill with
Evans, keeping abreast of us. Then you return to me." I would then say,
"Fiske, look in that house and around the barn and orchard and then
rejoin me down this road (pointing east)." I would have the civilian
join me and walk down the road with me while I questioned him.

Captain: Do you think you have made careful arrangements for
searching the house, etc., by leaving only one man to do the work?

Sergeant Adams: I have not sufficient men nor time enough to
do much more. I simply want to make sure things are reasonably safe
and I thought that a couple of men from the main body of the advance
guard would do any careful searching, questioning, etc., that might be
deemed necessary. I must not delay the march.

Captain: That is right. You learn nothing from the civilian and
he does not arouse any suspicion on your part. You continue along the
road. The fields to the north of the road are in wheat stubble; the
ground to the south, between your road and the railroad, is rough, rocky
grass land with frequent clumps of bushes. Davis and Evans, your right
flankers on the railroad fill, are just approaching the cut; Fiske has
rejoined; Corporal Jones and his men are about 200 yards from the road
forks at Brown's, and you and your four men are 200 yards in their
rear, at the turn of the road. At this moment a half dozen shots are fired
down the road in your direction from behind the wall along the edge of
the orchard on the Brown farm. This firing continues and your two
leading men are lying down at the roadside returning the fire. Tell me
quickly just what you are going to do?

Sergeant Adams: I order my four men to deploy as skirmishers
in that field (pointing to the rough ground south of the road); I go under

the fence with the men and lead them forward at a fast run, unless the fire is very heavy.

Captain (interrupting the Sergeant): Davis, you had just reached the cut on the railroad when this happened. What do you do?

Private Davis: I take Evans forward with me at a run through the cut. What do I see along the Chester Pike or Sandy Creek?

Captain: You see no sign of the enemy any place, except the firing over the wall.

Private Davis: I run down the south side of the fill and along towards the road with Evans to open fire on the enemy from their flank, and also to see what is in the orchard. I will probably cross the road so that I can see behind the stone wall.

Captain: That's fine and shows how you should go ahead at such a time without any orders. There is usually no time or opportunity at such a moment for sending instructions and you must use common sense and do something. Generally it would have been better to have tried to signal or send word back that there was nothing in sight along the road or in the valley, but in this particular case you could probably do more good by going quickly around in rear as you did, to discover what was there and assist in quickly dislodging whatever it was. If there had been no nose of the ridge to hide you as you came up and a convenient railroad fill to hurry along behind as you made for the road, your solution might have been quite different.

Sergeant, continue with your movements.

Sergeant Adams: I would attempt to rush the wall. If the fire were too heavy, I would open fire (at will) with all my men, and, if I seemed to get a little heavier fire than the enemy's, I would start half of my men forward on a rush while the others fired. I would try to rush in on the enemy with as little delay as possible, until it developed that he had more than a small detachment there. I assumed it was a delaying patrol in front of me, and as my mission requires me to secure the uninterrupted march of the main body, I must not permit any small detachment to delay me. If, however, it proves to be a larger force, for instance, the head of an advanced guard, I will lose some men by plunging in, but as I understand it, that is the duty of the point. Then again, if it be the head of a hostile advance guard, I will want to rush them out of their favorable position under cover of the stone wall, buildings and orchard, before any more of their force can come up. This would give the favorable position to our force; by acting too cautiously we would lose the valuable moments in which the enemy's reënforcements (next element of the advance guard) were coming up, with this desirable position bing weakly held by a small part of the enemy.

Captain: That is all correct. What messages would you have sent?

Sergeant Adams: Up to the present time I would not have sent any. I could not have sent any. I could not afford to take the time to send a man back, nor could I spare the man. Besides, all I could say was that we were fired on, and you should be able to see and hear that from where the company is.

Captain: About the time you reached the position of Corporal Baker the firing ceases, and when you reach the wall you see five mounted men galloping northeast up Farm Lane. The Brown farm appears to be deserted.

Sergeant Adams: I would turn to one of the men and say, "Run back to the Captain and tell him we were fired on from this orchard by a mounted patrol of five men who are galloping off up a lane to the northeast. I am going south." When he had repeated the message I would start south down the Chester Pike, directing Corporal Baker to follow this road south and to tell Davis to follow the high ridge west of the road, going through the clump of woods just ahead. I would send one man as a right flanker to follow the west bank of Sandy Creek. This would leave me with two men, one watching for signals from the front and along Sandy Creek, the other from Davis and from the rear. I would expect to see a patrol from the company moving across towards Boling Woods. Had I not been mixed up in a fight as I approached the Brown farm I would have sent two men as left flankers across country to the cut on the Chester Pike on the western edge of the Boling Woods.

Captain: Very good. That is sufficient for this problem. All of you should have caught the idea of the principal duties of the point and flankers of an advance guard. You must watch the country to prevent being surprised and you must at the same time manage to push ahead with the least possible delay. The point cannot be very cautious so far as concerns its own safety, for this would mean frequent halts which would delay the troops in rear, but it must be cautious about reconnoitering all parts of the ground near the road which might conceal large bodies of the enemy.

The leader of the point must be careful in using his men or he will get them so scattered that they will become entirely separated and he will lose all control of them. As soon as the necessity for flankers on one side of the line of march no longer exists, signal for them to rejoin and do not send them out again so long as you can see from the road all the country you should cover.

Problem No. 2. (Infantry)

Captain (to one platoon of his company): Let us assume that this platoon is the advance party of an advance guard, marching through Salem along the Chester Pike [Par. 511b (b)]. One squad is 350 yards in front, acting as the point. The enemy is thought to be very near, but only two mounted patrols have been seen during the day. The command is marching for Chester. The day is hot, the roads are good but dusty, and the crops are about to be harvested.

Sergeant Adams, explain how you would conduct the march of the advance party, beginning with your arrival at the cross roads in Salem.

Sergeant Adams: The platoon would be marching in column of squads and I would be at the head. Two pairs of connecting files would keep me in touch with the point. (Par. 511a.) I would now give this order: "Corporal Smith, take two men from your squad and patrol north along this road (pointing up the Tracy-Maxey road) for a mile and then rejoin the column on this road (Chester Pike), to the west of you." I would then say to Private Barker, "Take Carter and cut across to that railroad fill and go along the top of that (Sandy) ridge, rejoining the column beyond the ridge. Corporal Smith with a patrol is going up this road. Keep a lookout for him." When we reached the point where the road crosses the south nose of Sandy Ridge and I saw the valley in front of me with the long high ridge west of Sandy Creek, running parallel to the Chester Pike and about 800 yards west of it, I would give this order: "Corporal Davis, take the three remaining men in Corporal Smith's squad, cross the creek there (pointing in the direction of the Barton farm) go by that orchard, and move north along that high ridge, keeping the column in sight. Make an effort to keep abreast of the advance guard, which will continue along this road."

I gave Corporal Davis the remaining men out of Corporal Smith's squad because I did not want to break up another squad and as this is, in my opinion, a very important patrol, I wanted a noncommissioned officer in charge of it. Unless something else occurs this will be all the patrols I intend sending out until we pass the steel railroad trestle over Sandy Creek.

Captain: Your point about not breaking up a squad when you could avoid it by using the men remaining in an already broken squad, is a very important one. Take this particular case. You first sent out two pairs of connecting files between the advance party and your point—four men. This leaves a corporal and three men in that squad. If we assume that no patrols were out when we passed through Salem, this corporal and two of his men could have been sent up the Tracy-Maxey road, leaving one man to be temporarily attached to some squad. From the last mentioned squad you would pick your two men for the Sandy Ridge patrol and also the corporal and three men for the Barton farm, etc., patrol. This would leave three men in this squad and you would have under your immediate command two complete squads and three men. As the patrols return, organize new squads immediately and constantly endeavor to have every man attached to a squad. This is one of your most improtant duties, as it prevents diorder when some serious situation suddenly arises. Also it is one of the duties of the detachment commander that is generally overlooked until too late.

The direction you sent your three patrols was good and their orders clear, covering the essential points, but as you have in a very short space of time, detached nine men, almost a third of your advance party, don't you think you should have economized more on men?

Sergeant Adams: The Sandy Ridge patrol is as small as you can make it—two men. I thought the other two patrols were going to be detached so far from the column that they should be large enough to send a message or two and still remain out. I suppose it would be better to send but two men with Corporal Davis, but I think Corporal Smith should have two with him.

Captain: Yes, I agree with you, for you are entering a valley which is, in effect, a defile, and the Tracy-Maxey road is a very dangerous avenue of approach to your main body. But you must always bear in mind that it is a mistake to use one more man than is needed to accomplish the object in view. The more you send away from your advance party, the more scattered and weaker your command becomes, and

[445]

this is dispersion, which constitutes one of the gravest, and at the same time, most frequent tactical errors.

To continue the problem, we will suppose you have reached the stone bridge over Sandy Creek; the point is at the cross roads by the Smith house; you can see the two men moving along Sandy Ridge; and Corporal Davis' patrol is just entering the orchard by the Barton farm. Firing suddenly commences well to the front and you hear your point reply to it.

Sergeant Adams: I halt to await information from the point.

Captain: That is absolutely wrong. You command the advance party of an advance guard; your mission requires you to secure the uninterrupted march of the main body; and at the first contact you halt, thus interrupting the march (Par. 511). The sooner you reach the point, the better are your chances for driving off the enemy if he is not too strong, or the quicker you find out his strength and give your commander in the rear the much desired information.

Sergeant Adams: Then I push ahead with the advance party, sending back the following message—

Captain (interrupting): It is not time to send a message. You know too little and in a few minutes you will be up with the point where you can hear what has happened and see the situation for yourself. Then you can send back a valuable message. When but a few moments' delay will probably permit you to secure much more detailed information, it is generally best to wait for that short time and thus avoid using two messengers. When you reach the cross roads you find six men of the point deployed behind the fence, under cover of the trees along the County Road, just west of the Chester Pike, firing at the stone wall along the Mills' farm lane. The enemy appears to be deployed behind this stone wall, from the Chester Pike west for a distance of fifty yards, and his fire is much heavier than that of your point. You think he has at least twenty rifles there. You cannot see down the Chester Pike beyond the enemy's position. Your patrol on Sandy Ridge is midway between the 68 and 66 knolls, moving north. The ground in your front, west of the road, is a potato field; that east of the road as far as the swamp, is rough grass land.

Sergeant Adams: I give order, "Corporal Gibbs, deploy your squad to the right of the Pike and push forward between the Pike and the swamp. Corporal Hall (commands the point), continue a heavy fire. Here are six more men for your squad." I give him the four connecting files and two of the three men in the advance party whose squad is on patrol duty. "Corporal Jackson, get your squad under cover here. Lacey, run back to the major and tell him the point has been stopped by what appears to be twenty of the enemy deployed behind a stone wall across the valley 500 yards in our front. I am attacking with the advance party."

Captain: Corporal Davis (commands patrol near Barton farm), you can hear the firing and see that the advance is stopped. What do you do?

Corporal Davis: I would head straight across for the clump of woods on the ridge just above the Mills' farm, moving as rapidly as possible.

Captain: That is all right. Sergeant, Corporal Hall's squad keeps up a heavy fire; Corporal Gibbs' squad deploys to the right of the pike, rushes forward about 75 yards, but is forced to lie down by the enemy's fire, and opens fire. Corporal Gibbs, what would your command for firing be?

Corporal Gibbs: AT THE BOTTOM OF THAT WALL. BATTLE SIGHT. CLIP FIRING.

Captain: Why at the bottom of the wall?

Corporal Gibbs: The men are winded and excited and will probable fire high, so I gave them the bottom of the wall as an objective.

Captain: The enemy's fire seems as heavy as yours. Sergeant, what do you do?

Sergeant Adams: I give this order, "Corporal Jackson, deploy your squad as skirmishers on the left of Corporal Hall's squad and open fire." What effect does this additional fire have on the enemy?

Captain: His bullets seem to go higher and wilder. You appear to be getting fire superiority over him.

Sergeant Adams: If I do not see any signs of the enemy being reënforced, dust in the road behind his position, etc., I take immediate command of the squads of Corporals Hall and Jackson, and lead them forward on a rush across the potato field.

Captain: Corporal Gibbs, what do you do when you see the other two squads rush?

Corporal Gibbs: I order, **FIRE AT WILL**, and urge the men to shoot rapidly in order to cover the advance.

Captain: Sergeant Adams' squads are forced to halt after advancing about 150 yards.

Corporal Gibbs: I keep up a hot fire until they can resume their firing, when I lead my squad forward in a rush.

Captain: What do you do, Sergeant?

Sergeant Adams: I would have the Corporals keep up a heavy fire. By this time I should think the support would be up to the cross roads.

Captain: It is, but have you given up your attack?

Sergeant Adams: If it looks as if I could drive the enemy out on my next rush, I do so, but otherwise I remain where I am, as I have no reserve under my control and the action has gotten too serious for me to risk anything more when my chief is practically on the ground to make the next decision. He should have heard something about what is on the Pike behind the enemy, from the patrol on Sandy Ridge.

Captain: Your solution seems correct to me. Why did you send Corporal Gibbs' squad up between the pike and the swamp?

Sergeant Adams: It looked as if he would strike the enemy from a better quarter; there appeared to be better cover that way, afforded by the turn in the road, which must have some weeds, etc., along it, and the swamp would prevent him from getting too far separated from the remainder of the advance party.

Captain: The Sergeant's orders for the attack were very good. He gave his squad leaders some authority and attached his extra men to a squad. He did not attempt to assume direct control of individual men,

but managed the three squads and made the squad leaders manage the
individual men. This is the secret of successful troop leading. His
orders were short, plain and given in proper sequence.

Problem No. 3 (Infantry)

(See Fort Leavenworth map in pocket at back of book.)

Situation.

A Blue battalion, in hostile country, is in camp for the night,
August 5-6, at Sprong (ja'). At 9:00 P. M., August 5th, Lieutenant A,
Adjutant gives a copy of the following order to Sergeant B:

1st Battalion, 1st Infantry,
Sprong, Kansas,
5 Aug., '09.

Field Orders No. 5.

1. The enemy's infantry is six miles east of FORT LEAVENWORTH.
His cavalry patrols were seen at F (qg') today.

Our regiment will reach FRENCHMAN'S (oc') at noon tomorrow.

2. The battalion wi'l march tomorrow to seize the ROCK ISLAND
BRIDGE (q) at FORT LEAVENWORTH.

3. (a) The advance guard, consisting of 1st platoon Co. A and
mounted orderlies B,. C, and D, under Sergeant B, will precede the main
body at 400 yards.

(b) The head of the main body will march at 6:30 A. M., from 19,
via the 17 (jc')—15 (jg') 1—5 (lm')—FORT LEAVENWORTH (om')
road.

4. The baggage will follow close behind the main body under escort
of Corporal D and one squad, Co. B.

5. Send reports to head of main body.

C,
Major, Comdg.

Copies to the company commanders, to Sergeant B and Corporal D.

A. **Required,** 1. Give Sergeant B's estimate of the situation. (The
estimate of the military situation includes the following points:

1. His orders or mission and how much discretion he is allowed.
2. The ground as it influences his duty.
3. The position, strength and probable intentions of the enemy.
4. Sergeant B's decision.)

Answer. 1. The size of the advance guard, its route and the distance it is to move in front of the main body are prescribed by Major C.
Sergeant B is free to divide up the advance as he sees fit, to use the
various parts so as to best keep open the way of the main body, maintain the distance of 400 yards in front of it, and protect it from surprise
by the enemy.

2. The ground may be such as to make easy or to hinder reconnaissance, such as hills or woods; to impede or hasten the march, such as
roads, streams, defiles; to offer good or poor defensive positions; to offer
good or poor opportunities for an attack. Sergeant B sees from his map
that the ground is rolling and open as far as Kern (ji') with good
positions for reconnaissance and for defense or attack. There is a bridge
over Salt Creek (ig') which has steep banks and will be a considerable
obstacle if the bridge has been destroyed. From this creek to Kern the

advance would be under effective fire from Hancock Hill (ki'), so that these heights must be seized before the main body reaches 15 (jg').

Beyond Kern the heavy woods make reconnaissance difficult and must be treated somewhat like a defile by the point. (Par. 483.)

3. There is little to fear from the main body of the enemy which is 1½ miles farther from the Rock Island bridge than we are, but we know the enemy has cavalry. The size of the cavalry force is not known, and may be sufficient to cause us considerable delay, especially in the woods. The enemy's evident intention is to keep us from seizing the bridge.

4. Having considered all these points, Sergeant B comes to the following decision: * * * (Before reading the decision as contained in the following paragraph, make one of your own)

Answer: To have only an advance party with which to throw forward a point of 5 men 200 yards to the front and send out flankers, as needed (Par. 483); to send the three mounted orderlies well to the front of the point to gain early information of the enemy, especially on Hancock Hill (ji') and the ridge to the north of 11 (jj').

Required, 2. Sergeant B's order. (Par. 459.)

Answer. Given verbally to the platoon and mounted orderlies, at 9:30 P. M.

"The enemy's cavalry patrols were seen at F (qh') today; no hostile infantry is on this side of the Missouri river. The battalion will move tomorrow to Fort Leavenworth, leaving 19 (ja') at 6:30 A. M.

"This platoon and orderlies B, C, and D will form the advance guard, and will start from the hedge 400 yards east of 19 at 6:30 A. M. via the 17 (jc')—15 (jg')—5 (lm') road.

"The point, Corporal Smith and 4 men of his squad, will precede the remainder of the advance guard at 200 yards.

"I will be with the advance party. Private X and Y will act as connecting files with the main body."

The flankers will be sent out from time to time by Sergeant B as necessary.

Required, 3. The flankers sent out by Sergeant B between 19 (ja') and 15 (jg').

Answer. A patrol of 3 men is sent to Hill 900 southeast of 19 (ja'), thence by Moss (kc') and Taylor (lc') houses to Hill 840 east of Taylor, thence to join at 15 (jg').

Two men are sent from the advance party as it passes Hill 875.5 (ie') to the top of this hill to reconnoiter to the front and northeast. These men return to the road and join after the advance party has reached Salt Creek. Two men are sent ahead of the advance party at a double time to take position on 'Hill 875 northeast of J. E. Daniels' place (jf') and reconnoiter to the northeast and east.

Reasons. The patrol sent out on the south moves out far enough to get a good view from the hills which an enemy could observe or fire into the column. There is no necessity of sending out flankers north of the road at first, because from the road itself a good view is obtained. Hills 875.5 and 875 give splendid points for observing all the ground to the north and east. (Don't send flankers out unless they are necessary.)

512 (contd.)

Required, 4. When the advance party reaches J. E. Daniels' house (je') a civilian leaves the house and starts toward 15. What action does Sergeant B take?

Required, 5. When the advance party reaches Salt Creek bridge (jg') the point signals "enemy in sight," and Private H reports that he saw about 6 or 8 mounted men ride up to the edge of the woods at Kern, halt a moment, and disappear. What action does Sergeant B take?

Answer. He at once sends a message back by Private H stating the facts. He then orders the advance party to move forward, hastens up to the point and directs it to continue the march, seeking cover of fences and ravines and hill top.

Required, 6. When the point reaches Schroeder (jh') it receives fire from the orchard at Kern. What action is taken?

Answer. The men in the point are moved rapidly down the hill and gain shelter in the ravines leading toward Kern. Two squads are rapidly placed in line along the ridge west of Schroeder and under cover of their fire the remainder of the advance party run down the hill at 10 yards distance to join the point. A squad of this force is then hurried forward to the Kern house. Here the squad is stopped by fire and Sergeant B deploys two more squads which advance by rushes and drive out the enemy, found to be 10 cavalrymen. The squads left at Schroeder now join at double time and the advance party moves forward, without having delayed the march of the main body.

Problem No. 4 (Infantry)

Situation:

A Blue force of one regiment of infantry has outposts facing south on the line Pope Hill (sm')—National cemetery (pk')—E (qh'). A Red force is reported to have reached Soldiers' Home (3 miles south of Leavenworth) from the south at 7:00 o'clock this morning. Corporal A is directed by Sergeant B, in command of the left support at Rabbit Point (tn'), to take out a patrol toward the waterworks and south along the Esplanade (xo') to the Terminal bridge

Required, 1. Give Sergeant B's orders to Corporal A.

Answer. "The enemy, strength unknown, was at Soldiers' Home at 7:00 o'clock this morning. Another patrol will advance along Grant avenue (tm').

"Our outposts will remain here for the day.

"Select from the first section a patrol and reconnoiter this road (Farragut avenue) as far as the waterworks (vn'), thence by Esplanade to the Terminal bridge, and report on the ground in our front. When you reach the Terminal bridge return if no enemy is seen.

"Send reports here."

Required, 2. How many men does Corporal A select, and why? (Par. 456.)

Answer. Five men are taken because the patrol is to reconnoiter, not to fight, and on account of the distance to go and lack of information of the enemy, 2 or 3 messages may have to be sent.

Required, 3. What equipment should Corporal A have? (Par. 457.)

Required, 4. State the points to be noted by Corporal A in selecting his patrol and what inspection does he make? (Par. 460.)

Answer. He selects Privates C, D, E, F and G, on acount of their bravery, attention to duty and discretion. He directs them to carry one meal in their haversacks, full canteen and fifty rounds of ammunition. He then inspects them as to their physical condition, sees that they have proper equipment and that nothing to rattle or glisten is carried.

Required, 5. What does Corporal A next do? (Par. 461.)

Answer. He gives them their instructions as follows: "The enemy, strength unknown, was at Soldiers' Home (about three miles south of Leavenworth) at 7 o'clock this morning. There will be a friendly patrol along that road (pointing to Grant avenue). We are to reconnoiter along this road and down toward that bridge (pointing). Be very careful not to be seen, take advantage of all cover, and keep touch with C and myself on this road at the point of the patrol. In case we get separated meet at the waterworks (vn')."

He then explains the signals to be used, and moves the patrol in close order out along the road until it passes the sentinel at the bridge XV (un'), to whom he gives the direction to be taken by the patrol.

Required, 6. Upon leaving XV, what formation would the patrol take, and reasons for same. (Par. 463.)

Answer. Corporal A and Private C form the point on the road leading southwest of the waterworks; Private D moves on the left overlooking the railroad; Private E moves promptly up Corral creek (um') to the top of Grant Hill (um') to observe the country toward the southwest; Private F moves about 50 yards in rear of the point, followed at 50 yards by Private G.

Corporal A forms his patrol as stated because of the necessity of getting a view from the hill on each side. Only one man is sent out on each side because they can be plainly seen by the patrol on the road, and no connecting file is necessary. The distances taken along the road assure at least one man's escape, and Corporal A is in front to get a good view and to signal the flankers.

Problem No. 5 (Infantry)

Situation:

The head of the patrol is now at the bridge, XVI (un') northwest of the waterworks.

Private E has reached the top of Grant Hill and signals the enemy in sight; the patrol halts and Corporal A moves out to meet Private E who is coming down toward the patrol. He says he saw three mounted men ride up to Grant and Metropolitan avenues (wm') from the south and after looking north a moment move west.

Required, 1. Corporal A's action. (Pars. 474 and 474a.)

Answer. Corporal A at once writes the following message and sends it back by Private E:

"No. 1.

Patrol, Company B,
Farragut Avenue,
Northwest of Waterworks,
10 May, '09, 8:30 A. M.

To Commander Blue Left Support,
Rabbit Point.

Three mounted Reds, seen by Private E, just now reconnoitered at

[451]

Grant and Metropolitan avenues; they are moving west on Metropolitan avenue; the patrol will continue toward the Terminal bridge.

A,
Corporal.''

Reasons. The message is sent because this is the first time the enemy has been seen, and they have not been reported north of Soldiers' Home before. The message should state who saw the enemy, and the man seeing them should always carry the message telling of the facts. The patrol would not allow this small hostile patrol to stop its advance, but would proceed on its route cautiously to avoid being seen, and to see if the Red cavalrymen are followed by others of the enemy.

Required, 2. Give the method of reconnoitering the buildings at the waterworks and coal mine. (Par. 488.)

Answer. Private D carefully examines the east side of the enclosures and buildings, while Private C examines the west side. The remainder of the patrol halts concealed in the cut west of the north enclosure, until C and D signal no enemy in sight, whereupon the patrol moves forward along the road (XV—3rd St.), C and D advancing rapidly between the buildings to the town where they join the patrol.

Required, 3. Give the route followed by E from Grant Hill to edge of Leavenworth.

Answer. He moves down the east slope of Grant Hill to the ravine just east of the old R. R. bed (um'), being careful to keep concealed from the direction of Leavenworth. He moves up the ravine, keeping a sharp lookout to the front, and moving rapidly until abreast, if he has fallen behind. He takes the branch ravine lying just west of Circus Hill (vm'), and moves up to its end. Here he halts and makes careful inspection of Metropolitan avenue and the street south into the city. Being sure the coast is clear, he darts across the narrow ridge south of Circus Hill to the ravine to the east and then joins the patrol. He reports to Corporal A any indication of the enemy he may have seen.

Problem No. 6 (Infantry)

Situation:

A Blue force holds Fort Leavenworth (om') in hostile country. Outposts occupy the line Salt Creek Hill (gh')—13 (ij')—Sheridan's Drive, (mi') against the Reds advancing from the northwest.

At 4:30 P. M., June 25th, Sergeant A is given the following orders by Captain B, commanding the support:

"The enemy will probably reach Kickapoo late today. Our outposts extend as far north as Salt Creek Hill. There were six of our men prisoners at 45 (dc') this afternoon at 1 o'clock, being held by 15 home guards at Kickapoo. Take ... men from the company and move to Kickapoo recapture the prisoners and gain all the information you can of the enemy north of there."

Required, 1. How many men does Captain B name, and why? (Par. 456.)

Answer. Thirty men are assigned.

Reason. This is twice as many as the enemy holding the prisoners, and to secure secrecy no larger force than is absolutely necessary should

[452]

be taken. This force will allow men to surround the enemy while the remainder rush them.

Required, 2. Give the order of Sergeant A to his patrol. (See 6th requirement, Problem 4.)

Required, 3. What route will the patrol take?

Answer, 11 (jj')—13 (1j')—Salt Creek Hill (gh')—and along the edge of the woods east of the M. P. R. R. (fg') as far as the bridge opposite Kickapoo Hill—thence up Kickapoo Hill toward 45 (dc').

Reasons. S nce the patrol's orders do not require any reconnaissance before reaching Kickapoo the shortest and most practical route is chosen. The route as far as Salt Creek Hill lies behind our outpost line and is thus protected. The main roads are avoided because they will be carefully watched by the enemy. The edge of the woods east of the M. P. Ry. (beginning about ff') gives good cover and by moving to the bridge the patrol can probably sneak close in on the enemy and capture them by surprise.

Problem No. 7 (Infantry)

Situation:

The patrol reaches the top of Kickapoo Hill (cd'). Sergeant A and Private C move cautiously to the top and see the six prisoners in the cemetery (cd') just west of Kickapoo Hill, and a Red sentinel at each corner. Just west of the cemetery are about 10 more Reds. No others are visible.

Required, 1. What decision does Sergeant A make and what does he do?

Answer. He decides to capture the enemy by surprise. He leaves Private C to watch and, moving cautiously back to his patrol, makes the following dispositions: Corporal D with 10 men to move up to Private C and cover the enemy, remaining concealed. He takes the remainder of the patrol with fixed bayonets around the northeast slope of Kickapoo Hill in the woods and moves up the ravine toward 29. When his detachment arrives within about 100 yards of the enemy, they charge bayonet and rush them. Corporal D's party at the same time rush in from the opposite side. (Note: The enemy are demoralized by the surprise and are captured without a shot being fired.)

Required, 2. What action does Sergeant A now take?

Answer. He causes the enemy to be kept apart while he and his noncommissioned officers question them separately. He then questions the Blue prisoners, and furnishing them the guns taken from the Reds, sends them and the captured Reds back to our line under Corporal D, with a written message giving the information secured from his questions. (Par. 476.)

Required, 3. What does he then do?

Answer. Places his main body in concealment at the Cemetery (cd') and sends a patrol under Corporal H via 35—41—43, and one under Corporal F via 29—27—23 west to learn further of the enemy in execution of the second part of his orders.

The patrol under Corporal H sends back the following message:
"No 1.   Patrol Company A, 1st Infantry,
                                21 June, '09; 5:30 P. M.
Commander Expeditionary Patrol at 45:

A column of infantry is moving east about 1 mile west of
Schweizer (aa'); about 800 yards in front of this body is another small
body with 8 to 10 men 300 yards still farther east. It took thè main
body 2 min., 45 sec, to pass a point on the road. I remain in observation.
                                                    H,
                                                Corporal.''

Required, 3. The size of the command reported by Corporal H
and its formation. (Par. 475b.)

Answer. One battalion infantry (512 men), preceded by 1 section
at advance guard. The advance guard having only advance party and
point, 2¾ minutes x 175 = 481 men in the main body, leaving about
32 men for the advance men for the advance guard.

Problem No. 8 (Infantry)
General Situation:

A Blue force of one regiment of infantry has outposts facing
south on the line Pope Hill (sm'), National Cemetery (qk')—E (qi'). A
Red force moving north reached Soldiers' Home at 7 o'clock this morning.

Special Situation:

Corporal B is chosen by Sergeant A, commander of the right support at the National Cemetery, to take a patrol south as far as 20th street
(yf') and Metropolitan avenue (wh'), to report on the ground along the
route, and to reconnoiter the enemy. A friendly patrol moves along
Sheridan's Drive (i)—Atchison Hill (rg')—Southwest Hill (ue'), and
one on Prison Lane (rk').

Required, 1. Sergeant A's orders, verbatim (that is, word for
word).

2.   Give the various details attended to by Corporal B
before he moves out with his patrol.

3.   What is the formation of the patrol when its point
is at E (qh').

4.   When the patrol reaches 14 (ug'), how are the inter-
secting roads reconnoitered?

5.   Four mounted men are seen riding west at a walk at
64 (wh'). What action does Corporal A take?

6.   Describe the ground passed over by the patrol.

Problem No. 9 (Infantry).
Situation:

The enemy is moving east toward Frenchman (oc') and is
expected to reach there early tomorrow. A company at 72 (uj') forms
the left support of an outpost in hostile country, on the line 70 (vj')—
National Cemetery (qj'). At 4 P. M. Sergeant A is ordered to take a
patrol of 12 men and go to Frenchman and destroy the bridge there, and
remain in observation in that vicinity all night.

Required, 1. His orders to the patrol.

2.   The route the patrol will follow, and its formation
crossing the Atchison Hill—Government Hill
ridge.

3. Give the conduct of the patrol from Atchison Hill (rg')—Government Hill (tf') to its position at the bridge at Frenchman.

General Situation:

A Blue squadron is camped for the night at Waterworks (vn'), Fort Leavenworth, and has outposts on the line XIV (un')—Grant Hill (um')—Pr:son Hill (wk'). A Red force is reported to be advancing from the north on Kickapoo (cb').

Problem No. 10 (Cavalry)

Special Situation:

Lieutenant A, commanding the left support on Prison Hill, at 5 P. M., directs Sergeant Jones to take a patrol of 5 men from his platoon and move via Atchison Cross (ug') to the vicinity of Kickapoo and secure information of any enemy that may be in that locality. Another patrol is to go via Fort Leavenworth (ol').

Required, 1. The order given by Lieutenant A, verbatim. (Pars. 459 and 461.)

Answer. "Sergeant Jones, the enemy is north of Kickapoo, moving on that place. The squadron will remain here tonight; Sergeant B will take a patrol through Fort Leavenworth.

"Select a patrol of 5 men from your platoon and move out via Frenchman's (oc') toward Kickapoo.

"Secure any information you can of the enemy in that locality.

"Report on the condition of the bridges between here and 47 (fd').

"You may have to stay out over night.

"Send messages here."

Sergeant Jones selects five good men, directs them to take one cooked ration each and canteen full of water. He inspects the men and horses carefully; sees that no horse of conspicuous color or that neighs is taken. Explains the orders to his men, etc., as was done in the infantry patrol.

Required, 2. What route does the patrol take, and why?

Answer. Metropolitan avenue (w)—70 (vj')—72 (vj')—14 (ug')—Frenchman (oc')—17 (jc')—47 (ec').

Reasons. The enemy is distant and Kickapoo, the objective of the patrol, is seen from the map, which Sergeant Jones has, to be over an hour's ride at a walk and trot. It is not at all probable that the enemy will be met until the patrol reaches the vicinity of Kickapoo and Sergeant Jones decides to take the shortest and best road though it is a main highway, instead of Sheridan's Drive (j) of the F (qg')—15 (jg') lane.

It is always well for a patrol to avoid main highways when the enemy is near, especially in hostile country, but here the time saved more than justifies the use of the direct route.

Problem No. 11 (Cavalry)

Same situation as Problem 1.

Required, 1. The formation and conduct of the patrol as far as Frenchman's.

Answer. Sergeant Jones determines to move at a walk and trot (5 miles per hour) in order to reach the vicinity of Kickapoo and take up a position of observation before night. Sergeant Jones and Private B are in the lead, 2 men about 100 yards to the rear, the remaining 2 men about 75 yards in the rear of these. They move out at a trot along the road until Atchison Cross is reached. The two cross roads are reconnoitered without halting the patrol, inasmuch as from the cross roads a good view is had north and south.

From Atchison Cross to 16 (sf') the patrol moves at a walk, being up a slope from 4 to 6 degrees. Usually such a place would be rushed through, but the distance of the enemy makes this unnecessary. No scouting is done off the road through the woods, because of the distance of the enemy. On reaching the top of the hill the patrol is halted while Sergeant Jones moves up to the high ground south of the road at the crest, and in concealment searches with his glasses the road as far as Frenchman's, especially the village beyond G (qf'). Seeing no signs of the enemy he moves the patrol down the hill at a walk until the cut is passed and there takes a fast trot, so as to avoid being long in a position where they could be seen from the direction of Kickapoo. The same formation and gait are maintained as far as Gauss' (pd'), where a walk is taken to rest the horses and to gain opportunity to see if any enemy are holding the bridge at Frenchman's.

Situation:

Just as the patrol comes to a walk Sergeant Jones sees what appears to be a dismounted patrol moving south over the ridge about 650 yards north of Frenchman's. He can see three men.

Required, 2. Action taken by Sergeant Jones.

Answer. The patrol is moved into the orchard just off the road, while Sergeant Jones moves quickly to the top of the hill and, concealed by the trees, examines the road north to see if the 3 men are followed by others forming a part of a larger patrol or of a column. He finds the three men are not followed.

Required, 3. What does he do next?

Answer. He determines to capture the patrol by surprise. He has the horses led over south of the orchard hill so as not to be visible to the enemy. He then distributes his men along the north edge of the orchard, himself nearest the bridge, 2 men 75 yards back along the road toward G (qf'), then 2 men 75 yards farther along toward G. As the third man comes opposite him, Sergeant Jones cries "Halt", which is the signal for the other parties to similarly hold up their men.

Reasons. Sergeant Jones might either capture the hostile patrol or let it pass, and then proceed on his road. Since they are the first enemy seen and there is such a good chance to capture them, and as they may furnish definite information of the enemy's main force, he decides as stated. There is an objection in capturing them that he will have to send one or two men to take them to camp. The patrol is placed as described above so as to have the two men opposite each of the enemy, except for Sergeant Jones, who is alone. By thus covering each man of the hostile patrol by two of our men, they will at once see the folly of an effort to escape and no shot need be fired. One man is holding the horses.

Problem No. 12 (Cavalry)

Same situation as Problem 10.

Required:

1. What action does Sergeant Jones take before leaving the vicinity of Frenchman's?

2. Give the formation and conduct of the patrol after leaving here.

3. Give the report submitted by Sergeant Jones under his instructions in regard to bridges. (Par. 492.)

At 6:30 P. M. (it is dark at 7:30) the patrol reaches 17 (jc').

4. Give the route followed from here and the disposition of the patrol made for the night.

Problem No. 13 (Cavalry)

Situation

The Missouri river is the boundary between hostile countries.

A Blue separate brigade (3 regiments infantry, 1 squadron cavalry, 1 battery field artillery) is moving from Winchester (19 miles west of Leavenworth) to seize the Rock Island bridge (q) across the Missouri river at Fort Leavenworth. The cavalry squadron is camped at Lowemont, 8 miles west of Leavenworth, for night June 4-5. At 3 P. M. Sergeant Jones is directed to take a patrol of six men and move via the Rock Island bridge into Missouri and gain information of the enemy reported to be now just east of the river.

Required, 1. Give the formation of the patrol when it first comes on the map.

Required, 2. Give the conduct of the patrol from Mottin's (oa') to G (qf').

At Frenchman's Sergeant Jones met a farmer coming from Fort Leavenworth, who said about 200 hostile cavalry were seen just east of the Missouri about 2 P. M., moving towards the Terminal Bridge (z).

Required, 3. Action of Sergeant Jones. (Does he hold the man? Does he send a message? Does he change his plans or direction of march?)

The patrol reaches the top of the hill, Sheridan's Drive—Government Hill (tf').

Required, 4. What action does Sergeant Jones take before proceeding east?

FLANK GUARDS

The flanks of a column are ordinarily protected by the advance guard, which sends out patrols to carefully examine the country on both sides of the line of march. In some cases, however, the direction of march of the column is such that there is a great danger of the enemy's striking it in flank and some special provision is necessary to furnish additional security on the threatened flank. This is done by having a detachment, called a flank guard, march off the exposed flank. The flank guard usually follows a road, parallel to the one on which the column is marching and at least 1,000 yards (effective rifle range) beyond it. If hostile artillery is feared this distance is much greater.

The flank guard regulates its march so as to continue abreast of the advance guard of the main column. It takes a formation similar to an advance guard, does most of its patrolling to the front and on the exposed flank, and keeps in constant touch with the main column by means of mounted or dismounted messengers.

In case the enemy is encountered the flank guard drives him off if practicable or takes up a defensive position, protecting the march of the main column, and preventing the enemy from disturbing the latter's march.

REAR GUARD

513. Definition and Duties. A rear guard is a detachment of a marching column following in rear to protect the main column from being surprised and to prevent the march from being delayed or interrupted.

When the main column is marching toward the enemy the rear guard is very small and its duties relatively unimportant. It is principally occupied in gathering up stragglers.

When the main column is marching away from the enemy (retreating) the rear guard is all important. It covers the retreat of the main body, preventing the enemy from harassing or delaying its march.

Strength. The strength of a rear guard is slightly greater than that of an advance guard, as it cannot expect, like the latter, to be reinforced in case it is attacked, as the main column is marching away from it and avoiding a fight.

Form of Order. The rear guard commander, on the receipt of the retreat order, issues a rear guard order, according to the following general form:

Field Orders
No.. — .. .. (Title)
                (Place)   (Date and hour)
    Troops   1 (Information of enemy and of our supporting troops.)
                  2 (Plan of commander—duty of rear guard)

(a) Reserve—in   3. (a) (Instructions for reserve—place and time of departure, or approximate distance from main body—reconnaissance.)
     order of march
       (Troops)
(b) Support: (b) (Instructions for support—place and time of departure or distance from reserve—any special reconnaissance.)
     (Commander)
       (Troops)
(c) Rear Cavalry: (c) (Instructions for rear cavalry—place and time of departure, road or country to be covered—special mission.)
     (Commander)
       (Troops)
(d) Right (left)   (d) (Instructions for flank guard—place and time of departure, route, special mission.)
     Flank Guard:
     (Commander
       (Troops)

    4. (Instructions for field train when necessary—usually to join train of main body.)
    5. (Place of commander or where messages may be sent.)
(How and to whom issued.)   (Signature.)

The distance of a rear guard from the main body and its formation are similar to those of an advance guard. The elements corresponding to the advance cavalry, the point, and the advance party of an advance guard are termed the rear cavalry, rear point and rear party, respectively. The support and reserve retain the same designations.

A rear guard formed during an engagement to cover the withdrawal or retreat of the main body, may first be compelled to take up a defensive position behind which the main body forms up and moves off. It may be forced to withdraw from this position by successive skirmish lines, gradually forming up in column on the road as it clears itself from fighting contact with the enemy.

The rate of march of the rear guard depends upon that of the main body. The main body may be much disorganized and fatigued, necessitating long halts and a slow marching rate.

Action of the Rear Guard. The withdrawal of defeated troops is delayed, if possible, until night. If it becomes necessary to begin a retreat while an engagement is in progress, the rear guard is organized and takes up a defensive position generally behind the fighting line; the latter then falls back and assembles under cover of the rear guard.

The rear cavalry gives away before the enemy's pursuit only when absolutely necessary, maintains communication with and sends information to the rear guard commander, and pays special attention to the weak points in the retreat, namely, the flanks. It makes use of every kind of action of which it is capable, according to the situation, and unless greatly outnumbered by hostile cavalry, it causes considerable delay to the enemy.

When the enemy is conducting an energetic pursuit the rear guard effects its withdrawal by taking up a succession of defensive positions (that is, where the nature of the ground enables the rear guard to defend itself well) and compelling the enemy to attack or turn them. (It should be understood that these successive defensive positions must, in the case of a large force, be from two to four miles apart and in the case of a small force at least one-half mile apart—not a few hundred yards as is frequently attempted in peace maneuvers.)

When the enemy's dispositions for attack are nearly completed, the rear guard begins to fall back, the cavalry on the flanks being usually the last to leave. The commander designates a part of the rear guard to cover the withdrawal of the remainder; the latter then falls back to a new position in rear, and in turn covers the withdrawal of the troops in front. These operations compel the enemy continually to deploy or make turning movements, and constantly retard his advance.

The pursuit may be further delayed by obstacles placed in the enemy's path; bridges are burned or blow up; boats removed or destroyed; fords and roads obstructed; tracks torn up; telegraph lines cut, and houses, villages, woods and fields fired. Demolitions and obstructions are prepared by engineers, assisted, if necessary, by other troops detailed from the reserve, and are completed by the mounted engineers of the rear party at the last moment.

The instructions of the supreme commander govern in the demolition of important structures.

OUTPOSTS

514. Definition and Duties. Outposts are detachments thrown out to the front and flanks of a force that is in camp or bivouac, to protect the main body from being surprised and to insure its undisturbed rest. In fact, an outpost is merely a stationary advance guard. Its duties, in general, are to *observe* and *resist*—to observe the enemy, and to resist him in case of attack. Specifically its duties are:

(a) To observe toward the front and flanks by means of stationary sentinels and patrols, in order to locate the enemy's whereabouts and learn promptly of his movements, thus making it impossible for him to surprise us.

(b) To prevent the main body from being observed or disturbed.

(c) In case of attack, to check the enemy long enough to enable the main body to prepare for action and make the necessary dispositions.

Size. The size of the outpost will depend upon many circumstances, such as the size of the whole command, the nearness of the enemy, the nature of the ground, etc. A suitable strength for an outpost may vary from a very small fraction to one-third of the whole force. However, in practice it seldom exceeds one-sixth of the whole force—as a rule, if it be greater, the efficiency of the troops will be impaired. For a single company in bivouac a few sentinels and patrols will suffice; for a large command, a more elaborate outpost system must be provided. The most economical form of outpost is furnished by keeping close contact with the enemy by means of outpost patrols, in conjunction with resisting detachments on the avenues of approach.

Troops at a halt are supposed to be resting, night or day, and the fewer on outpost the more troops will there be resting, and thus husbanding their strength for approaching marches and encounters with the enemy. Outpost duty is about the most exhausting and fatiguing work a soldier performs. It is, therefore, evident that not a man or horse more than is absolutely necessary should be employed, and that the commander should use careful judgment in determining the strength of the outpost, and the chiefs of the various outpost subdivisions should be equally careful in disposing their men so as to permit the greatest possible number to rest and sleep undisturbed, *but at the same time always considering the safety of the main body as the chief duty.*

515. Composition. The composition of the outpost will, as a rule, depend upon the size and composition of the command, but a mixed outpost is composed principally of infantry, which is charged with the duty of local observation, especially at night, and with resisting the enemy, in case of attack, long enough for the main body to prepare for action.

The cavalry is charged with the duty of reconnaissance, and is very useful in open country during the day.

Artillery is useful to outposts when its fire can sweep defiles or large open spaces and when it commands positions that might be occupied by hostile artillery.

Machine guns are useful to command approaches and check sudden advances of the enemy

Engineers are attached to an outpost to assist in constructing entrenchments, clearing the field of fire, opening communication laterally and to the rear. The outpost should be composed of complete organizations. For example, if the outpost is to consist of one company, do not have some of the platoons from one company and the others from another, and if it is to consist of one battalion, do not have some of the companies from one battalion and others from another, etc.

FORMATION OF OUTPOSTS

516. Subdivisions. As in the case of an advance guard, the outpost of a large force is divided into elements or parts, that gradually increase in size from front to rear. These, in order from the main body, are the reserve, the line of supports, the line of outguards, and the advance cavalry, and their formation, as shown by the drawing below, may be likened to an open hand, with the fingers apart and extended, the wrist representing the main body, the knuckles the line of supports,

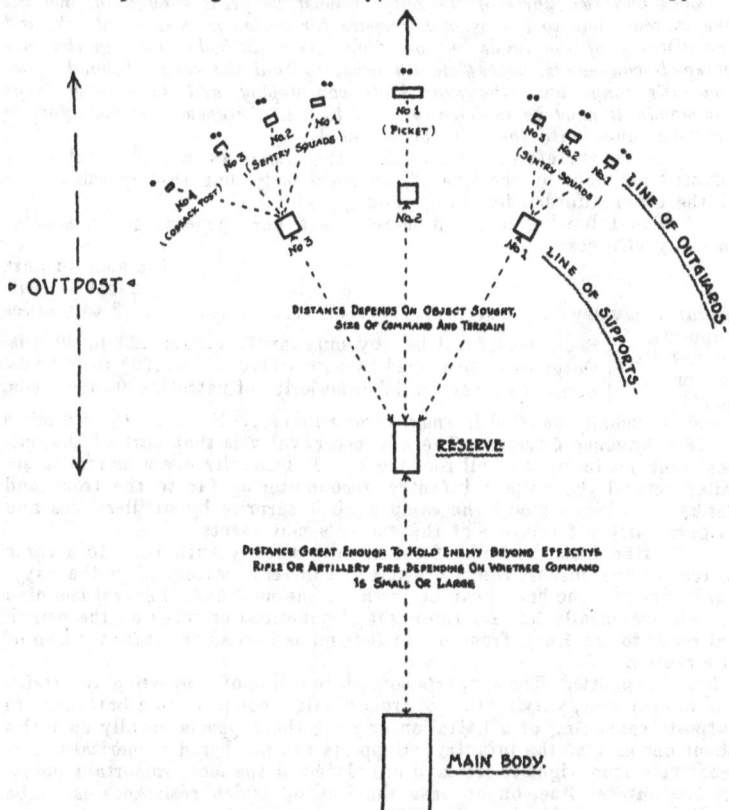

DISTANCE DEPENDS ON OBJECT SOUGHT, SIZE OF COMMAND AND TERRAIN

DISTANCE GREAT ENOUGH TO HOLD ENEMY BEYOND EFFECTIVE RIFLE OR ARTILLERY FIRE, DEPENDING ON WHETHER COMMAND IS SMALL OR LARGE

the first joints the line of outguards, the second joints the line of sentinels and the finger tips the advance cavalry.

In case of attack each part is charged with holding the enemy in check until the larger element, next in rear, has time to deploy and prepare for action.

517. Distances Between the Subdivisions. The distances separating the main body, the line of supports, the l ne of outguards, the line of sentries and the advance cavalry, will depend upon circumstances. There can be no uniformity in the distance between supports and reserves, nor between outguards and supports, even in the same outpost. The avenues of approach and the important features of the ground will largely control the exact positions of the different parts of the outposts. The basic principle upon which the distances are based, is: *The distance between any two parts of the outpost must be great enough to give the one in rear time to deploy and prepare for action in case of attack, and the distance of the whole outpost from the main body must, in the case of small commands, be sufficiently great to hold the enemy beyond effective rifle range until the main body can deploy, and, in case of large commands, it must be sufficiently great to hold the enemy beyond effective artillery range until the main body can deploy.*

It is, therefore, evident that the distances will be materially affected not only by the size of the main body, but also by the nature of the cover afforded by the ground.

The following is given merely as a very general guide, subject to many changes:

Distance to next
element in rear.

Advance cavalry ....2 to 6 miles

Supports (Generally two or more)
{ Sentinels (furnished by outguard).........20 to 40 yds.
{ Outguards (furnished by support).........200 to 500 yds.
{ Support proper furnishes majority of patrols. 400 to 800 yds.

Reserve (usually omitted in small commands)..............½ to 2 miles

518. Advance Cavalry. The advance cavalry is that part of the outpost sent out in front of all foot troops. It generally operates two to six miles beyond the outpost infantry, reconnoitering far to the front and flanks in order to guard the camp against surprise by artillery fire and to give early information of the enemy's movements.

After dusk the bulk of the cavalry usually withdraws to a camp in rear of the outpost reserve, where it can rest securely after the day's hard work and the horses can be fresh for the next day. Several mounted patrols are usually left for the night at junctions or forks on the principal roads to the front, from one to four miles beyond the infantry line of observation.

519. Supports. The *supports* constitute a line of *supporting* and *resisting* detachments, varying in size from a half a company to a battalion. In outposts consisting of a battalion or more the supports usually comprise about one-half of the infantry. Supports are numbered numerically consecutively from right to left and are placed at the more important points on the outpost line, on or near the line on which resistance is to be made in case of attack.

As a rule, roads exercise the greatest influence on the location of supports, and a support will generally be placed on or near a road.

Each support has assigned to it a definite, clearly-defined section of front that it is to cover, and the support should be located as centrally as possible thereto.

520. Outguards. The outguards constitute the line of small detachments farthest to the front and nearest to the enemy, and their duty is to maintain uninterrupted observation of the ground in front and on the flanks; to report promptly hostile movements and other information relating to the enemy; to prevent unauthorized persons from crossing the line of observation; to drive off small parties of the enemy, and to make temporary resistance to larger bodies. For convenience outguards are classified as pickets, sentry squads, and cossack posts. They are numbered consecutively from right to left in each support.

521. *A picket* is a group consisting of two or more squads, ordinarily not exceeding half a company, posted in the line of outguards to cover a given sector. It furnishes patrols and one or more sentinels, double sentinels, sentry, squads, or cossack posts for observation.

Pickets are placed at the more important points in the line of outguards, such as road forks. The strength of each depends upon the number of small groups required to observe properly its sector.

522. *A sentry squad* is a squad posted in observation at an indicated point. It posts a double sentinel in observation, the remaining men resting near by and furnishing the reliefs of sentinels. In some cases it may be required to furnish a patrol.

523. *A cossack post* consists of four men. It is an observation group similar to a sentry squad, but employs a single sentinel.

At night, it will sometimes be advisable to place some of the outguards or their sentinels in a position different from that which they occupy in the daytime. In such case the ground should be carefully studied before dark and the change made at dusk. However, a change in the position of the outguard will be exceptional.

524. *Sentinels* are generally used singly. in daytime, but at night double sentinels will be required in most cases. Sentinels furnished by cossack posts or sentry squads are kept near their group. Those furnished by pickets may·be as far as 100 yards away.

Every sentinel should be able to communicate readily with the body to which he belongs.

Sentinel posts are numbered consecutively from right to left in each outguard. Sentry squads and cossack posts furnished by pickets are counted as sentinel posts.

If practicable, troops on outpost duty are concealed and all movements made so as to avoid observation by the enemy; sentinels are posted so as to have a clear view to the front and, if practicable (though it is rarely possible), so as to be able, by day, to see the sentinels of the adjoining outguards. Double sentinels are posted near enough to each other to be able to communicate easily in ordinary voice.

Sentinels are generally on duty two hours out of six. For every sentinel and for every patrol there should be at least three reliefs; therefore, one-third the strength of the outguards gives the greatest number of men that should be on duty as sentinels and patrols at one time.

Skillful selection of the posts of sentinels increases their field of observation. High points, under cover, are advantageous by night as well as by day; they increase the range of vision and afford greater facilities for seeing lights and hearing noises. Observers with good field glasses may be placed on high buildings, on church steeples or in high trees.

Glittering objects on uniform or equipment should be concealed. It is seldom necessary to fix bayonets, except at night, in dense fog, or in very close country.

Reliefs, visiting patrols, and inspecting officers, approach sentinels from the rear, remaining under cover if possible.

For the usual orders of a sentry on the line of observation, see Par. 699.

525. Reserve. The reserve forms a general support for the line of resistance. It is, therefore, centrally located near the junction of roads coming form the direction of the enemy, and in concealment if practicable.

Of the troops detailed for outpost duty, about one-half of the infantry, generally all of the artillery, and the cavalry not otherwise employed, are assigned to the reserve. If the outpost consists of less than two companies the reserve may be omitted altogether.

The arms are stacked and the equipments (except cartridge belts) may be removed. Roads communicating with the supports are opened.

When necessary, the outpost order states what is to be done in case of attack, designates places of assembly and provides for interior guards. Interior guards are posted in the camp of the reserve or main body to maintain order, and furnish additional security. Additional instructions may be given for messing, feeding, watering, etc. In the vicinity of the enemy or at night a portion of the infantry may be required to remain under arms, the cavalry to hold their horses (cinches loosened), and the artillery to remain in harness, or take up a combat position.

In case of alarm, the reserve prepares for action without delay, and word is sent to the main body. In combat, the reserve reinforces the line of resistance, and if unable to check the enemy until the arrival of the main body, delays him as much as possible.

The distance of the reserve from the line of resistance varies, but is generally about half a mile; in outposts of four companies or less this distance may be as small as 400 yards.

526. Patrols. Instead of using outguards along the entire front of observation, part of this front may be covered by patrols only. These should be used to cover such sections of the front as can be crossed by the enemy only with difficulty and over which he is not likely to attempt a crossing after dark.

In daylight much of the local patrolling may be dispensed with if the country can be seen from the posts of the sentinels. However, patrols should frequently be pushed well to the front unless the ground in that direction is exceptionally open.

Patrols must be used to keep up connection between the parts of the outpost except when, during daylight, certain fractions or groups are mutually visible. After dark this connection must be maintained throughout the outpost except where the larger subdivisions are provided with wire communication.

The following patrols are usually sent out from the main bodies of the supports:

(a) Patrols of from three men to a squad are sent along the roads and trails in the direction of the enemy, for a distance of from one to five miles, depending on how close the enemy is supposed to be, whether or not there is any advance cavalry out, and how long the outpost has been in position. The extreme right and left supports send patrols well out on the roads to the flanks. These patrols generally operate continuously; as soon as one returns from the front, or possibly even before it returns, another goes out in the same general direction to cover the same country. Frequently a patrol is sent out along a road to the front for two or three miles with orders to remain out until some stated time—for example, 4 P. M., dusk or dawn. It sends in important information, and remains out near the extremity of its route, keeping a close watch on the surrounding country.

An effort should always be made to secure and maintain contact with the enemy, if within a reasonable distance, in order that his movements or lack of movement may be constantly watched and reported on. The usual tendency is towards a failure to send these patrols far enough to the front and for the patrol leader to overestimate the distance he has traveled. A mile through strange country with the ever-present possibility of encountering the enemy seems three miles to the novice.

At night the patrols generally confine their movements to the roads, usually remaining quietly on the alert near the most advanced point of their route to the front.

The majority of such patrols are sent out to secure information of the enemy—reconnoitering patrols—and they avoid fighting and hostile patrols, endeavoring to get in touch with the enemy's main force. Other patrols are sometimes sent out to prevent hostile detachments from approaching the outposts; they endeavor to locate the hostile patrols, drive them back, preventing them from gaining any vantage point from which they can observe the outpost line. These are called combat patrols and have an entirely different mission from reconnoitering patrols.

(b) Patrols of from two men to a squad, usually two men, are sent from the support around the line of its outguards, connecting with the outguards of the adjacent supports, if practicable. These are "visiting patrols," and they serve to keep the outguards of a support in touch with it and with each other; to keep the commander of a support in touch with his outguards and the adjacent supports; and to reconnoiter the ground between the outguards. Since a hostile force of any size is practically forced to keep to the roads, there are rarely ever any supports and very few outguards posted off the roads, the intervals being covered by patrols, as just described.

When going out a patrol will always inform the nearest sentinel of the direction it will take and its probable route and hour of return.

Detail for Patrols. Since for every patrol of four men, twelve are required (3 reliefs of 4 men each), the importance of sending out just enough men and not one more than is actually needed, can readily be understood. As fast as one visiting patrol completes its round, another should usually be sent out, possibly going the rounds by a slightly different route or in the reverse direction. The same generally applies to the reconnoitering and combat patrols, though frequently they are sent out for the entire day, afternoon or night, and no 2d and 3d relief is required. Three reliefs are required for the sentinel or sentinels at the post of the supports, so care should be taken to establish but one post, if it can do all that is required. It should not be considered that every man in the support should be on duty or on a relief for an outguard, a patrol or sentinel post. There should be as many men as possible in the main body of a support (this term is used to distinguish this body from the support proper, which includes the outguards and their sentinels) who only have no duty other than being instantly available in case of attack.

527. Flags of Truce. Upon the approach of a flag of truce, the sentry will at once notify the commander of the outguard, who will in turn send word to the commander of the outpost and ask for instructions. One or more men will advance to the front and halt the party at such distance as to prevent any of them from overlooking the outposts. As soon as halted, the party will be ordered to face in the opposite direction. If permission is given to pass the party through the outpost line, they will be blindfolded and led under escort to the commander of the outpost. No conversation, except by permission of the outpost commander, is to be allowed on any subject, under any pretext, with the persons bearing the flag of truce.

528. Entrenchments and Obstacles. The positions held by the subdivisions of an outpost should generally be strengthened by the construction of entrenchments and obstacles, but conditions may render this unnecessary.

529. Concealment. Troops on outpost must keep concealed as much as is consistent with the proper performance of their duties; especially should they avoid the sky line.

530. Detached Posts. In addition to ordinary outguards, the outpost commander may detail from the reserve one or more detached posts to cover roads or areas not in general line assigned to the supports.

In like manner the commander of the whole force may order detached posts to be sent from the main body to cover important roads or localities not included in the outpost line.

Detached posts may be sent out to hold points which are of importance to the outpost cavalry, such as a ford or a junction of roads; or to occupy positions especially favorable for observation, but too far to the front to be included in the line of observation; or to protect flanks of the outpost position. Such posts are generally established by the outpost commander, but a support commander might find it necessary to establish a post practically detached from the rest of his command. They usually vary in strength from a squad to a platoon. The number and strength of detached posts are reduced to the absolute needs of the situation.

531. Examining Posts. An examining post is a small detachment, under the command of an officer or a noncommissioned officer, stationed at some convenient point to examine strangers and to receive bearers of flags of truce brought in by the outguards or patrols.

Though the employment of examining posts is not general in field operations, there are many occasions when their use is important; for example: When the outguards do not speak the language of the country or of the enemy; when preparations are being made for a movement and strict scrutiny at the outguards is ordered; at sieges, whether in attack or defense. When such posts are used, strangers approaching the line of observation are passed along the line to an examining post.

No one except the commander is allowed to speak to persons brought to an examining post. Prisoners and deserters are at once sent under guard to the rear.

532. Cavalry Outpost. Independent cavalry covering a command or on special missions, and occasionally the advance cavalry of a mixed command, bivouac when night overtakes them, and in such cases furnish their own outposts. The outposts are established, in the main, in accordance with the foregoing principles, care being taken to confine outpost work to the lowest limits consistent with safety. No precaution, however, should be omitted, as the cavalry is generally in close proximity to the enemy, and often in territory where the inhabitants are hostile.

The line of resistance is occupied by the supports the latter sending out the necessary outguards and patrols. Each outguard furnishes its own vedettes (mounted sentinels), or sentinels. Due to the mobility of cavalry, the distances are generally greater than in an outpost for a mixed command. An outguard of four troopers is convenient for the day time, but should be doubled at night, and at important points made even stronger. The sentinels are generally dismounted, their horses being left with those of the outguards.

Mounted cavalry at night can offer little resistance; the supports and outguards are therefore generally dismounted, the horses being under cover in rear, and the positions are strengthened by intrenchments and obstacles. By holding villages, bridges, defiles, etc., with dismounted rifle fire, cavalry can greatly delay a superior force.

There should always be easy communication along the line of resistance to enable the cavalry to concentrate at a threatened point.

A support of one squadron covers with its outposts a section rarely longer than two miles.

As such a line is of necessity weak, the principal reliance is placed on distant patrolling. If threatened by infantry, timely information enables the threatened point to be reinforced, or the cavalry to withdraw to a place of safety. If there is danger from hostile cavalry, the roads in front are blocked at suitable points, such as bridges. fords, defiles, etc., by a succession of obstacles and are defended by a few dismounted men. When compelled to fall back these men mount and ride rapidly to the next obstacle in rear and there take up a new position. As the march of cavalry at night is, as a rule, confined to roads, such tactics seriously delay its advance.

In accordance with the situation and the orders they have received, the support commanders arrange for feeding, watering, cooking, resting and patrolling. During the night the horses of the outguards remain saddled and bridled. During the day time cinches may be loosened, one-third of the horses at a time. Feeding and watering are done by reliefs. Horses being fed are removed a short distance from the others.

Independent cavalry generally remains in outpost position for the night only, its advance being resumed on the following day; if stopped by the enemy, it is drawn off to the flanks upon the approach of its own infantry.

ESTABLISHING THE OUTPOST

533. The outpost is posted as quickly as possible, so that the troops can the sooner obtain rest. Until the leading outpost troops are able to assume their duties, temporary protection, known as the *march outpost*, is furnished by the nearest available troops.

Upon receipt of the *halt order* from the commander of the main column, the outpost commander issues the *outpost order* with the least practicable delay.

The *halt order,* besides giving the necessary information and assigning camp sites to the parts of the command, details the troops to constitute the outpost, assigns a commander therefor, designates the general line to be occupied, and, when practicable, points out the position to be held in case of attack.

The *outpost order* gives such available information of the situation as is necessary to the complete and proper guidance of subordinates; designates the troops to constitute the supports; assigns their location and the sector each is to cover; provides for the necessary detached posts; indicates any special reconnaissance that is to be made; orders the location and disposition of the reserve; disposes of the train if the same is ordered to join the outpost; and informs subordinates where information will be sent In large commands it may often be necessary to give the order from the map, but usually the outpost commander will have to make some preliminary reconnaissance, unless he has an accurate and detailed map.

Generally it is preferable for the outpost commander to give verbal orders to his support commanders from some locality which overlooks the terrain. The time and locality should be so selected that the support commanders may join their commands and conduct them to their positions without causing unnecessary delay to their troops. The reserve commander should, if possible, receive his orders at the same time as the support commanders. Subordinates to whom he gives orders separately should be informed of the location of other parts of the outpost.

In large outposts written orders, which are issued in the following form, are frequently most convenient:

534.

	Field Orders	(Title)
	No.	(Place)
	Troops	(Date and hour)
(a)	Advance Cavalry: (Commander) (Troops)	1 (Information of the enemy and of our supporting troops.)
(b)	The Support · No. 1 (Commander) (Troops)	2. (Plan of commander—to establish outpost, approximate line of resistance.)
	No. 2 (Commander) (Troops)	3 (a) (Instructions for advance cavalry—contact with enemy, roads or country to be specially watched, special mission)
	No. 3 (Commander) (Troops)	(b) (Instructions for support—positions they are to occupy, and sections of line of resistance, which they are to hold, intrenching, etc.)
(c)	Detached Post: (Commander) (Troops)	(c) (Instructions for detached posts—position to be occupied, duties, amount of resistance.)
(d)	Reserve: (Commander) (Troops)	(d) (Instruction for reserves—location, observation of flanks, conduct in case of attack, duties of special troops.)
	No. 4 (Instruction for field train if it has accompanied the outpost)	
	No 5 (Place of commander or where messages may be sent)	
		(Signature)

(How and to whom issued)

(Note: In the case of a small outpost the order is usually verbal.)

After issuing the initial orders, the outpost commander inspects the outpost, orders the necessary changes or additions, and sends his superior a report of his dispositions.

The reserve is marched to its post by its commander, who then sends out such detachments as have been ordered and places the rest in camp or bivouac, over which at least one sentinel should be posted. Connection must be maintained with the main body, the supports, and nearby detached posts.

The supports march to their posts, using the necessary covering detachments when in advance of the march outpost. A support commander's order should fully explain the situation to subordinates, or to the entire command, if it be small. It should detail the troops for the different outguards and, when necessary, define the sector each is to cover. It should provide the necessary sentinels at the post of the support, the patrols to be sent therefrom, and should arrange for the necessary intrenching.

In posting his command the support commander must seek to cover his sector (the front that he is to look after) in such manner that the enemy can not reach, in dangerous numbers and unobserved, the position of the support or pass by it within the sector intrusted to the support. On the other hand, he must economize men on observation and patrol duty, for these duties are unusually fatiguing. He must practice the greatest economy of men consistent with the requirements of practical security.

As soon as the posting of the support is completed, its commander carefully inspects the dispositions and corrects defects, if any, and reports the disposition of his support, including the patrolling ordered, to the outpost commander. This report is preferably made by means of a sketch.

By day the outpost will stack arms and the articles of equipment, except the cartridge belt and canteen, will be placed by the arms. At night the men will invariably sleep with their arms and equipment near them.

In addition to the sentinel posted over the support, a part of the support, say one-third or one-fourth, should always be awake at night.

Each outguard is marched by its commander to its assigned station, and especially in the case of a picket, is covered by the necessary patrolling to prevent surprise.

Having reached the position, the commander explains the situation to his men and establishes reliefs for each sentinel, and, if possible, for each patrol to be furnished. Besides these sentinels and patrols, a picket must have a sentinel at its post.

The commander then posts the sentinels and points out to them the principal features, such as towns, roads, and streams, and gives their names. He gives the direction and location of the enemy, if known, and of adjoining parts of the outpost.

He gives to patrols the same information and the necessary orders as to their routes and the frequency with which the same shall be covered. Each patrol should go over its route once before dark.

Each picket should maintain connection by patrols with the outguards on its right and left.

535. Intercommunication. It is most important that communication should be maintained at all times between all parts of the outpost, and between the outpost and the main body. This may be done by patrols, messengers, wire or signal.

The commander of the outpost is responsible that proper communication be maintained with the main body, and the support commanders keep up communication with the outguards, with the adjoining supports and with the reserve. The commander of a detached post will maintain communication with the nearest outguard.

536. Changes for the Night. In civilized warfare, it is seldom necessary to draw the outpost closer to the main body at night in order to diminish the front; nor is it necessary to strengthen the line of observation, as the enemy's advance in force must be confined to the roads. The latter are therefore strongly occupied, the intervening ground being diligently patroled.

In very open country or in war with savage or semi-civilized people familiar with the terrain, special precautions are necessary.

537. Relieving the Outpost. Ordinarily outposts are not kept on duty longer than twenty-four hours. In temporary camps or bivouac they are generally relieved every morning. After a day's advance the outpost for the night is usually relieved the following morning when the support of the new advance guard passes the line of resistance. In retreat the outpost for the night usually forms the rear guard for the following day, and is relieved when it passes the line of observation of the new outpost.

Outguards that have become familiar with the country during the day time should remain on duty that night. Sentinels are relieved once in two hours, or oftener, depending on the weather. The work of patrols is regulated by the support commander.

Commanders of the various fractions of an outpost turn over their instructions and special orders, written and verbal, to their successors, together with the latest information of the enemy, and a description of the important features of the country. When practicable the first patrols sent out by the new outposts are accompanied by members of the old outpost who are familiar with the terrain. When relieved the old outguards return to their supports, the supports to the reserve and the latter to the main body; or, if more convenient, the supports and reserves return to the main body independently, each by the shortest route.

When relieved by an advance guard, the outpost troops ordinarily join their units as the column passes.

Evening and shortly before dawn are hours of special danger. The enemy may attack late in the day in order to establish himself on captured ground by intrenching during the night; or he may send forward troops under cover of darkness in order to make a strong attack at early dawn. Special precaution is therefore taken at those hours by holding the outpost in readiness, and by sending patrols in advance of the line of observation. If a new outpost is to be established in the morning, it should arrive at the outpost position at daybreak, thus doubling the outpost strength at that hour.

538.   OUTPOST PROBLEMS

Problem No. 1 (Infantry)

Lieutenant (to two squads of his company): Two battalions of our regiment have camped by Baker's Pond (Elementary Map) for the night. It is now 3 P. M. on a rainy day in August. The enemy is thought to be about five miles to the south of us. Our platoon is the left support of the outpost and is stationed at the road fork on the Chester Pike, by the Mason house. The Twin Hills-Lone Hill ridge is taken care of by other troops. Corporal Baker, where do you think I should place outguards?

Corporal Baker: One at the junction of the Mills farm lane and the Chester Pike, and one at the steel railroad trestle over Sandy Creek.

Lieutenant: Those positions are both too far from the support, almost a half mile, but they cover the two main avenues of approach and there is no good place for a position nearer the support. A position farther north of the Mills farm lane would have its view obstructed by the wall and trees along the lane and the wall would be a bad thing to leave unoccupied such a short distance to your front. So in this case, in spite of the excessive distances from the support, I think the two positions are well chosen. Each should be an outguard of a squad, for in the day time, in addition to furnishing a sentinel to observe to the front, they should have some power of resistance, particularly at the trestle. At night they should each have one double sentinel post. This requires three reliefs of two men each, which, with the corporal, only leaves one extra man, who can be used as a messenger.

Corporal Baker, I order you to take your squad and post it as Outguard No. 1, at the junction of this (Chester) pike and that farm lane (Mills farm) in front. Corporal Davis' squad will be Outguard No. 2, at the railroad trestle over there (pointing). Friendly troops will be on the ridge to the east of your position. Your meals will be cooked here and sent to you.

Explain how you post your squad.

Corporal Baker: I order Smith to double time 150 yards to the front and act as point for the squad. I then march the squad down to its position, keeping Smith about 200 yards in front until I have arranged everything. I then post Brown under cover of the trees along the lane where he can look down the road as far as possible and I tell him, "Brown, you are to take post here, keeping a sharp lookout to the front and flanks. The enemy is thought to be about five miles south (pointing) of us. This is the Chester Pike. That creek over there is Sandy Creek. Salem is about a mile and three-quarters down this pike in that (S. E.) direction. York is a mile and a half in that (S. W.) direction. Our troops are on that ridge (Twin Hills) and a squad is at the trestle over there. It is Outguard No. 2. You are in Outguard No. 1. You know where we left our platoon. It is our support. Signal Smith to come in." I then have the squad pitch their shelter tents along the northern side of the wall, where they will be hidden to view from the front by the trees along the lane and the wall. I want the men to get shelter from the rain as soon as possible. I then instruct the men of the squad, in the same manner that I did Brown; I notice the time, and detail Davis as second relief and Carter as third relief for Brown's post.

I then direct two men to take all the canteens and go over to that farm (Mills) and fill them, first questioning the people about the enemy and about the country around here. I also direct these two men to get some straw or hay for bedding in the shelter tents, and instruct them to return with as little delay as possible.

I wait until they return and order two other men to go down to the cross roads, question the people there, look the ground over and return here. I caution them not to give any information about our force or the outguard. I would see that the sentinel's position was the best available and that the men had as comfortable quarters as possible, without being unduly exposed to view and without interfering with their movements in case of attack. They would keep their rifles at their sides at all times and not remove their equipments. After dark I put two men on post at the same time. To do this I arrange three reliefs of two men each. They are posted in pairs for two hours at a time.

If no patrol from the support appeared within a half hour after I first took position I would send a messenger back to you to see if everything was all right and tell you what I had done.

Lieutenant: I think the two men sent to the crossroads should have been started out before sending anyone to the Mills house as this was a more important point. The Field Service Regulations state that outguards do not patrol to the front, but what you did was entirely correct. You were securing yourself in your position and should be familiar with your immediate surroundings. You should have told the crossroads patrol to determine how much of an obstacle Sandy Creek was. I suppose you assumed the swamp was impassable.

The sentinel in this case is, I suppose, across the lane from the outguard about ten or fifteen yards in advance. After dark the double sentinel post should be posted on the pike about thirty yards in advance of the outguard.

Very frequently it would not be wise to put up your shelter tents on outguard. But here, considering the rain and the protection the trees and wall furnish, it was wise to do so.

The noncommissioned officer in charge of an outguard should be very precise in giving his orders and in making his arrangements, details, etc. The discipline must be strict; that is, the men must be kept under absolute control, so that in case of sudden attack there will be no chance of confusion and the outguard commander will have his men absolutely in hand and not permit any independent action on their part. This is often not the case, owing to the familiar relations that usually exist in our army between a corporal and the members of his squad.

We will not have time to go into the arrangements for Outguard No. 2 other than to say that the conditions there are somewhat different from those Corporal Baker has had to deal with. The outguard should be posted on the west bank of Sandy Creek and the sentinel at the southeastern end of the trestle. A skirmish trench should be dug down the western slope of the fill west of the creek, and extended across the track by throwing up a parapet about two and one-half feet high, slightly bent back towards the northeast so as to furnish cover from fire from the east bank of the creek, north of the trestle. The shelter tents could be pitched as "lean-tos" against the western slope of the fill, and hidden by bushes and branches of trees.

(Note: The details of commanding this outguard, its action in case of attack, what should be done with a passing countryman, etc., can be profitably worked out in great detail.)

Problem No. 2

Lieutenant (to six squads): We will take the same situation as we had in Problem 1, with squad outguards as before.

Sergeant Adams, you have command of the platoon and have sent out the two outguards. Explain your arrangements for the support.

Sergeant Adams: I have the men fall out by squads and rest on the side of the road while I look the ground over. I then tell Sergeant Barnes, "You will have immediate charge of the guard, cooking, visiting patrols, etc., here at the support. Detail three men from Corporal Evan's squad as first, second and third relief for the sentinel over the support. Post your sentinel at the road fork and give him the necessary instructions as to the outguards, the adjacent support which is on this road (pointing west) on top of that ridge, etc. I will give you further instructions later." I then fall in the remainder of the support (one sergeant, one cook, four corporals and twenty-seven privates, three squads being intact and one man on duty as sentinel) and have shelter tents pitched under cover of the orchard and Mason house. While this is being completed I select a line for a trench, about thirty-five yards long, behind the fence on the east and west road and extending east of the Chester Pike about fifteen yards, slightly bent back towards the northeast. No trench in the road. I then say to Sergeant Foss, "Take Graves' squad and construct a shelter trench along this line (indicating), having the parapet concealed. Cut the fences so as to furnish easy access."

I then say to Corporal Evans, "Take three men from your squad and, as a reconnoitering patrol, cross the trestle there (pointing), and follow that road (pointing to the Boling-Salem road) into Salem, reconnoitering that village. Then take up a position on that ridge (pointing to Sandy Ridge) and remain out until dusk. Send me a message from Sandy Ridge with a sketch and description of the country."

I assume that Corporal Evans is familiar with the information about the enemy, the location of our outguards, etc.

Selecting five men from Corporal Geary's squad and the remaining man of Corporal Evans' squad (three having been detailed for sentinel duty, and three sent out on patrol duty with Corporal Evans), I turn them over to Sergeant Barnes, saying, "Here are six men to furnish three reliefs for a visiting patrol of two men. Have this patrol visit Outguard No. 2 and cross the trestle, going south down the east bank of the creek; thence recross the creek at the road bridge, visiting Outguard No. 1; thence across to the adjacent outguard of the support on our left, which is somewhere on that ridge (pointing to the Twin Hills-Lone Hill Ridge); and thence to the starting point. Have them locate that support on their first trip. You can reverse the route and make such minor changes from time to time as you think best. Report to me after they have completed the first round. Make arrangements for sending supper to the outguards. Take two men from Corporal Jackson's squad to carry it out. Be careful that the cook fire is not visible. I am going out to visit Outguard No. 1 and then No. 2. You will have charge until I return.''

The men have stacked arms in front of the tents and have removed all equipment but their belts.

I would now visit the outguards, taking a man with me, and see if they are properly located. I would instruct the outguard commanders as to what to do in case of attack, in case strangers approach, point out their line of retreat in case of necessity, etc. I would make a sketch of the position and send it, with a description of my dispositions, to the commander of the outpost.

Lieutenant: Your arrangements and dispositions appear satisfactory. You should have been more prompt in sending Corporal Evans out with his patrol. Why didn't you send a patrol towards York, or south along the Chester Pike?

Sergeant Adams: I considered that the support on my right would cover that ridge (Twin Hills-Lone Hill), and that the route I laid out for Corporal Evans would cover the Chester Pike and the country east of Sandy Creek at the same time, thus avoiding the necessity for two patrols.

Lieutenant: That seems reasonable, but you should have given some specific orders about reporting on the width, depth, etc., of Sandy Creek, which might prove a very valuable or dangerous obstacle. You can readily see how quickly a command becomes broken up and depleted in strength, and how important it is to make only such detachments as are necessary. It looks as if your outguards might have been made smaller considering the size of your platoon (6 squads), but I think the squad outpost is so much better than one not composed of a complete unit, that it is correct in this case. With Corporal Evans' patrol of three men, the visiting patrol requiring six men, the sentinel post requiring three men, Sergeant Barnes, and the two outguards, you have thirty men actually on duty or detailed for duty, out of fifty-one. Of course, the men constituting the outguards, the man detailed for the visiting patrol and support sentinel, have approximately two hours on duty and four hours off duty, so they get some rest. Furthermore, you should have a three-man patrol watching the crossroads at Salem during the night, Corporal Evans' patrol having returned. This patrol should be relieved once during the night, at a previously stated hour, which means six more men who do not get a complete night's rest.

Sergeant Adams: Isn't Salem rather far to the front to send a patrol at night?

Lieutenant: Yes, it is, but unless you touch the crossroads there you would have to have two patrols out, one near Maxey's farm and one on the Chester Pike. As it is you are leaving the road from York to the crossroads in front of Outguard No. 1, uncovered, but you should find that this is covered by a patrol from the adjacent support. The crossroads in front of Outguard No. 1 is the natural place for a stationary, night patrol, but it is so close to the outguard that the benefit derived from a patrol there would be too small to justify the effort.

(Note: Further details of the duties of this support can be gone into. The messages should be written, and patrols carried through their tour of duty with the resulting situations to be dealt with; the sentinels tested as to their knowledge of their duties, etc. Also note carefully the manner in which the support commander uses his noncommissioned officers for carrying out his intentions, and thus avoids the most objectionable and inefficient practice of dealing directly with the privates.)

Problem No. 3   (Infantry)

(See Fort Leavenworth map in pocket at back of book.)

Situation:

A Blue force, Companies A and B, 1st Infantry, under Captain A, in hostile country, is covering the Rock Island Bridge and camped for the night, April 20-21, on the south slope of Devin ridge (rm'). The enemy is moving northward from Kansas City (30 miles south of Leavenworth). At 3:30 P. M. Captain A receives a message from Colonel X at Beverly (2 miles east of Rock Island Bridge, (qo'), stating that two or three companies of hostile infantry are reported five miles south of Leavenworth at 2:30 P. M. No enemy is west of Leavenworth. Captain A decides to place one platoon on outpost.

Required, 1. Captain A's order.

Answer. Verbally: "Two or three Red companies were five miles south of Leavenworth at 2:30 P. M. today. No enemy is west of Leavenworth. We will camp here. 1st Platoon, 'A' company, under Sergeant A, will form the outpost, relieving the advance guard (2d Platoon Co. A). The line, Pope Hill (sm')—Rabbit Point (tn') will be held. Detached posts will be placed on Hill 880, west of Merritt Hill (rl'), and on Engineer Hill (ql'). In case of attack the outpost line will be held.

"The baggage will be at the main camp.

"Messages will reach me on Devin Ridge (rm')."

Issued verbally to officers and Sergeant A.

Required, 2. Give verbatim (word for word) the order issued by Sergeant A.

Answer. "Two companies of the enemy were five miles south of Leavenworth at 2:30 P. M. today. Our camp is to be here. This platoon will be the outpost on the line, Rabbit Point (im')—Pope Hill (sm').

"The right support, 1st section, less 1 squad, under Sergeant B, will take position north of Pope Hill and cover the following front: the ravine (XIX—Merritt Hill) west of Grant avenue to the ravine about midway between Grant Avenue and Rabbit Point (tn').

"The left support, 2d section, less 1 squad under Sergeant H, will take position on north slope of Rabbit Point and will cover the following front: The ravine midway between Grant Avenue and Rabbit Point to Missouri River.

"Corporal D, you will take the eight men of your squad and form a detached post on Engineer Hill (qk').

"Corporal E, take your squad and form a detached post on Hill 880 west of Merritt Hill (rl').

538 (contd.)

"If attacked hold your front. Each support and detached post will entrench.

"Send messages to me at right support."

The outpost moves out, each support and detached post separately, without throwing out covering patrols, because the advance guard is now holding the front. There is no reserve.

Required, 3. What does Sergeant A do now?

Required, 4. What does Sergeant B do as soon as he reaches Pope Hill?

(Note: During the remainder of the afternoon one man up in a tree on Grant Avenue will be the only observing post necessary for this support. At night an outguard would be placed on Grant Avenue with continuous patrols along the front, because the open ground furnishes easy approach to the enemy. A post of four men might also be placed on the bridge over Corral Creek (um').

Required, 5. The location of supports and the main body of detached post on Engineer Hill.

Required, 6. What patrolling would be done from the left support?

CHAPTER VIII

MAP READING

INTRODUCTION

539. This chapter on map reading presents two phases of the subject. In order that the beginner may grasp the rudimentary principles without difficulty, the subject is first considered in the most elementary manner, and later, for the benefit of the student who has already acquired a slight knowledge of the subject, the same ground is covered in a less elementary way.

For some unknown reason, military map reading has always been considered a very difficult subject to master, and the beginner, starting out with this idea, tries to find it difficult. Therefore, it is not strange that he finds the subject hard to understand and laborious to study. As a matter of fact, it is far easier to learn to read a map than it is to learn to patrol, write messages, give proper orders, etc.

The most ignorant tourist easily uses the complicated maps in the guide books to find his way about; men, women and children, riding in automobiles, use the road maps of the country without a thought of having acquired the knowledge of some difficult art; but as soon as the military student considers map reading, he decides he has a most difficult subject to master, and he proceeds, unconsciously, to make it difficult.

PART I

A **military map** is a drawing, made to represent some section of country, showing the things that are of military importance, such as roads, streams, bridges, houses, and hills. The map must be so drawn that you can tell the distance between any two points, the heights of hills, and the relative positions of everything shown.

Map Reading

By **map reading** is meant the ability to get a clear idea of the ground represented by the map with the same ease one reads a book or newspaper. This means to grasp at once the distance on the ground corresponding to a given distance on the map, to get a correct idea of the network of streams and roads, heights, slopes, depressions, and all forms of military cover and obstacles. The first thing necessary in map reading, therefore, is to have a thorough knowledge of the scale of maps.

The Scale of a Map

540. In order that you may be able to tell the distance between any two points, for instance, between Salem and Boling on the Elementary Map (in back of book), the map must be so drawn that a certain distance, say one inch, on the map always represents a certain distance on the ground, say one mile.

Suppose Boling is five miles from Salem, and one inch on the map is to represent one mile on the ground; then on the map Boling would be shown five inches from Salem, and any person knowing the scale of

540 (contd.)

that map (1 inch=1 mile), could at once determine the distance from Salem to Boling. He would measure the number of inches between the two towns and know that the actual distance on the ground was as many miles as he had measured inches. Suppose, for example, he found it was three and one-half inches to Boling; then he would know that the two places were three and one-half miles apart.

Another example, suppose the scale of your map reads 6 inches=1 mile and you wish to know the distance you have to march from a farm to a certain crossroads. You measure the distance between these two points on the map and find it to be nine inches. You at once know that the actual distance, ground distance, is one and one-half miles. For, if six inches on the map equals one mile on the ground, nine inches must equal one and a half miles—just as twelve inches would equal two miles, and twenty-four inches would represent four miles on the ground.

Instead of writing the scale on the map thus, 6 inches=1 mile or 6 inches to the mile (which means the same thing), you may find the scale represented by a long line or pair of lines divided into numbered lengths. Thus:

100   50   0   100   200   300   400   500   600 yards

Figure 1.

In fact, this is even a simpler method of indicating the scale of a map. You do not have to convert measurements of inches into miles, yards, feet, or whatever inches on the map represent on the ground. No ruler of inches is necessary. From the 0 on the scale shown in Figure 1 to the 600, is 600 yards. This means that this length on the map represents 600 yards on the ground, and as the scale has seven subdivisions of 100 yards each, and one of these again subdivided into four parts of twenty-five yards each, you can quickly find out the ground distance between points on the map.

It must be noted here that each subdivision of the scale is marked, not with its actual length, but with the distance which it represents on the ground. This is ordinarily known as a graphical scale, and is the most frequent method of indicating the scale of a map.

Example. If you wish to determine the distance from A to B, along the road shown in Figure 2,

A| B|

100   50   0   100   200   300   400   500   600 yards

Figure 2.

take a piece of paper and lay its edge along the road; mark the edge opposite A and opposite B; lay the edge of the paper along the scale (shown in Figure 2), and it shows the distance to be 675 yards. When you have to determine several distances from the map, it is more convenient first to lay the edge of the paper along the scale and mark off divisions like those of the scale. You then have a copy of the scale which you can quickly apply to any portion of the map and read off the ground distances instantly. This has supposedly been done in Figure 2, where a rough copy of the actual scale is shown applied to the road between A and B.

Example. Along the lower border of the Elementary Map is a graphical scale of miles and a second graphical scale, reading in yards. How far is it from York to Oxford by the most direct road (Valley Pike)? Take a piece of paper and lay its edge along the scale of miles; mark off similar divisions on the edge of the paper. This gives you a length of one mile. Apply this along the Valley Pike, starting at York. You will find the distance is four times the length of the paper and about one-half mile over. Therefore, as the scale length you used represents one mile, the distance is four and one-half miles. If you desired to know the distance in yards from York to the Cemetery on the County Road, you would repeat the same process, taking your distances from the graphical scale reading yards, and you would find the distance to be 750 yards.

Using the scale of miles, you can find the distance from York to Salem by road (three miles), from Salem to Boling by road (four miles), and so on.

The distances between points on a map represent corresponding Ground distances measured on the level (horizontal distances). For example, the distance from York to Boling that you would walk would include the extra distance required in ascending and descending the hill and valley between these two points. The distance on the map, however, represents the distance between the towns, measured as though the intervening ground were absolutely level.

540a. Direction. In order that the map may correctly show the positions of the different roads, streams, hills, houses, etc., with respect to one another, they must be given the same relative locations on the map that they occupy on the ground. The map must also show the points of the compass.

541. Meridians. If you look along the upper left hand border of the Elementary Map you will see two arrows pointing towards the top of the map, thus:

Figure 3.

They are pointing in the direction that is north on this map. The arrow with a full barb points toward the north pole of the earth. The arrow with but half a barb points toward what is known as the magnetic pole of the earth. This magnetic pole is a point up in the arctic regions near the geographical or true north pole, which, on account of its magnetic qualities, attracts one end of all compass needles and causes them to point towards it. As it is near the true north pole, this serves to indicate the north direction to a person using a compass. The arrow with the full barb is called the true meridian, as it points to the true north pole. The arrow with the half barb is called the magnetic meridian as it points, not accurately towards the true north pole, but towards the magnetic pole. When you are using a compass, the needle points toward the magnetic pole, which is close enough to the true north for your purposes.

You now know from the meridians that in going from York to Oxford you travel north; from Boling to Salem you must travel south; going from Salem to York requires you to travel west; and from York to Salem you travel east. Suppose you are in command of a patrol at York and are told to go to Salem by the most direct line across country. You look at your map and see that Salem is exactly east of York. Next you take out your field compass (Figure 16, Par. 551), raise the lid, hold the box level, allow the needle to settle and see in what direction the north end of the needle points (it would point towards Oxford). You then know the direction of north from York, and you can turn your right and go due east towards Salem.

Having once discovered the direction of north on the ground, you can go to any point shown on your map without other assistance. If you stand at York, facing north and refer to your map, you need no guide to tell you that Salem lies directly to your right; Oxford straight in front of you; Boling in a direction about halfway between the directions of Salem and Oxford, and so on.

542. **Contours.** So far we have only dealt with the methods of. representing the horizontal (level) distances between points, and the directions with regard to the different points of the compass (north, south, east and west). In order to show on a map a correct representation of ground, the differences in elevation (vertical distances) of the terrain (ground)—the hills, valleys, ravines and flat land—must be indicated. This is usually done by means of contours.

A contour is a line on a map which shows the route one might follow on the ground and walk on the absolute level. If you went half way up the side of a hill and, starting there, walked entirely around the hill, neither going up any higher nor down any lower, and you made a sketch of the route you had followed, the line representing your path around the hill would be, in effect, a contour. By means of these contour lines at different vertical (up and down) elevations, the hills, valleys, etc., can be graphically shown on a map.

For example, on the Elementary Map, Sandy Ridge and Long Ridge are two isolated hills: Suppose this country were flooded with water twenty feet above the lowest point (Sandy Creek), the lines (contours) around Sandy Ridge and Long Ridge marked, 20, would then mark the edge of the water (water line) around the lower slopes of these hills. The "20" means that the contours so marked show the lines of the same elevation on the ground that are just twenty feet vertically above the lowest point in the section of country shown by this map. Now suppose the water rose twenty feet more, that is, to a height of forty feet. Then the contours marked 40 would indicate the water line around these hills—all the ground towards the 20-foot contour and below would be inundated. If the water rose to sixty feet, then all of Long Ridge would be under water, but the two small knolls on Sandy Ridge outlined by the sixty-foot contour, would still be out of water, as their crests are shown to be slightly higher (66, 68 and 65). Contours are taken at a fixed vertical distance apart. In this case it is twenty feet.

542a. An excellent idea of what is meant by contours and contourlines can be gotten from Figures 4 and 5. Let us suppose that formerly the island represented in Figure 4 was entirely under water and that by a sudden disturbance the water of the lake fell until the island stood twenty feet above the water, and that later several other sudden falls of the water, twenty feet each time, occurred, until now the island stands 100 feet out of the lake, and at each of the twenty feet elevations a distinct water line is left. These water lines are perfect contour-lines measured from the surface of the lake as a reference (or datum) plane. Figure 5 shows the contour-lines in Figure 4 projected, or shot down, on a horizontal (level) surface. It will be observed that on the gentle slopes, such as F-H (Fig. 4), the contours (20, 40) are far apart. But on the steep slopes, as R-O, the contours (20, 40, 60, 80, 100) are close together. Hence, it is seen that contours far apart on a map indicate gentle slopes, and contours close together, steep slopes. It is also seen that the shape of the contours gives an accurate idea of the form of the island. The contours in Fig. 5 give an exact representation not only of

Figure 5.

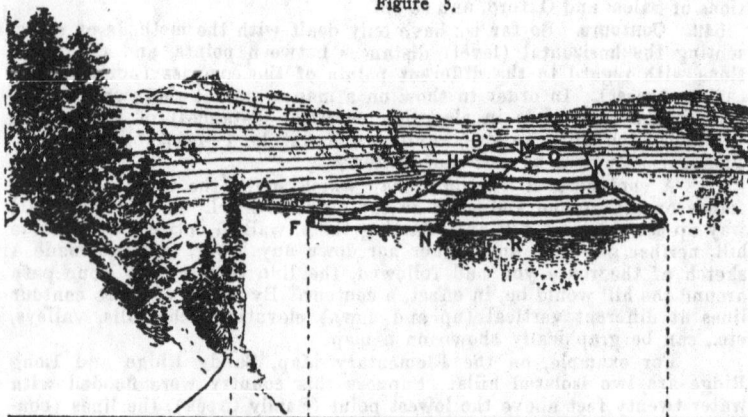

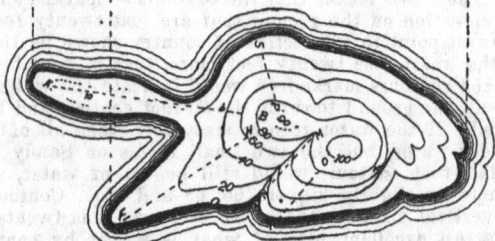

Figure 4

the general form of the island, the two peaks, O and B, the stream, M-N, the Saddle, M, the water shed from F to H, and steep bluff at K, but they also give the slopes of the ground at all points. From this we see that the slopes are directly proportional to the nearness of the contours— that is, the nearer the contours on a map are to one another, the steeper is the slope, and the farther the contours on a map are from one another, the gentler is the slope. A wide space between contours, therefore, represents level ground.

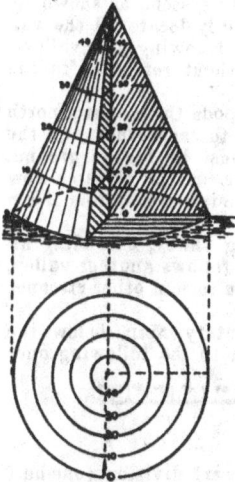 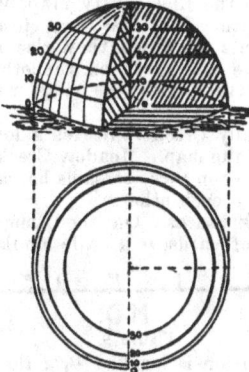 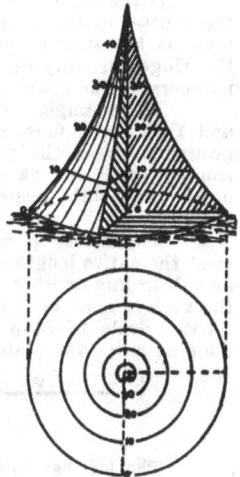

Figure 6.   Figure 7.   Figure 8.

The contours of a cone (Figure 6) are circles of different sizes, one within another, and the same distance apart, because the slope of a cone is at all points the same.

The contours of a half sphere (Figure 7), are a series of circles, far apart near the center (top), and near together at the outside (bottom), showing that the slope of a hemisphere varies at all points, being nearly flat on top and increasing in steepness toward the bottom.

The contours of a concave (hollowed out) cone (Figure 8) are close together at the center (top) and far apart at the outside (bottom).

The following additional points about contours should be remembered:

(a) A Water Shed or Spur, along with rain water divides, flowing away from it on both sides, is indicated by the higher contours bulging out toward the lower ones (F-H, Fig. 5).

(b) A Water Course or Valley, along which rain falling on both sides of it joins in one stream, is indicated by the lower contours curving in toward the higher ones (M-N, Fig. 5).

(c) The contours of different heights which unite and become a single line, represent a vertical cliff (K, Fig. 5).

(d) Two contours which cross each other represent an overhanging cliff.

(e) A closed contour without another contour in it, represents either a hill top (figure cone) or a depression (a volcano) depending on whether its reference number is greater or smaller than that of the outer contour. A hilltop is shown when the closed contour is higher than the contour next to it; a depression is shown when the closed contour is lower than the one next to it.

If the student will first examine the drainage system, as shown by the courses of the streams on the map, he can readily locate all the valleys, as the streams must flow through valleys. Knowing the valleys, the ridges or hills can easily be placed, even without reference to the numbers on the contours.

For example: On the Elementary Map, Woods Creek flows north and York Creek flows south. They rise very close to each other, and the ground between the points at which they rise must be higher ground, sloping north on one side and south on the other, as the streams flow north and south, respectively, (see the ridge running west from Twin Hills.)

The course of Sandy Creek indicates a long valley, extending almost the entire length of the map. Meadow Creek follows another valley, and Deep Run another. When these streams happen to join other streams, the valleys must open into each other.

543. Scale of Map Distances. On the Elementary Map, below the scale of miles and scale of yards, is a scale similar to the following one:

M.D.

Figure 9.

The left-hand division is marked ½°; the next division (one-half as long) 1°; the next division (one-half the length of the 1° division) 2°, and so on. The ½° division means that where adjacent contours on the map are just that distance apart, the ground has a slope of ½ a degree between these two contours, and slopes up toward the contour with the higher reference number; a space between adjacent contours equal to the 1° space shown on the scale means a 1° slope, and so on.

What is a slope of 1°? By a slope of 1° we mean that the surface of the ground makes an angle of 1° with the horizontal (a level surface. See Figure 14, Par. 550). The student should find out the slope of some hill or street and thus get a concrete idea of what the different degrees of slope mean. A road having a 5° slope is very steep.

By means of this scale of M. D.'s on the map, the map reader can determine the slope of any portion of the ground represented, that is, as steep as ½° or steeper. Ground having a slope of less than ½° is practically level.

544. Conventional Signs. In order that the person using a map may be able to tell what are roads, houses, woods, etc., each of these features are represented by particular signs, called conventional signs. On the Elementary Map the conventional signs are all labeled with the name of what they represent. By examining this map the student can quickly learn to distinguish the conventional signs of most of the ordinary features shown on maps. These conventional signs are usually graphical representations of the ground features they represent, and, therefore, can usually be recognized without explanation.

For example, the roads on the Elementary Map can be easily distinguished. They are represented by parallel lines (=====). The student should be able to trace out the route of the Valley Pike, the Chester Pike, the County Road, and the direct road from Salem to Boling.

Private or farm lanes, and unimproved roads are represented by broken lines (====). Such a road or lane can be seen running from the Barton farm to the Chester Pike. Another lane runs from the Mills farm to the same Pike. The small crossmarks on the road lines indicate barbed wire fences; the round circles indicate smooth wire; the small, connected ovals (as shown around the cemetery) indicate stone walls, and the zigzag lines (as shown one mile south of Boling) represent wooden fences.

Near the center of the map, by the Chester Pike, is an orchard. The small circles, regularly placed, give the idea of trees planted in regular rows. Each circle does not indicate a tree, but the area covered by the small circles does indicate accurately the area covered by the orchard on the ground.

Just southwest of Boling a large woods (Boling Woods) is shown. Other clumps of woods, of varying extent, are indicated on the map.

The course of Sandy Creek can be readily traced, and the arrows placed along it, indicate the direction in which it flows. Its steep banks are indicated by successive dashes, termed *hachures*. A few trees are shown strung along its banks. Baker's Pond receives its water from the little creek which rises in the small clump of timber just south of the pond, and the hachures along the northern end represent the steep banks of a dam. Meadow Creek flows northeast from the dam and then northwest toward Oxford, joining Woods Creek just south of that town. York Creek rises in the woods 1¼ miles north of York, and flows south through York. It has a west branch which rises in the valleys south of Twin Hills.

A railroad is shown running southeast from Oxford to Salem. The hachures, unconnected at their outer extremities, indicate the fills or embankments over which the track runs. Notice the fills or embankments on which the railroad runs just northwest of Salem; near the crossing of Sandy Creek; north of Baker's Pond; and where it approaches the outskirts of Oxford. The hachures, connected along their outer extremities, represent the cut through which the railroad passes. There is only one railroad cut shown on the Elementary Map—about one-quarter of a mile northeast of Baker's Pond—where it cuts through the northern extremity of the long range of hills, starting just east of York. The wagon roads pass through numerous cuts—west of Twin Hills, northern end of Sandy Ridge, southeastern end of Long Ridge, and so on. The small T's along the railroad and some of the wagon roads, indicate telegraph or telephone lines.

The conventional sign for a bridge is shown where the railroad crosses Sandy Creek on a trestle. Other bridges are shown at the points the wagon roads cross this creek. Houses or buildings are shown in Oxford, Salem, York and Boling. They are also shown in the case of a number of farms represented—Barton farm, Wells farm, Mason's, Brown's, Baker's and others. The houses shown in solid black are substantial structures of brick or stone; the buildings indicated by rectangular outlines are "out buildings," barns, sheds, etc.

545. **Example of Method Followed in Reading a Map.** Suppose you are out in the field in a campaign and are ordered to march to the sec-

tion of country represented by the Elementary Map and take military control of it. You are given a copy of this map to study over the situation and familiarize yourself with the country. How would you go about reading this map?

You would first look at the scale at the bottom of the map and see about how much distance on the map represented a mile on the ground. Then you would look for the meridian and see which direction was north.

Oxford, Boling, Salem and York are the only towns or villages — all small. Oxford is about four miles due north of York and about two and one-half miles west of Boling, and Boling is about four miles north of Salem. A direct road connects Salem and Boling. The Chester Pike runs northwest out of Salem, and then due north, furnishing, with the crossroads, means of communication between Oxford and Boling, Oxford and Salem, York and Boling, and York and Salem. A railroad passes through Oxford and Salem.

There are numerous streams in the country, but Sandy Creek, five feet deep and sixty feet wide, is the only one of any size. It passes about halfway between Salem and York, flows north for about three miles, turns east, and disappears off the map about a mile south of Boling. The course of this creek and the smaller ones, mark the valleys in the district. Baker's Pond, two miles southeast of Oxford, is the only large body of water.

There are several prominent hills or ridges. Just east of York a range of hills commences, and runs north about three miles, with several east and west spurs. It reaches a height of eighty feet in several places, and completely commands the Chester and Valley Pikes and the valley through which Sandy Creek flows. East of Sandy Creek and nearly a mile northwest of Salem, is a long, "hog-backed" hill (Sandy Ridge), a little over a mile long, rising sixty feet out of the valley, and running north and south. From its crest an extensive view of the valley through which Sandy Creek, the Chester Pike and the railroad run can be obtained. Between Oxford and Boling is a similar hill (Long Ridge) about forty feet high. About two-thirds of the way between Salem and Boiling, the western extremity of a high ridge is shown, with Bald Knob rising to a height of 100 feet above the surrounding low ground. It is the highest point shown on the map. A quarry has been cut into its southern face, to which a switch runs from the railroad. There are several small knolls, notably the one between Bald Knob and Salem.

The country is dotted with farms and orchards, but is lightly timbered, except for the extensive Boling Woods. Some of the roads and lanes are bordered by lines of trees, and the majority of the fences along the roads are of barbed or smooth wire. Telegraph or telephone lines follow the railroad and the principal highways. There are a few stone walls and two swamps.

Three highway bridges span Sandy Creek, one stone, one steel and one of wood. The railroad crosses this creek on a steel trestle about 200 yards long; the track approaching each end of the trestle on a high fill or embankment. The width, depth and steep banks of this creek make these bridges of considerable importance.

PART II

(Note: Part II presents the subject of Map Reading in a more comprehensive manner than Part I.)

A Map is a representation on paper of a certain portion of the earth's surface.

A Military Map is one which shows the relative distances, directions and elevations of all features of military importance on the ground represented.

546. Scale of Maps. A map is drawn to scale, that is, each unit of distance on the map must bear a fixed proportion to the corresponding distance on the ground. If one inch on the map, for instance, equals one mile (63,360 inches) on the ground, then one-third inch equals one-third of a mile, or $^{63360}/_3 = 21,120$ inches on the ground, etc. The term distance in this book means horizontal distance; vertical distance is called elevation or depression, depending on whether the point spoken of is higher or lower than another.

For example (see Fort Leavenworth map in back of book), the distance from Frenchman's (oc') in a straight line to McGuire (qh') is 2,075 yards, but to walk this distance would require the ascent and descent of Sentinel Hill, so that the actual length of travel would be considerably greater than the horizontal distance between the two points. In speaking of distance between towns, boundaries, etc., horizontal distance is always meant. The fixed relation between map distances and corresponding ground distances must be constantly kept in mind.

Methods of Representing Scales. There are three ways in which the scale of the map may be represented:

1st. By words and figures, as 3 inches=1 mile; 1 inch=200 feet.

2d. By Representative Fraction (abbreviated R.F.), which is a fraction whose numerator represents units of distance on the map and whose denominator, units of distance on the ground.

For example, R. F. $= \dfrac{1 \text{ inch (on map)}}{1 \text{ mile (on ground)}}$ which is equivalent to R. F.= 1/63360, since I mile=63,360 inches. So the expression, "R. F. 1/63360" on a map merely means that 1 inch on the map represents 63,360 inches (or 1 mile) on the ground. This fraction is usually written with a numerator 1, as above, no definite unit of inches or miles being specified in either the numerator or denominator. In this case the expression means that one unit of distance on the map equals as many of the same units on the ground as are in the denominator. Thus, 1/63360 means that 1 inch on the map=63,360 inches on the ground, 1 foot on the map= 63,360 feet on the ground; 1 yard on the map=63,360 yards on the ground, etc.

3d. By Graphical Scale, that is, a drawn scale. A graphical scale is a line drawn on the map, divided into equal parts, each part being marked not with its actual length, but with the distance which it represents on the ground. Thus, in Figure 1, page 26, the distance from 0 to 50 represents 50 yards on the ground; the distance from 0 to 100, 100 yards on the grounds, etc. And if the scale were applied to road running from A to B (Figure 2, Par. 540), it would show that the length of the road is 675 yards.

It will readily be seen that a map scale must be known by the student in order that he may have a correct idea of the distances between objects represented on the map. This is necessary in determining lengths of marches, ranges of small arms and artillery, relative lengths of roads to a given point, etc. Therefore, if under service conditions one should have only a map without a scale, or one with only an R. F. on it, he would first of all be compelled to construct a graphical scale to read yards, miles, etc., or one showing how many miles one inch represents. Fortunately, almost every map has a graphical scale, and there will be but few occasions on which it will be necessary to construct a graphical scale.

546a. Construction of Scales. The following are the most usual problems that arise:

1. Having given the R. F. on a map, to find how many miles on the ground are represented by one inch on the map. Let us suppose that the R. F. is $\frac{1}{21120}$.

Solution

Now, as previously explained, $\frac{1}{21120}$ simply means that one inch on the map represents 21,120 inches on the ground. There are 63,360 inches in one mile. 21,120 goes into 63,360 three times—that is to say, 21,120 is $\frac{1}{3}$ of 63,360, and we, therefore, see from this that one inch on the map represents $\frac{1}{3}$ of a mile on the ground, and consequently it would take three inches on the map to represent one whole mile on the ground. So, we have this general rule: To find out how many miles one inch on the map represents on the ground, divide the denominator of the R. F. by 63,360.

2. Being given the R. F. to construct a graphical scale to read yards. Let us assume that $\frac{1}{21120}$ is the R. F. given—that is to say, one inch on the map represents 21,120 inches on the ground, but, as there are 36 inches in one yard, 21,120 inches $= \frac{21120}{36}$ yds. $= 586.66$ yds.—that is, one inch on the map represents 586.66 yds. on the ground. Now, suppose about a 6-inch scale is desired. Since one inch on the map $= 586.66$ yards on the ground, 6 inches (map) $= 586.66 \times 6 = 3\,519.96$ yards (ground). In order to get as nearly a 6-inch scale as possible to represent even hundreds of yards, let us assume 3,500 yards to be the total number to be represented by the scale. The question then resolves itself into this: How many inches on the map are necessary to represent 3,500 yards on the ground. Since, as we have seen, one inch (map) $= 586.66$ yards (ground), as many inches are necessary to show 3,500 yards as 586.66 is contained in 3,500; or $\frac{3500}{586.66} = 5.96$ inches.

Figure 10

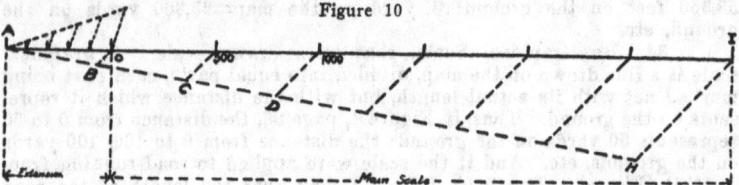

[488]

Now lay off with a scale of equal parts the distance A-I (Figure 10) = 5.96 inches (about 5 and 9½ tenths), and divide it into 7 equal parts by the construction shown in figure, as follows: Draw a line A-H, making any convenient angle with A-I, and lay off 7 equal convenient lengths (A-B, B-C, C-D, etc.), so as to bring H about opposite to I. Join H and I and draw the intermediate lines through B, C, etc., parallel to H-I. These lines divide A-I into 7 equal parts, each 500 yards long. The left part, called the Extension, is similarly divided into 5 equal parts, each representing 100 yards.

3. **To construct a scale for a map with no scale.** In this case, measure the distance between any two definite points on the ground represented, by pacing or otherwise, and scale off the corresponding map distance. Then see how the distance thus measured corresponds with the distance on the map between the two points. For example, let us suppose that the distance on the ground between two given points is one mile and that the distance between the corresponding points on the map is ¾ inch. We would, therefore, see that ¾ inch on the map = one mile on the ground. Hence ¼ inch would represent ⅓ of a mile, and 4-4, or one inch, would represent $4 \times \frac{1}{3} = 4\text{-}3 = 1\frac{1}{3}$ miles.

The R. F. is found as follows:

R. F. $\dfrac{1 \text{ inch}}{1\frac{1}{3} \text{ mile}} = \dfrac{63{,}360 \times 1\frac{1}{3} \text{ inches}}{1 \text{ inch}} = \frac{1}{84480}.$

From this a scale of yards is constructed as above (2).

4. To construct a graphical scale from a scale expressed in unfamiliar units. There remains one more problem, which occurs when there is a scale on the map in words and figures, but it is expressed in unfamiliar units, such as the meter (= 39.37 inches), strides of a man or horse, rate of travel of column, etc. If a noncommissioned officer should come into possession of such a map, it would be impossible for him to have a correct idea of the d'stances on the map. If the scale were in inches to miles or yards, he would estimate the distance between any two points on the map to be so many inches and at once know the corresponding distance on the ground in miles or yards. But suppose the scale found on the map to be one inch = 100 strides (ground), then estimates could not be intelligently made by one unfamiliar with the length of the stride used. However, suppose the stride was 60 inches long; we would then have this: Since 1 stride = 60 inches, 100 strides = 6,000 inches. But according to our supposition, 1 inch on the map=100 strides on the ground; hence 1 inch on the map=6,000 inches on the ground, and we have as our R. F.,

$\dfrac{1 \text{ inch (map)}}{6{,}000 \text{ inches (ground)}} = \frac{1}{6000}.$ A graphical scale can now be constructed as in (2).

Problems in Scales

The following problems should be solved to become familiar with the construction of scales:

Problem No. 1. The R. F. of a map is $\frac{1}{21000}$. Required: 1. The distance in miles shown by one inch on the map; 2. To construct a graphical scale of yards; also one to read miles.

that a contour line is a line that joins points on the surface of the earth, which are the same height--that is, which are in the same level plane. The projection of a contour line on a horizontal surface (a map) is called a contour. Elevations and depressions may, therefore, be represented on maps by imagining the surface of the ground being cut by a number of horizontal planes that are the same distance apart, and then projecting (or shooting) on a horizontal plane the lines so cut on the earth's surface.

(Note: Read over Par. 542a, before studying what follows.)

549. **Map Distances.** The horizontal distance between contours on a map (called map distance, or M. D.) is proportional to the slope of the ground represented—that is to say, the greater the slope of the ground, the less is the horizontal distance between the contours; the less the slope of the ground represented, the greater is the horizontal distance between the contours.

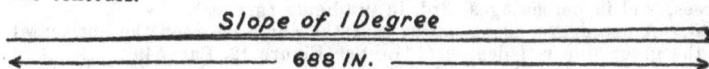

Slope of 1 Degree

← ———————— 688 IN. ———————— →

Figure 12.

Slope (degrees)	Rise (feet)	Horizontal Distance (inches)
1 deg.	1	688
2 deg.	1	$\dfrac{688}{2} = 344$
3 deg.	1	$\dfrac{688}{3} = 229$
4 deg.	1	$\dfrac{688}{4} = 172$
5 deg.	1	$\dfrac{688}{5} = 138$

Figure 13.

It is a fact that 688 inches horizontally on a 1 degree slope gives a vertical rise of one foot; 1376 inches, two feet, 2064 inches, three feet, etc., from which we see that on a slope of 1 degree, 688 inches multiplied by vertical rises of 1 foot, 2 feet, etc., gives us the corresponding horizontal distance in inches. For example, if the contour interval (Vertical Interval, V. I.) of a map is 10 feet, then 688 inches x 10 equals 6880 inches, gives the horizontal ground distance corresponding to a rise of 10 feet on a 1 degree slope. To reduce this horizontal ground distance to horizontal map distance, we would, for example, proceed as follows:

Let us assume the R. F. to be 1/15840—that is to say, 15,840 inches on the ground equals 1 inch on the map, consequently, 6880 inches on the ground equals 6880/15840, equals .44 inch on the map. And in the case of 2 degrees, 3 degrees, etc., we would have:

[491]

550

$$\text{M. D. for } 2° = \frac{6880}{15840 \times 2} = .22 \text{ inch;}$$

$$\text{M. D. for } 3° = \frac{6880}{15840 \times 3} = .15 \text{ inch, etc.}$$

From the above, we have this rule:

To construct a scale of M. D. for a map, multiply 688 by the contour interval (in feet) and the R. F. of the map, and divide the results by 1, 2, 3, 4, etc., and then lay off these distances as shown in Figure 9, Par. 543.

FORMULA

$$\text{M. D. (inches)} = \frac{688 \times \text{V. I. (feet)} \times \text{R. F.}}{\text{Degrees (1, 2, 3, 4, etc.)}}$$

550. Slopes. Slopes are usually given in one of three ways: 1st, in degrees; 2nd in percentages; 3rd, in gradients (grades).

1st. A one degree slope means that the angle between the horizontal and the given line is 1 degree (1°). See Figure 12, Par. 549.

2d. A slope is said to be 1, 2, 3, etc., per cent, when 100 units horizontally correspond to a rise of 1, 2, 3, etc., units vertically.

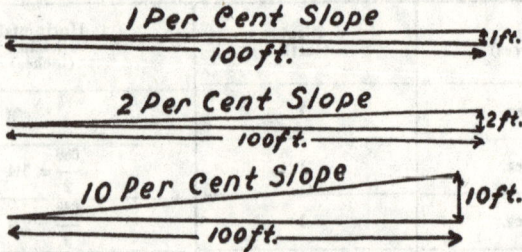

Figure 14.

3d. A slope is said to be on one (⅟₁), two on three, (⅔), etc., when one unit horizontal corresponds to 1 vertical; three horizontal correspond to two vertical, etc. The numerator usually refers to the vertical distance, and the denominator to the horizontal distance.

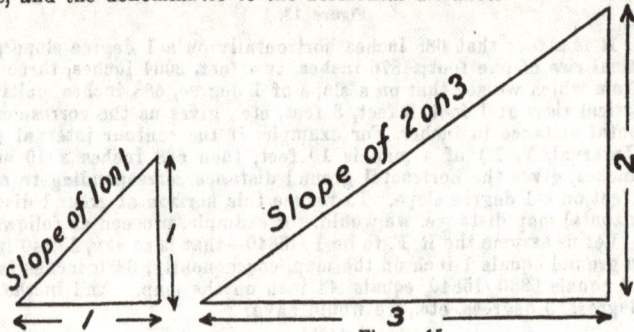

Figure 15.

Degrees of slope are usually used in military matters; percentages are often used for roads, almost always of railroads; gradients are used of steep slopes, and usually of dimensions of trenches.

Effect of Slope on Movements

60 degrees or 7/4 inaccessible for infantry;
45 degrees or 1/1 difficult for infantry;
30 degrees or 4/7 inaccessible for cavalry;
15 degrees or 1/4 inacessible for artillery;
5 degrees or 1/12 accessible for wagons.

The normal system of scales prescribed for U. S. Army field sketches is as follows: For road sketches, 3 inches = 1 mile, vertical interval between contours (V. I.) = 20 ft.; for position sketches, 6 inches = 1 mile, V. I. = 10 ft.; for fortification sketches, 12 inches = 1 mile, V. I. = 5 ft. On this system any given length of M. D. corresponds to the same slope on each of the scales. For instance, .15 inch between contours represents a 5° slope on the 3-inch, 6-inch and 12-inch maps of the normal system. Figure 9, Par. 543, gives the normal scale of M. D.'s for slopes up to 8 degrees. A scale of M. D.'s is usually printed on the margin of maps, near the geographical scale.

Directions on Maps

551. Having given the means used for determining horizontal distances and relative elevations represented on a map, the next step is the determination of horizontal directions. When these three facts (distance, height and direction) are known of any point with respect to any other point, its position is then fully determined. For instance (see map in pocket at back of book), Pope Hill (sm') is 800 yards from Grant Hill (um') (using graphical scale), and it is 30 feet higher than Grant Hill, since it is on contour 870 and Grant Hill is on contour 840; Pope Hill is also due north of Grant Hill, that is, the north and south line through Grant Hill passes through Pope Hill. Therefore, the position of Pope Hill is fully determined with respect to Grant Hill.

The direction line from which other directions are measured is usually the true north and south line (known as the True Meridian) or the plane of the magnetic needle, called the Magnetic Meridian. These two lines do not usually have the same positions, because at all points of the earth's surface the true meridian is the straight line joining the observer's position and the North Pole of the earth, whereas the direction of the magnetic needle varies at different points of the earth, at some places pointing east of and at others west of, the True Pole. At the present time the angle which the magnetic needle (called Magnetic Declination) makes with the True Meridian, is at Fort Leavenworth, 8° 23' east of north.

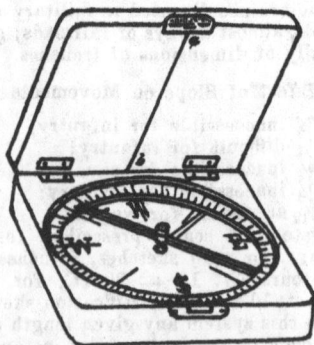

Figure 16.

It is important to know this relation because maps usually show the True Meridian and an observer is generally supplied with a magnetic compass. Figure 16 shows the usual type of Box Compass. It has 4 cardinal points, N, E, S and W marked, as well as a circle graduated in degrees from zero to 360°, clockwise around the circle. To read the magnetic angle (called magnetic azimuth) of any point from the observer's position the north point of the compass circle is pointed toward the object and the angle indicated by the north end of the needle is read.

Orientation

552. In order that directions on the map and on the ground shall correspond, it is necessary for the map to be oriented, that is, the true meridian of the map must lie in the same direction as the true meridian through the observer's position on the ground, which is only another way of saying that the lines that run north and south on the map must run in the same direction as the lines north and south on the ground. Every road, stream or other feature on the map will then run in the same direction as the road, stream or other feature itself on the ground, and all the objects shown on the map can be quickly identified and picked out on the ground.

Methods of Orientating a Map

1st. By magnetic needle: If the map has a magnetic meridian marked on it as is on the Leavenworth map (in pocket at back of book), place the sighting line, a-b, of the compass (Fig. 16) on the magnetic